BIOCHEMISTRY RESEARCH TRENDS

RADON

GEOLOGY, ENVIRONMENTAL IMPACT AND TOXICITY CONCERNS

BIOCHEMISTRY RESEARCH TRENDS

Additional books in this series can be found on Nova's website
under the Series tab.

Additional e-books in this series can be found on Nova's website
under the e-book tab.

BIOCHEMISTRY RESEARCH TRENDS

RADON

GEOLOGY, ENVIRONMENTAL IMPACT AND TOXICITY CONCERNS

AUDREY M. STACKS
EDITOR

Copyright © 2015 by Nova Science Publishers, Inc.

All rights reserved. No part of this book may be reproduced, stored in a retrieval system or transmitted in any form or by any means: electronic, electrostatic, magnetic, tape, mechanical photocopying, recording or otherwise without the written permission of the Publisher.

We have partnered with Copyright Clearance Center to make it easy for you to obtain permissions to reuse content from this publication. Simply navigate to this publication's page on Nova's website and locate the "Get Permission" button below the title description. This button is linked directly to the title's permission page on copyright.com. Alternatively, you can visit copyright.com and search by title, ISBN, or ISSN.

For further questions about using the service on copyright.com, please contact:
Copyright Clearance Center
Phone: +1-(978) 750-8400   Fax: +1-(978) 750-4470   E-mail: info@copyright.com.

NOTICE TO THE READER

The Publisher has taken reasonable care in the preparation of this book, but makes no expressed or implied warranty of any kind and assumes no responsibility for any errors or omissions. No liability is assumed for incidental or consequential damages in connection with or arising out of information contained in this book. The Publisher shall not be liable for any special, consequential, or exemplary damages resulting, in whole or in part, from the readers' use of, or reliance upon, this material. Any parts of this book based on government reports are so indicated and copyright is claimed for those parts to the extent applicable to compilations of such works.

Independent verification should be sought for any data, advice or recommendations contained in this book. In addition, no responsibility is assumed by the publisher for any injury and/or damage to persons or property arising from any methods, products, instructions, ideas or otherwise contained in this publication.

This publication is designed to provide accurate and authoritative information with regard to the subject matter covered herein. It is sold with the clear understanding that the Publisher is not engaged in rendering legal or any other professional services. If legal or any other expert assistance is required, the services of a competent person should be sought. FROM A DECLARATION OF PARTICIPANTS JOINTLY ADOPTED BY A COMMITTEE OF THE AMERICAN BAR ASSOCIATION AND A COMMITTEE OF PUBLISHERS.

Additional color graphics may be available in the e-book version of this book.

Library of Congress Cataloging-in-Publication Data

ISBN: 978-1-63463-742-8
Library of Congress Control Number: 2014958460

Published by Nova Science Publishers, Inc. † New York

CONTENTS

Preface		vii
Chapter 1	Electret Ion Chambers for Characterizing Indoor, Outdoor, Geologic and Other Sources of Radon *Payasada Kotrappa*	1
Chapter 2	Space-Time Distribution of Radon-222 from Groundwater-Streamflow-Atmosphere Interactions in the Karst Systems of the Campania Region (Southern Italy) *Michele Guida, Domenico Guidaand Albina Cuomo*	43
Chapter 3	History-Radon-Geology Connections *I. Burian, J. Merta and P. Otahal*	99
Chapter 4	Efficiency of Four Different Methods of Ventilation for Radon Mitigation in Houses *Lydia Leleyter, Benoît Riffault, Benoît Basset,* *Mélanie Lemoine, Hakim Hamdoun and Fabienne Baraud*	109
Chapter 5	The Use of Radon and Thoron in Balneotherapy *Fábio Tadeu Lazzerini and Daniel Marcos Bonotto*	117
Chapter 6	The Possible Applications of Radon Inhalation Treatment as Antioxidant Therapy for Hepatopathy *Takahiro Kataoka, Akihiro Sakoda, Reo Etani, Yuu Ishimori,* *Fumihiro Mitsunobu and Kiyonori Yamaoka*	133
Chapter 7	Radon in Tap Water in the Territory of Tbilisi City *Nodar Kekelidze, Teimuraz Jakhutashvili, Eremia Tulashvili,* *Manana Chkhaidze, Zaur Berishvili, Lela Mtsariashvili* *and Master Irina Ambokadze*	149
Chapter 8	Radon in Water — Hydrogeology and Health Implication *N. Todorović, J. Nikolov, T. Petrović Pantić, J. Kovačević,* *I. Stojković and M. Krmar*	163

Chapter 9	Indoor Radon Activity Concentration Measurement Using Charcoal Canister *G. Pantelić, M. Živanović, J. Krneta Nikolić, M. Eremić Savković, M. Rajačić and D. Todorović*	**189**
Chapter 10	Methods of Radon Measurement *J. Nikolov, N. Todorović, S. Forkapić, I. Bikit, M. Vesković, M. Krmar, D. Mrđa and K. Bikit*	**209**
Chapter 11	Radon Buildup in Dwellings, Spas and Caves: Facts and Interpretations *I. Bikit, S. Forkapic, D. Mrdja, K. Bikit, N. Todorovic and J. Nikolov*	**227**
Chapter 12	The Analysis of Radon Diffusion Emanation and Adsorption on Different Types of Materials *Sofija Forkapić, Dušan Mrđa, Ištvan Bikit, Kristina Bikit, Selena Grujić and Uranija Kozmidis-Luburić*	**251**
Index		**269**

PREFACE

Radon is a naturally occurring volatile gas formed from the alpha radioactive decay of radium. It is colorless, odorless, tasteless, chemically inert, and radioactive. Of all the radioisotopes that contribute to natural background radiation, radon presents the largest risk to human health. This book discusses the geology, the environmental impact and the toxicity concerns of radon. Some of the topics included are the use of radon in balneotherapy; the health implications of radon in tap water; methods of radon measurement; and facts and interpretations of radon buildup in dwellings, spas and caves.

Chapter 1 – Electret Ion Chambers (EICs) are portable, passive, accurate integrating ionization chambers that do not require a battery or any external source of power. An EIC consists of an electret, a charged Teflon® disk, enclosed inside an electrically conducting plastic enclosure. The electret serves both as a source of electrostatic field for ion collection and also as a sensor for quantifying the ions collected. The passive EICs, also popularly known as E-PERM® (electret passive environmental radon/radiation monitors) are widely used passive radon detectors in US, Europe, Canada and other countries for the measurement of indoor and outdoor radon concentrations and other applications. Figure 1 illustrates how a radon measuring EIC works. The radon gas passively diffuses into the chamber through small filtered holes into the volume of the chamber, and the alpha particles emitted by the decay process ionize the air molecules inside the chamber. Negative ions produced inside the chamber are collected on the positively charged electret, causing a reduction of its surface charge. The reduction in charge (initial charge minus the final charge of the electret) is a function of the radon concentration, the test duration, and the chamber volume. The charge on the electret surface is measured by using a specially designed portable electret reader. The collected data is analyzed by software using algorithms obtained by appropriate calibrations. The EICs are used not only for indoor and outdoor measurements, but also for characterizing a number of geologically important radon related parameters. Such parameters include, radon in water, radon flux from ground and other surfaces, radon progeny concentration in air, and for geophysical prospecting for uranium. This article further provides historic development and standardization of EIC system. The article also provides the theory and practice of measuring geologically important parameters of radon. Bibliography provides a list of publications for those who wish to pursue the technology further.

Chapter 2 – Karst systems provides 25% of drinking water resources to the world's population and sustains aquatic life in most fluvial eco-systems. In contrast, the singular process of aquifer recharge, the particular mechanism of subterranean pathway and the complex interactions between surface and groundwater make these systems highly variable in space-time hydrological behaviour and vulnerable to contamination and pollution. In order to provide a useful approach to integrate a traditional approach at the above problem resolution, this chapter describes the findings from Radon-222 activity concentration monitoring data from stream-flow and in-stream springs measurement in typical Mediterranean karst landscapes. The study areas of concern are located in the protected area of the Campania region (Southern Italy), primarily in the Cilento and Vallo di Diano National Park-European Geopark and Regional Park of Picentini Mnt, and surrounding area. In these protected areas, the management of the relevant water resources requires adequate groundwater assessment by performing hydro-geomorphological and hydrological modelling supporting planning tasks in water protection for domestic drinking use, riverine wildlife preservation and water quality maintenance in application of the European Water Framework Directive (EWFD). In the framework of an interdisclinary research program carried out at a regional scale, a Regional River Monitoring System has been realized, at basin, segment and reach scale to collect experimental data about ^{222}Rn activity concentration, in addition to physical-chemical and streamflow rate measurements. Montly campaigns were performed from 2007 to 2010 by means field and laboratory Rn222 activity concentration measurements in the streamflow and inflow spring waters, using RAD7-H2O and Water Probe (Durridge, Inc., USA). Appropriate sampling procedures and measurement protocols have been tested, taking into account the different local hydrogeological and hydrological situations occurring along the karst systems. At segment scale, data elaborations from selected segment of the monitored rivers provide location and downstream influences of the groundwater inflows from river banks and bed, also in absence of valuable streamflow discharge increments. At river reach scale, more detailed data enable to improve a preliminary model in the ^{222}Rn degassing rate and provide a first contribution to surficial-groundwater seasonal hydrograph separation. The analysis of the seasonal monitoring data trends from karst springs confirms the the utility and reliability of the Radon222 as tracers in a general hydrogeological conceptual model, highlighting the complex behaviour of the multilevel groundwater circuits, the uppermost in caves, the middle in conduits and the lowermost in fracture network, corresponding to the differentiated recharge types in the fluvial-karst hydro-geomorphological system. Ultimaly, these results are useful in the assessment of the high-level of Radon-in-air atmosphere diffusion enriched by Radon222 degassing from differen karst springs feeding surficial water bodies. Preliminary esperimental results from two study cases are provided in order to planning geosite fruition and irrigation water uses.

Chapter 3 – The history of Central Europe (Bohemia) is connected with natural radioactivity. In the first part of the 20th century, a first maximum allowed concentration of radon for uranium miners was implemented. Recently we measured a very high radon concentration in soil air at depths 0.8 m (3 MBq·m^{-3}) in the center of the town Jachymov (Joachimsthal) – where Mme Curie discovered the element radium. In the second half of the 20th century, uranium became a strategic raw material and prospecting for uranium deposits became very intense. One of the prospecting methods was placing a deposition foil at the depth of about one meter and then measuring the surface activity caused mostly by ^{218}Po (first radon decay product) deposited on it. Other sources of information about radium concentration can be obtained using gamma dose rate measurements, accomplished either by walking, using cars, or flying in airplanes. A map of gamma dose rates of the whole country was published in 1977. There is a rough correlation between uranium and radium concentrations, and of radon, in soil. This can be demonstrated based on large number of measurement. In the 1980s the uranium mining boom was decreasing and all over the world the radon industry was born. In Bohemia (Czech Republic) it is very justified, due to high radon concentration in soil, and to bad insulation of building foundations. At present the evaluation of radon risk index in building sites is compulsory. This evaluation is based on at least 15 measurements of radon-in-soil at the depth 0.8 m. There are two more necessary parameters utilized: the permeability of soil and estimation of geologic structure (possibility of change of parameters in vertical profile). The reason of this detailed local evaluation is that, due to complicated geologic structure, within a distance of the order of meters the concentration of radon in soil air could be dramatically different. For quality control of results the measuring companies are certified by SONS (State Office for Nuclear Safety). One of the necessary requirements is the accuracy of a measuring device here the checking is delegated to our Authorized Metrologic Center, which is traceable to PTB Braunschweig, Germany). There exists a correlation between geologic characteristics – uranium (radium) content in soil (gamma dose) - and radon in soil (influenced by permeability and the presence or absence of geological faults). There is also a correlation between global estimation of radon risk emission (map of risk 1:200 000) and radon in underground water. These relationships will be shown here using results of one of the authors' research tasks.

Chapter 4 – Calvados (France) is a region that is naturally rich in radon. The radon contents in a nursery school and an individual house, located in a same town in Calvados, are measured. These two buildings do not have either basement nor underfloor space and are directly on the elevation. It turned out that important radon concentration, with volume activity that could even exceeds 1000 Bq/m^3, were observed in both cases. In France, the public authorities distinguish three levels of exposure (below 400 Bq/m^3, between 400 and 1000 Bq/m^3 and over 1000 Bq/m^3) with respect to the management of the risk associated with radon in places accessible to the public (according to the "Arrêté du 22 juillet 2004, J.O. 11 août 2004). In case of volume activities exceeding 1000 Bq/m^3 the French public authorities recommend that "important corrective actions must necessarily and rapidly be undertaken".

To decrease the radon concentration in a confined atmosphere, two natural ventilation methods and two forced ventilation techniques were tested. The four studied techniques are: Punctual Natural Ventilation (*PNV*), which consists in renewing the internal air by simple aeration twice a day for 15 minutes. Continue Natural Ventilation (*CNV*) which consists in creating additional aeration in the first floor (10 cm diameter) and/or aeration of the basement thanks to some cavities linked to the outside by a pipe. Insufflating Mechanical Ventilation (*IMV*) which consists in setting off the air through the house roof and injecting fresh air into the house (at a temperature over or equal to 15°C). Mechanical Ventilation with Double flow (*MVD*) which assures the air renewal by insufflating fresh air into the living areas and extraction of the used air from the wet rooms. The PVN efficiency is evidenced by an important reduction in radon levels (closed to 0 Bq/m^3), when the windows are opened. However, as soon as the windows are closed, at dusk, when the outside temperatures decrease, as a consequence of the stack effect, the convective transfer of radon, hence the radon concentration increases quickly. Thus this technique was not satisfactory. The other 3 tested techniques (CNV, IMV and MVD) were perfectly adapted and completely satisfactory. Indeed they allowed a very important decrease (from 80 to 95%) of the radon average concentrations. Moreover, the radon concentrations peaks are consistently below 300 Bq/m^3, even during the cold periods which are characterized by strong convective transfer of radon in the buildings as a consequence of the stack effect.

Chapter 5 – Radon (^{222}Rn, half-life 3.8 days) is a naturally occurring volatile noble gas formed from the normal radioactive decay series of ^{238}U, according to the following decay sequence: ^{238}U (4.49 Ga, α) → ^{234}Th (24.1 d, β⁻) → ^{234}Pa (1.18 min, β⁻) → ^{234}U (0.248 Ma, α) → ^{230}Th (75.2 ka, α) → ^{226}Ra (1622 a, α) → ^{222}Rn (3.83 d, α) → ... Thoron (^{220}Rn, half-life 56 seconds) is another naturally radioactive volatile noble gas formed in the ^{232}Th decay series according to the sequence: ^{232}Th (14.0 Ga, α) → ^{228}Ra (5.8 a, β⁻) → ^{228}Ac (6.2 h, β⁻) → ^{228}Th (1.9 y, α) → ^{224}Ra (3.7 d, α) → ^{220}Rn (55.6 s, α) → … Radon and thoron are colorless, odorless, tasteless, chemically inert and radioactive gases produced continuously in rocks and soils through α-decay of ^{226}Ra and ^{224}Ra, respectively, with some atoms escaping to the surrounding fluid phase, such as groundwater and air. They are subjected to recoil at "birth", with the emanated fraction relatively to that produced in the solid phase being dependent on factors such as total surface area of solids and concentration/distribution of ^{238}U (^{226}Ra) in the minerals. ^{222}Rn decays to stable lead according to the sequence: ^{222}Rn (3.83 d, α) → ^{218}Po (3.05 min, α) → ^{214}Pb (26.8 min, β⁻) → ^{214}Bi (19.7 min, β⁻) → ^{214}Po (0.16 ms, α) → ^{210}Pb (22.3 a, β⁻) → ^{210}Bi (5 d, β⁻) → ^{210}Po (138.4 d, α) → ^{206}Pb. ^{220}Rn decays to stable lead according to the sequence: ^{220}Rn (55.6 s, α) → ^{216}Po (0.14 s, α) → ^{212}Pb (10.6 h, β⁻) → ^{212}Bi (60.6 min, β⁻-64.1% or α-35.9%) → ^{212}Po (0.3 μs, α) or ^{208}Tl (3.0 min, β⁻) → ^{208}Pb. High ^{222}Rn concentrations occur in groundwaters in many areas where wells are used for domestic water supply, inclusive in small rural water supplies. Some natural processes related to high concentration of radon in groundwater are: low transmissivity zones, uranium content of the source rock, severe chemical weathering, hydrothermal solution, deposition, extensive fracturing and variations in stress in rocks associated with seismicity. Potential health hazards from radon in consuming water have been considered worldwide, especially when groundwaters are utilized for public water supplies, because ^{222}Rn concentrations in surface waters are often less than 3.7 Bq/L, while in groundwater the ^{222}Rn concentrations commonly are 10-100 times higher. Despite the concerns coupled to the health risks due to ingestion of dissolved radon and thoron in drinking water, these radioactive gases have been sometimes used in balneotherapy in view of attributed benefic physiological effects to human health. This chapter reports the results of investigations held elsewhere focusing some of these aspects.

Chapter 6 – The possibility of antioxidant therapy has been reported for several diseases such as ischemic stroke. The therapy could also be applied for diseases caused by reactive oxygen species (ROS). It has been reported that ROS or free radicals may cause various types of hepatopathy, including alcoholic liver disease. Low dose (0.5 Gy) X- or γ-irradiation activates the antioxidative functions of the mouse liver and inhibits ROS- or free radical-induced hepatopathy. Radon therapy is performed mainly for pain-related diseases in Japan and Europe. Several clinical studies have been reported, but the possible mechanisms of the beneficial effects remain unknown. Recently, the authors have reported that the possible mechanism of radon therapy is the activation of antioxidative functions following radon inhalation. For example, radon inhalation inhibits and alleviates chronic constriction injury induced pain or inflammatory pain in mice due to the activation of antioxidative functions. In addition, although hepatopathy is not the main indication for radon therapy, their recent studies suggested that radon inhalation inhibits hepatopathy caused by ROS or free radicals. In this chapter, based on experiments with mice, the authors reviewed the possible applications of radon inhalation as an antioxidant therapy for hepatopathy from the viewpoints of recent antioxidant therapy, hepatopathy induced by ROS or free radicals, and the beneficial effects of radon inhalation for hepatopathy.

Chapter 7 – Content of radioactive gas radon – Rn-222 in tap water of municipal water-supply system in various territorial sites of Tbilisi city – capital of Georgia – has been investigated. Within the framework of study there were analyzed the water resources used for supply of urban population by drinking water. It is shown, that now water supply is made from 11 sources of natural water which can be divided on two essentially various groups - sources in which underground waters (basically, from artesian wells) are used, and sources in which surface waters (river and from water reservoirs) are used. Water samples were selected in the residential buildings located in the main territorial sites of the city – in total 52 territorial entities have been allocated. Researches were carried out in the period January-December, 2013. Total amount of control points has made 118 points. Samples in nearby settlements (10 control points) were selected for comparison. In many control points sampling and the control of radon content was carried out monthly. Modern radon detector RAD7 was used for determination of radon content. It was established that radon content in water considerably changes depending on sampling time (that connects with possible changes of specific conditions of water transport to the consumer – distance from intermediate storage reservoirs, duration of stay in water mains, etc.) as well as on location of control point (that connects with primary prevalence in certain territories of water transport from surface sources – in this case activity of samples corresponded to group with very low radon content (<0.3 Bq/L) and low (0.3 - 1.0 Bq/L), or from underground sources – in this case activity of samples corresponded to group with typical radon content (1.0 - 3.0 Bq/L) and above typical (3.0 - 10.0 Bq/L)). Based on the received data (more than 700 results) there was issued radon map of tap water in the city territory. Comparison with literary data has been carried out, in particular it is noticed, that the received values do not exceed recommended reference levels and are not dangerous for the population.

Chapter 8 – Radon presence in the environment is associated mainly with trace amounts of uranium and radium in rocks and soil. Underground rock containing natural uranium continuously releases radon into water in contact with it (groundwater). When groundwaters reach the surface, in spas, wells or springs, the radon concentrations decrease sharply with the water movement and with purification treatment. But if the water is consumed directly from the point of emergence, as is habitual in rural sites, the time is often not long enough to prevent the health risks associated with its short-lived daughters. Hence, there is a need to determine the radon activity concentrations in groundwaters used directly (or indirectly through irrigation) and to estimate the doses received by the public consuming these waters. The risk due to radon in drinking-water derived from groundwater is typically low compared with that due to total inhaled radon but is distinct, as exposure occurs through both consumption of dissolved gas and inhalation of released radon and its daughter radionuclides. Moreover, the use of radon-containing groundwater supplies not treated for radon removal (usually by aeration) for general domestic purposes will increase the levels of radon in the indoor air, thus increasing the dose from indoor inhalation. Radon analyses of groundwater samples can – beside the health implications – supply useful information for hydrogeological and hydrological purposes, groundwater quality (missing or existence of a protective soil cover), infiltration and exfiltration of groundwater, and age determinations of groundwater after seepage.

Chapter 9 – Active charcoal detectors are used for testing the concentration of radon in dwellings. The method of measurement is based on radon adsorption on coal and measurement of gamma radiation of radon daughters. Detectors used for the measurement were calibrated by ^{226}Ra standard of known activity in the same geometry. The contributions to the final measurement uncertainty are identified, based on the equation for radon activity concentration calculation. The quantities that contribute to the combined measurement uncertainty in charcoal canister method for radon concentration screening were identified as uncertainties of: counting statistics, efficiency, calibration factor for radon adsorption rate, decay factor, time of exposure and measurement. Different methods for setting the region of interest for gamma spectrometry of canisters were discussed and evaluated. The obtained radon activity concentration and uncertainties do not depend on peak area determination method. Standard and background canisters are used for QA&QC, as well as for the calibration of the measurement equipment. Standard canister is a sealed canister with the same matrix and geometry as the canisters used for measurements, but with the known activity of radon. Background canister is a regular radon measurement canister, which has never been exposed. Carbon filters were unsealed and exposed in closed rooms for 2 to 3 days. Detectors were placed at distance of 1 m from the floor and the walls. Upon closing the detectors, the measurement was carried out after achieving the equilibrium between radon and its daughters (at least 3 hours) using NaI or HP Ge detector. Radon concentration as well as measurement uncertainty was calculated according to US EPA protocol 520/5-87-005. Considering the measured concentration values of ^{222}Rn in dwelling units in Belgrade, as well as flaws of randomized sampling methods, the situation is not upsetting. Radon concentration in more than 80 % of apartments was lower than 200 Bq/m^3, which is within normal limits for apartments. Radon concentration in 6 % of apartments and in 4 % of schools was higher than 400 Bq/m^3 and intensive airing was recommended. For these dwellings additional measurements are required, followed by reparation of the facilities.

Chapter 10 – Radon-222 is a radioactive, noble gas. As a chemically inert gas, it is easily released from soil, building materials, and water, to emanate to the atmosphere. Research carried out in recent decades has shown that, under normal conditions, more than 70% of a total annual radioactive dose received by people originates from natural sources of ionizing radiation, whereby 40% is due to inhalation and ingestion of natural radioactive gas radon ^{222}Rn and its decay products. Many techniques have been developed over the years for measuring radon and radon progenies, because of its hazard effects on human health. Conceptually, measurement techniques can be divided into three board broad categories: (1) grab sampling, (2) continuous and active sampling, and (3) integrative sampling. In this section different techniques for radon measurement and comparison of those techniques will be presented. For radon in air measurement: RAD7, Alpha Guard, passive detectors and charcoal canisters. For activity concentration of radon in water: RAD7-H$_2$O, liquid scintillation counting. In general, the following generic guidelines should be followed when performing radon measurements during site investigations: The radon measurement method used should be well understood and documented; Long term measurements are used to determine the true mean radon concentration; The impact of variable environmental conditions (e.g., humidity, temperature, dust loading, and atmospheric pressure) on the measurement process should be accounted for when necessary. Consideration should be given to effects on both the sample collection process and the counting system; The background response of the detection system should be accounted for; If the quantity of interest is the working level, then the radon progeny concentrations should be evaluated. If this is not practical, then the progeny activities can be estimated by assuming they are 50% of the measured radon activity.

Chapter 11 – Radon as a natural radioactive gas, the daughter nucleus of the long lived ^{238}U is present in all parts of nature. Although its concentration in open air is very low and contributes negligibly to health risk, underground or in closed spaces radon might be a serious health risk problem. This was emphasized a long time ago, and series of radon mapping measurements and legislatives have been established in the meantime. Despite the fact that radon levels in dwellings are usually limited to about 300 Bq/m^3, lot of spas use much higher radon levels (about 10 000 Bq/m^3) for medical purposes. A lot of experimental techniques and methods have been adopted for radon measurements. For long time measurements (about 6 months) usually solid-state track detectors are used. Alpha spectroscopy combined with radon samplers is a method for determination of temporary radon concentrations in air, water and soil gas. Frequently, activated charcoal can be exploited for radon sampling from the air, followed by gamma spectroscopy determination of radon concentration. In the chapter the experimental results of the Novi Sad Nuclear Physics Group are presented and compared with worldwide published results. The associated health effects are estimated and discussed.

Chapter 12 – Since people spend most of the time inside the buildings it is of great importance to analyze the radon diffusion through buildings materials in order to prevent indoor radon build-up. Insulating properties of different types of materials against radon were studied by means of radon diffusion coefficient. A method has been developed in our laboratory by using closed air circulation system which includes RAD 7 radon detector connected by tubes to tightly closed glass chamber with radon source materials covered by well known thickness of insulating materials. This experimental setup has been upgraded for radon emanation coefficient measurements. The granulation effects on the radon adsorption and radon emanation rate of several building materials (ceramic plates, sand, red brick and siporex brick) with different radium Ra-226 content were investigated and discussed. The possibility of using natural zeolite for radon concentration reduction was also considered. On the other hand powder and liquid substances which are stored within closed rooms and exposed to higher radon concentrations can adsorb a certain radon amounts, depending on characteristics and granulation of powder, as well as radon solubility in liquids. This can lead to increase of dose received by general public, if such substances are used as human food, components for food or cosmetics. Research of radon adsorption by liquids and powders is also useful for correction of gamma spectrometry determination of Ra-226 concentrations in such samples mostly based on post-radon gamma lines. Radon adsorption by zeolite on various granulation was explored in this chapter. The radon adsorption coefficients were calculated based on gamma spectrometry measurements of materials and countinuous monitoring of radon inside the chamber with examined material and radon source.

In: Radon
Editor: Audrey M. Stacks

ISBN: 978-1-63463-742-8
© 2015 Nova Science Publishers, Inc.

Chapter 1

ELECTRET ION CHAMBERS FOR CHARACTERIZING INDOOR, OUTDOOR, GEOLOGIC AND OTHER SOURCES OF RADON

Payasada Kotrappa, PhD[*]
Rad Elec Inc, Industry Lane, Frederick, MD, US

ABSTRACT

Electret Ion Chambers (EICs) are portable, passive, accurate integrating ionization chambers that do not require a battery or any external source of power. An EIC consists of an electret, a charged Teflon[®1] disk, enclosed inside an electrically conducting plastic enclosure. The electret serves both as a source of electrostatic field for ion collection and also as a sensor for quantifying the ions collected. The passive EICs, also popularly known as E-PERM[®2] (electret passive environmental radon/radiation monitors) are widely used passive radon detectors in US, Europe, Canada and other countries for the measurement of indoor and outdoor radon concentrations and other applications. Figure 1 illustrates how a radon measuring EIC works. The radon gas passively diffuses into the chamber through small filtered holes into the volume of the chamber, and the alpha particles emitted by the decay process ionize the air molecules inside the chamber. Negative ions produced inside the chamber are collected on the positively charged electret, causing a reduction of its surface charge. The reduction in charge (initial charge minus the final charge of the electret) is a function of the radon concentration, the test duration, and the chamber volume. The charge on the electret surface is measured by using a specially designed portable electret reader. The collected data is analyzed by software using algorithms obtained by appropriate calibrations. The EICs are used not only for indoor and outdoor measurements, but also for characterizing a number of geologically important radon related parameters. Such parameters include, radon in water, radon flux from ground and other surfaces, radon progeny concentration in air, and for geophysical prospecting for uranium.

---

[*] Corresponding author: Payasada Kotrappa Ph.D. President. Rad Elec Inc, 5716-A, Industry Lane, Frederick, MD 21704 US, Tel: 301-694-0011 www.radelec.com, pkotrappa@radelec.com.
[1] Teflon® is the registered trademark for Teflon manufactured by Dupont.
[2] E-PERM® is the registered trademark EIC manufactured by Rad Elec Inc.

This article further provides historic development and standardization of EIC system. The article also provides the theory and practice of measuring geologically important parameters of radon. Bibliography provides a list of publications for those who wish to pursue the technology further.

PART 1. DEVELOPMENT AND STANDADIZATION OF THE ELECTRET ION CHAMBER (EIC) SYSTEMS FOR CHARECERIZING RADON

Introduction

An electret is an important part of the EIC. An electret as defined by Sessler [1] is a piece of dielectric material exhibiting a quasi-permanent electrical charge. The surface charge of the electret produces a strong electrostatic field capable of collecting ions of opposite signs. Until recently, electrets have been regarded as a curious analogue of magnets, worthy of academic interest. However with the development of high dielectric fluorocarbon polymers, such as Teflon, electrets have become reliable electronic components capable of maintaining high constant electrostatic fields even under high temperature and humidity conditions. Properly made electrets can have extraordinary stability with a discharge rate of 1 to 4 % per year when stored in storage cap, as shown by Kotrappa [10]. These are attractive due to the fact that it is possible to get high electrostatic fields without the use of batteries or high voltage units. These have found many applications such as self biased miniature microphones and other useful devices [1, 2]. Marvin [5] was the first to suggest that the reduction of charge on the electret was due to the collection of ions of the opposite sign from the surrounding air. He proposed the use of an electret in a closed chamber as a gamma dosimeter. His idea was not practical at that time because the charge on the electrets made in the early years was not stable because the non availability of materials such as Teflon. However, Bauser and Ranger [4] used a pair of thin Teflon electrets of opposite sign charges to collect and measure ions produced inside an ionization chamber. They showed that the radiation dose calculated from this measurement agreed well with the actual dose received by the chamber. They also showed that the performance was fairly insensitive to variations in humidities and temperatures in the range normally encountered in the environment. The dose information on their electrets was retained without loss for a period of more than one year. This work established the scientific basis for the performance of the EIC. The next innovation was the development of a single electret dosimeter [38]. In this case one side of the electret was coated with carbon. The electret was located at the bottom of the chamber with the conducting side in contact with the chamber. The electret provided the electric field with respect to the chamber. This combination worked similar to that of Bauser (4). Theoretical aspects of electrostatic fields in such ionization chambers were worked out by Fallone and his coworkers covered many basic aspects and several applications [27-35].

Further, basic development work and a number of applications was done by Pretzsch and his coworkers [14-22, 24]. Some of the earlier work was done by Kotrappa and his coworkers covering some aspects of radon and radon progeny concentrations in air [7, 8, 9].

First systematic study of the electret ion chambers, for use in passive mode was made by Kotrappa [11]. Such device was named an "electret passive environmental radon monitor based on ionization measurement" E-PERM®.

This study recognized that the environmental gamma radiation (terrestrial and cosmic) also contributes to the ionization due to radon during radon measurement. This signal needed to be subtracted during the radon measurement. It also depends upon the material used for the chamber construction. The study included different wall materials so as to determine the materials that provide minimal signal from gamma radiation. A plastic chamber coated with colloidal carbon provided the lowest response to gamma radiation when compared with aluminum or steel. Radon responses were independent of chamber material as long as those are electrically conducting. The radon responses were proportional to the volume of the chambers. Electrets made from thicker Teflon showed a higher response compared to electrets made from thinner Teflon. These observations were further used in building a family of E-PERMs required in the measurement of different concentrations and for different periods. Research in this study also included different methods of making electrets stable and with time, temperature and humidity variation. Further research included the development of the electret reader. This study was the basis for the design of the standardized EIC system shown in Figure 1 described by Kotrappa [12]. Introduction of standardized versions of electret ion chambers and the associated equipment led to rapid expansion of research and development leading to additional research and applications.

The EIC system consists of several components. These include an appropriately characterized and stabilized electret, an electret reader to read the surface charge of the electret in units of volts, the chambers made of electrically conducting polypropylene of different volumes with arrangement to remote closing and opening of electrets, the appropriate calibration factors and analyzing algorithms.

Electrets

It is important to discuss how the electret, being an important component of EIC, is made and how it is characterized. Sessler [1, 2], a pioneer in the field of electrets defines an electret as "a piece of dielectric material exhibiting quasi-permanent electrical charge". The term "quasi permanent" means that the time constants characteristic for the decay of the charge are much longer than the time periods over which studies are performed using the electret. The most practical material used for making electrets is fluorocarbon materials such as Teflon (PTFE or FEP) because of their very high surface and volume resistivity. This property prevents charges on the electrets from recombining, resulting in a long life time. Figure 2 provides schematics of an electret and methods of making an electret. The word "electret" comes from the word "magnet".

A magnet has both north and south poles, and an electret has both positive and negative charges. By grounding the appropriate side it is possible to get either positive or negative electrets. There are a large number of methods of introducing a charge into the Teflon disk [2, 3]. Methods which are relevant to practical electrets for use in an EIC, are "internal polarization by dipole orientation" and surface ion deposition by "breakdown of electrical fields". These are illustrated in items 1 and 2 in Figure 2.

In typical Teflon material there are randomly oriented dipoles. When the material is heated to a high temperature under an applied high electric field, the dipoles get oriented. Temperature needed for dipole orientation is called "glass temperature". At such temperatures the medium is semi fluid without compromising the properties of the material.

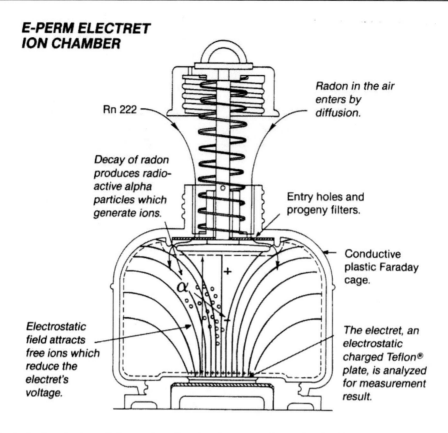

Figure 1. Basic functioning of electret ion chamber for measuring radon in air.

The appropriate temperature is in the range 150 to 200 °C and the applied field is about 10kV cm^{-1}. When the material is slowly cooled still under the electric field, the dipoles are frozen. Such electrets were made and used in an EIC by Kotrappa [11, 12]. The method of making electrets by application of breakdown field is illustrated in item 2 of Figure 2. This method is fully described by Sessler and West [3]. A piece of Teflon is sandwiched between two electrodes separated by low conductivity glass. Typically 2.5 mm soda glass with an electric field of about 40kV cm^{-1} provided electrets with appropriate surface charges for use with an EIC [11, 12]. The exploded view of the different parts of an electret is shown in Figure 4. Usually the negative side of the electret is coated with a thin layer of carbon. Further, these are loaded into a holder made of electrically conducting plastic. The electret holders are designed to be directly loaded into the charging jigs. After charging, it is loaded into a storage cap. Such electrets are further annealed to make them stable against varying temperatures and humidity and other environmental elements. The surface charge is further reduced by exposing the surface of the electret to ions of opposite sign to about 750 volts for electret ion chambers [11, 12]. Figure 4 shows the standardized version of electret assembly with the associated components. By removing the storage cap, the electret assembly can be loaded into the charging jig for charging the electret. Such an electret is shown in the left lower part of Figure 7. Electrets are characterized by their thicknesses, the surface charge density, the area and the dielectric constant of the electret material. Figure 3 illustrates the parameters which characterize the electret and it also illustrates how the surface potential of an electret is measured.

What Is An Electret?

1. It is an electrical analogue of a magnet.
2. It CARRIES a permanent electrical charge of either sign.
3. It is MADE from a dielectric material with high internal resistivity.

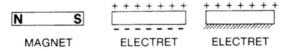

How Is An Electret Made?

There are several methods of making electrets, most popular are:

1. Internal polarization by dipole orientation.

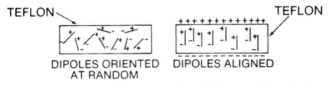

2. Surface ion deposition by breakdown of electrical fields.

Figure 2. What is an electret and how it is made.

The relationship in Figure 3 gives the relationship of these parameters with surface potential V_S. The surface potential V_S is measured by a method known as "capacitive probe method", fully described by Sessler [1], and is illustrated in the lower part Figure 3. When a movable shutter is moved out of the electret, the charge is induced on the probe, which in turn charges the capacitor C and the voltage on the capacitor is measured in a digital panel meter through an ultra high impedance operational amplifier. When the shutter is pushed back, the capacitor is discharged. When the shutter is pulled again the measurement is repeated. This electret surface voltmeter is available for measuring the surface voltage of commercially available electrets used in the EIC. Figure 7 shows a commercially available electret reader that works on this principle. For routine measurement, the electret is introduced into the receptacle. The lever is drawn to read the electret and when the lever is released the shutter goes back and is ready for another measurement. Being a non contact method, measurements can be repeated without affecting the surface voltage on the electret. The instrument is usually calibrated with a simulated electret with a known applied voltage. The manufacturer provides all the necessary information on the appropriate use of the electret voltage reader. Two types of electrets are manufactured and supplied by the manufacturer for use with commercially used EICs. The 1.588 mm thick electret (called thick or ST electret) is made from PTFE Teflon and the 0.127 mm thick electret (called thin or LT electret) is made from FEP Teflon.

How Is An Electret Characterized?

1. **By the sign of the charge.**
2. **By the surface charge density or equivalent surface potential.**
3. **By the thickness of the dielectric material used.**

Approximate relationship between the surface charge density (σ_s), the surface potential (V_s), and thickness of the electret (t).

$$V_s = \frac{\sigma_s T}{\varepsilon_0 \varepsilon}$$

How Is An Electret Measured?

An electret voltage reader works on the principle of induced voltage measurement.

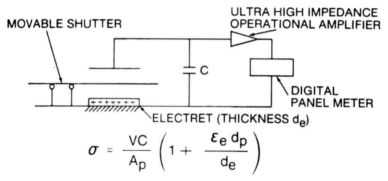

$$\sigma = \frac{VC}{A_p}\left(1 + \frac{\varepsilon_e d_p}{d_e}\right)$$

Figure 3. How is electret characterized and how is it measured.

These both have an electret surface area of 8.43 cm². Figure 5 illustrates a simple EIC used for characterizing alpha and beta radiations. Alpha or beta radiation ionizes air inside the chamber that also houses an electret. The electret, usually carrying positive charge, collects the negative ions. After a desired exposure period, the source is removed, electret is taken out and the surface potential is measured. The initial reading, final reading, and the exposure period are all used in calibration algorithms to calculate the desired characteristics of the source. Figure 5 illustrates the principle of electret ion chambers.

Conversion of the Electret Discharge in Volts to Collected Charge in Coulombs

Equation (1) gives the relationshipbetween various parameters.

$$\frac{Q}{V_S} = \frac{E E_0 A}{d} \tag{1}$$

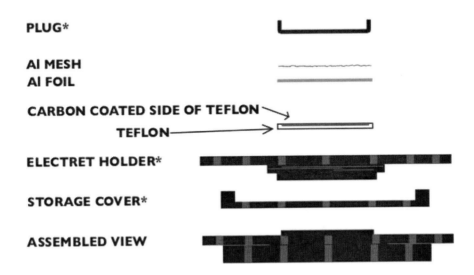

*Made of electrically conducting polypropylene.

Figure 4. Exploded view of components of electrets.

Using equation (1) it is possible to calculate the relationship between the surface voltage and the surface charge. Such a relationship is useful in the theoretical calculation of the responses of EICs.

Q is the total charge on the electret
E is the dielectric constant of the electret = 2 for Teflon
E_0 is the permittivity of space = 8.854 x 10^{-14} C.V^{-1} cm^{-1}
A is the area of the electret = 8.43 cm^2
d is the thickness of the electret
Vs is the surface voltage of the electret in volts
For d (thickness of electret) = 0.1588 cm
Q/Vs = 9.40 x 10^{-12} CV^{-1} (2)

1 volt change on electret corresponds to a collection of 9.40 x 10^{-12} coulombs
For d (thickness of electret) of 0.0.0127 cm

Q/Vs = 1.18 x 10^{-10} CV^{-1}

1 volt change on electret corresponds to a collection of 1.18 x 10^{-10} coulombs

What Is An Electret Ion Chamber?

It is a passive device
An electret loaded into an electrically conducting plastic chamber forms an electret ion chamber

Surface voltage of an electret is not affected by variation in humidity and temperatures. Any change is due to collection of ions.

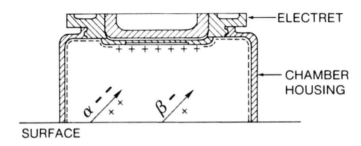

Electret Ion Chamber Configuration as a Windowless Counter for Measurement of Alpha and Beta Contaminations

Figure 5. What is an electret ion chamber and how alpha and beta radiation measured with EIC.

Example of Calculating the Response Factor for an EIC Gamma Monitors

The Bragg Gray equation for uniformly irradiated (by X and/or gamma radiation) air cavity the conversion rate (CR) is calculated and is given by equation (3).

$$CR = 33.97 \times 10^{-14} \text{ C.mrad}^{-1} \text{.ml}^{-1} \tag{3}$$

Dividing equation (3) by equation (2) gives the response in V mrad^{-1} ml^{-1}, multiplying by 58 (the volume of the chamber) gives the response factor in V mrad^{-1}.

The result is about 2.0 volt per mrad. This is very close to the experimental response factor [42].

Such methods are used for calculating theoretical responses of EICs for different radiations.

Practical EIC units are used with electrets in the range of 100 to 750 volts. This is also the range for operating conventional ionization chambers. The final voltage becomes the initial voltage for the next measurement.

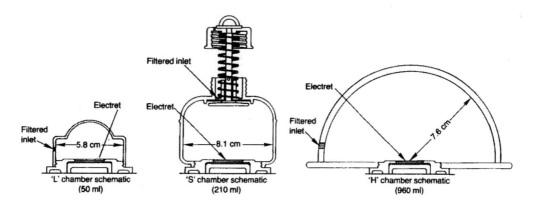

Figure 6A. Cross section ns of different EIC chambers used for radon measurement.

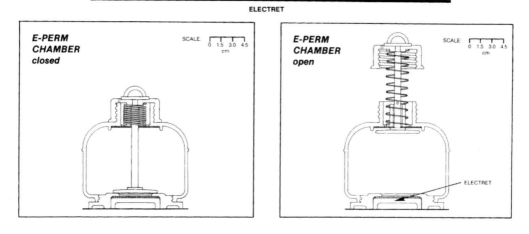

Figure 6B. Principle of operating EIC in on/off mode. Left is off and right is on.

Each electret gives a large number of measurements, adding to the economy of the technology. The response factor, also called as calibration factors (CF) that converts the data to the quantity that is being measured These are discussed later while discussing the use of EIC for different applications.

Comparison between Electret Ion Chambers and Conventional Ionization Chambers

The electret ion chamber (EIC) is a unique form of conventional of ionization chambers used for measuring ionizing radiations. In a typical ionization chamber, electric field is maintained between the two plates using a high voltage source. When ionizing radiation such as gamma radiation passes through the chamber volume, a measurable current is established in proportion to the intensity of radiation.

This current gives a measure of the ionizing radiation. In EIC, we have two readings, one is the initial volt on electret and the other is final volt on the electret.

In case of EIC, electret provides not only the high voltage needed for ionization chamber to function, but also serves as a sensor. The difference between the two readings is the sensor signal. The average reading of the electret, often called as mid-point voltage defines the average electric field. The rate of discharge of the electret defines the current, similar to the current in conventional ionization chambers. Since EIC can be used for extended periods (from days to months), the measurable current can be very small, and hence it is possible to measure very low levels ionization, not practical in conventional ionization chambers. Further, EIC performs well with highly humid atmospheres making it a very practical tool to measure ionizing gases, such as radon and thoron, tritium and other ionizing noble gases, in air in homes or in an atmosphere where humidity can be varying.

Scientific Basis for Electret Ion Chambers

Bauser and Range [4] were the first to use a pair of thin Teflon electrets of opposite charges to collect and measure ions produced inside an ionization chamber. They showed that the radiation dose calculated from this measurement agreed well with the actual dose received by the chamber. They also showed that the performance was fairly insensitive to variations in humidities and temperatures in the range normally encountered in the environment. The dose information on their electrets was retained without loss for a period of more than one year.

©C. Kurt Holter 1988.

Figure 7. Components of EIC for radon measurements; top left: radon EIC, top right: electret voltage reader, bottom left: electret in open mode.

This study laid a solid scientific basis for electret ion chambers (EIC) technology. These are also called as electret dosimeters.

Providing two electrets of opposite polarity served the purpose of demonstration of technology, but this was not practical. Next attempt was to build an EIC with one electret located at the bottom of a conducting chamber. If the electret carries positive charge, the entire conducting chamber is at negative potential (ground). Electret can be taken out and measured. Only one electret need to be measured, simplifying the operations.

This innovation of using single electret in EIC technology reported by Pretzsch and his coworkers [14 to 24] and later by Kotrappa and his coworkers [7, 8, 9]. These workers showed that the drop in surface potential of their single electret dosimeter also behaved according to established ion chamber theory, and they went on to demonstrate its use as personal dosimeters, and possible uses for radon measurement. The theoretical aspects of electrostatic fields in such ionization chamber were worked out by Fallone and his coworkers [27 to 35]. Kotrappa [7, 8, 9] and his coworkers were the first to report a rough correlation between the reductions of surface voltage and cumulative ^{222}Rn exposure in a passive chamber arrangement and further observed that this change in reduction did not appear to be sensitive to humidity change. This became a foundation for further development of practical EICs for measuring. ^{222}Rn (also called as radon).

R and D Leading to a Practical EIC for Indoor Radon Measurements

In the early 80s, there was a sudden large interest in measuring indoor radon concentration because of the discovery of homes with very high radon levels. This led to a systematic research and development of EIC aimed at for measuring radon concentrations in homes. Kotrappa and his coworkers [11] described in detail the conceptual design and performance of EIC for measuring radon concentration and various parameters that determine the performance. The study included the chambers of different volumes, ranging from 40 ml to 1250 ml and electrets of different thicknesses, ranging from 51 μm to 2.3 mm. The study concluded that the response was directly proportional to the volume of the chamber and proportional to the thickness of the electret. The study recognized the importance of response to gamma radiation level when measuring low concentrations of radon usually found in home, and further recognized the importance of the wall material in reducing the background due to gamma radiation. The gamma response successively decreased with the steel, Al and Aluminum lined plastic, respectively. It was also shown that electret produced and used in this study have high stability for practical use from a period of 2 days to one year, even at relatively high humidity. This study also provided the parameters needed to design an EIC required in meeting any requirement. This paper also provided the basic design parameters of an electret voltage reader suitable for routine use.

Standardization of EIC for Indoor Radon Measurement

It was important to move the R and D results to practical device. Next landmark work was to design a practical radon monitoring system using the parametric studies published earlier [11].

This led Kotrappa and associates [12] to standardization of practical electret passive environmental radon monitoring system abbreviated as E-PERM. The system consists of electrets, chambers and electret voltage readers and calculation procedures. The chamber and the electret holders were made of electrically conducting plastic to minimize the response of the gamma background to the radon measurement. The components were made by injection molding for high reproducibility. A spring loaded shutter mechanism is introduced to cover and uncover the electret from outside to serve as on-off mechanism. The optimized system components were: an electret voltage reader that is simple to use and accurate to ±1 volt of surface potential, chamber volume is 210 ml and electret thicknesses are 0.1.524 mm and 0.127 mm. These E-PERMS were designed for making radon measurements in 2 to 7 days with 1.52 mm thick electret and one month to 12 months using 0.127 mm thick electret. The paper [12] describes fully the calibration, performance, error analysis, and the lower limits of detection. The entire system was commercially produced so that it can be easily used. The system was evaluated by USEPA and was introduced as an approved method for measuring indoor radon concentration in air. This system became very popular and is used in many countries by radon and radiation specialists. The system has been evaluated by several investigators in many countries [25, 26, 46, 51, 74] and shown to meet the specifications [12] for measuring radon in air. Standardized version of EIC is often called as E-PERM, as such the words EIC and E-PERM means the same.

Description of Radon Measuring EIC and Related Components

Figure 1 shows the schematic of typical radon measuring EIC. Radon gas diffuses into the chamber through filtered openings. Radon decays inside the chamber leading to other radioactive decay products (^{214}Po and ^{218}Po). The decay products being particles diffuse to the interior surfaces of the chamber. Ionizing radiations emitted by the decay of radon and the decay products ionize air in the chamber. The negative ions are drawn to the surface of the positively charged electret located at the bottom of the chamber. The surface charge of the electret gets depleted due to the collection of ions. The depleted charge of the electret over a time period is equivalent to the time integrated charge produced by radon and the associated decay products. This in turn is proportional to the integrated radon concentration over that period. The technical basis for the measurement of indoor radon using the EIC has been fully described in three papers by Kotrappa and associates [11, 12, 13].

Components of EIC

The major components of EIC are: electret, chamber, mechanical system to cover the electret when EIC is not in use, the electret reader and the analysis tools to calculate radon concentration.

The Electret
The electret used in the EIC is a disk of Teflon® (Du Pont) which has been electrically charged and processed by special procedures so that the charge on the electret remains stable

even at high humidity or low/high temperatures. The production and processing of electrets is fully described elsewhere [11, 12, 13].

The Electret Reader

A method popularly known as shutter method or capacitative probe (Figure 3) method is adopted as the basis for the design of a dedicated electret reader to measure the surface voltage of the electrets. An electret is firmly positioned the electret receptacle and is positioned at a precise distance from the sensing plate. A metallic shutter separates electret from the sensor. When the shutter is pulled out, a charge is induced on sensor plate and generates a voltage across the capacitor. This voltage is measured by an ultra high impedance circuit to read out on the LCD meter. The meter is calibrated using a dummy electret with precise surface voltages. The electret reader reads out the electret volts correct to 1 volt.

It initializes to zero reading before registering a new reading. It has auto off set to 2 minutes of non-operation. The reader measures 1 to 1999 volts. It reads both positive and negative voltages. Figure 7 shows a photograph of the electret voltage reader.

The relationship between the surface volts to surface charge depends upon the thickness of the electret. Conversion factors can be used to convert reading in volts to charge on the electret in coulombs. For 1.542 cm thick electrets, the relationship between the surface charge and the surface potential is: Coulombs volt^{-1} = 9.69 x 10^{-12}.

For 0.0127 cm thick electret, the relationship between the surface charge and the surface potential is: Coulombs volt^{-1}= 1.175 x 10^{-10}. Open area is 8.43 cm^2. One side of the electret is electrically conducting and is held firmly in a holder that is also made of electrically conducting plastic. Electrets are produced and processed in the holder and are covered with a screw cap. When cap is unscrewed, electret can be lowered into the well of the electret reader to read the surface potential in volts, or can be screwed into the EIC chamber.

EIC Chambers

These are made of electrically conducting polypropylene, made by injection molding to maintain the required precision and reproducibility. This further helps minimize the response from gamma radiation [11]. Chambers of 3 different volumes are available to get different sensitivities. The volumes are 58 ml, 210 ml and 960 ml.

Figure 6A gives schematics each one of these chambers and the location of electrets. Figure 6B illustrates the method of switching the EIC on and off. In the off position electret is closed and EIC can be transported or stored without picking up any additional signal. Such on/off design of different design is also incorporated in 50 ml chamber [66].

Basic Electret Ion Chamber System for Measuring Radon Gas

Electret ion chambers (EICs) are portable, passive, accurately integrating ionization chambers that do not require a battery or any external source of power. Figure 1 shows the schematic of typical EIC used for measuring radon gas.

Detailed Method of Deriving Calibration Equations for a Radon EIC [12, 13]

The E-PERM® Electret Ion Chambers (EICs) have been widely used for research in indoor and outdoor radon measurements. Calibration factors are fitted to an equation that relates the calibration factors to the initial and final voltages. The calibration equations currently in use restrict the use of electrets to the initial readings of 750-250 volts. Recent research indicated that it is possible to derive the calibration equations applicable for wider ranges. A detailed procedure is described for calibrating SST EICs and deriving an appropriate equation, applicable over the range of 750 volts to 70 volts. Furthermore, the newly derived equation fits the experimental data with better precision, compared to the currently used equations.

Calibration Procedure

Because of the continuously decreasing nature of the electret voltages during a measurement, the calibration factors are not constants and depend upon the initial and final voltages of electrets. Calibration factor is related to midpoint voltage (MPV), the average between the initial and final voltages. Calibration factors are fitted to an equation that relates the calibration factors to the MPV). Equation (4) is used for calculating the radon concentration.

$$RnC = \frac{(IV - FV)}{(T) \times (CF)} - BG$$

(4)

where: RnC is the radon concentration in the radon test chamber (pCi/L or Bq/m3). T is the exposure period in days. IV and FV are the initial and final voltages respectively. CF is the calibration factor in volts per (pCi L^{-1}-days) or in volts per (Bqm^{-3}-days). BG is the radon equivalent per unit gamma radiation. The constant 0.087 is the radon concentration (pCi L^{-1}) equivalent for 1 μRh^{-1}. Similar unit can be used in SI units.

A set of EICs with staggered initial voltages (from 100 volts to 750 volts) are exposed in a standard radon test chamber for a known radon concentration for a known time period.

Using equation (4), CF is calculated for the respective MPV.

Fitting an Appropriate Equation to the Experimental Data

The data can be fitted to a linear regression equation (5) or equation (6)

$$CF=A+ B \times MPV \quad ----- \tag{5}$$

$$CF=C +D \, Ln(MPV) \quad ---- \tag{6}$$

A, B, C and D are constants.

In the earlier work[9] equation (5) was used. Linear regression equation was fitted between CF and MPV. Such equation was found to fit good only between the MPV of 700 volts and 200 volts only.

In recent work[10} the data was fitted between CF and natural logarithm of MPV. Such equation was found to fit good between the MPV of 700 volts and 100 volts.

The constants A and B; C and D are provided by the manufacturer for several combinations of chambers and two standard thicknesses of electret.

Table 2 gives experimental CF and the calculated CF using the fitted equation (6). Experimental data when fitted to natural logarithm of MPV appears to fit perfectly leading to an error of less than 2%.

When initial and final voltages are determined, equation (6) is used to calculate CF. Then equation (4) is used to calculate radon concentration.

The Performances of EIC for Measuring Radon

EICs are not affected by varying temperature and humidity found in the normal environment

EICs are not affected by the external electric field

EICs are not affected by the air draft

EICs are not affected by the presence of thoron in air

EICs are not affected by the ions generated by external ion sources

EICs are not affected by the external dust load

EICs are not affected by the magnetic fields or any radiofrequency sources

*The performance is affected by the presence of X or gamma radiation and elevations. Manufacturers recommendations are avaialble for the users. *Standard corrections are applicable for sampling at different elevations.

*Standard corrections are applicable for different gamma backgrounds.

Table 2. Percent deviation of CF relative to measured CF using equation

MPV	Measured CF	Fitted CF	% Dev(relative to measured)
731.7	2.0075	2.0484	2.0
697.5	2.0734	2.0345	1.9
649	2.0033	2.0136	0.5
549.8	1.9456	1.9654	1.0
450.5	1.9662	1.9075	3.0
354.9	1.8178	1.8382	1.1
259.2	1.7395	1.7468	0.4
163.6	1.6159	1.6131	0.2
70.1	1.3644	1.3668	0.2

Analysis procedures for sensitivity, error analysis and derivation of detectable limits at different concentrations is published and is provided by the manufacturer.

Inter-comparison of the performances compared to reference levels have been studied at national and international standard test chambers and in blind test.

The results have shown acceptable performances [25, 26, 46, 51, 74]. In none of these exercises there has been any adverse remarks and confirm the observation made in discussions on the performances. In many of these inter-comparison exercises, EIC is used as reference detector to assess other devises.

Advanced Researches on EIC

Most of the basic research on EIC is done by three groups: Kotrappa and associates in India and US, Pretzsch and associates in Germany and Fallone and associates in Canada. Relevant references are cited in the section on references. Some [2] researches of significance are mentioned in the following paragraphs.

Uncertainty Evaluation

Uncertainty evaluation is provided in detail [12]. This includes sensitivity analysis and the calculation of errors for each radon measurement. Further procedures are discussed (Manufacturers Manual) for calculating the lower method detection for different combinations of chambers and electrets in the (Manufacturers Manual).

Caresana and his associates [26] studied several commercially available EICs and evaluated the uncertainties associated with the measurement of radon and gamma radiation. In this work, attention is focused upon the electret ion chambers (EIC), widely used in radon concentration measurements. Measurements of gamma radiation sensitivity are performed in a secondary standard calibration laboratory and measurement of radon concentration sensitivity is performed in a radon chamber 0.8 m^3 in volume.

Raw data are analyzed to evaluate the calibration factors and the combined uncertainties are determined. The aim of the study is more rigorous than the study reported earlier by Kotrappa [12]. This study conforms to international standards (ISO 95) and some of this information is in draft ISO/ WD/ 766 (2007).

Basically results agree well with the simplified version given by Kotrappa [12].

Simplified version has an advantage for easy calculations without compromising the accuracy.

Advanced Method of Deriving Equation for Calibration Factors [12, 13]

EICs are designed to be used over the entire useful range (750 to 70 volts. Further these need two readings, initial and final volts for computation of results. This requires a calibration factors for any pair of readings characterized by mid- point voltage (MPV).

Because of continuously decreasing nature of electric field during the use of an EIC, from 750 volts down to 70 volts, calibration factor (CF) also varies depending upon the MPV.

These features are very different compared to other passive radon detectors such as alpha track detectors whose calibration factors remains the same for the entire range of use. These features of EIC make the calibration of the devices more complicated.

Detailed method of calibration is already described. The most recent research of fitting the CF with natural logarithm of MPV improved the fit and make it applicable over wider range of electret voltages compared to earlier practice [12].

The study also proved that CF decreases exponentially with MPV because of the increasing recombination of ions at the decreasing electric field.

Significant Observations in Advanced Researches Relating to Radon EIC

Reference [14] gives a novel method of the concept of reusable electrets. This is applicable to the measurement of X-rays using EIC not usable for radon. Reference [15] gives theoretical model for calculating the responses for radon EIC. This is of significance for radon EIC. The methods provide a theoretical basis for calculating approximate responses, finally trusting on the experimental results. Reference [16] is a very useful to calculate the response for different dimensions of the chambers.

It also analyzes the commercially available EIC chambers [12] in comparison with the optimized models. This concludes that commercially produced models are close to the optimized dimensions and can be marginally improved. Reference [15, 16] continues to analyze the optimized models for the effect of relative humidity and pressures. Experimentally determined effects for commercially produced EIC [37] are similar in most cases. Reference [19, 20, 21, 24] applies the theoretical models for designing EICs for photons and neutrons, not applicable for radon EIC.

Reference [22, 23] analyzes the surface charges on the electrets, both before the use and after the use of commercially produced electrets. The study concludes that surface charges tend to become more uniform after the usage of EIC. Recent experimental study [13] shows that the equation for calibration factor applies equally well down to the electret voltages of 70 volts, discounting any effect on surface charges of electrets.

Reference [27, 28, 29, 30, 31, 33, 34, 35] reports further studies on the charge distribution on electrets and electro static fields inside the EIC and possible applications for characterizing radioactive sources used in medical applications. These studies are fundamental in nature and are useful in further researches. Reference [38] discusses the feasibility of the use of EIC for radon measurement.

One particular work [32] is of some interest in further developments. This deals with the radiation induced conductivity (RIC) of Teflon in electret ion chambers. This is of a particular interest while using EIC for characterizing X-rays. In such EIC unit, electret not only collects the ions produced inside the EIC chambers, but also directly irradiated by X-rays.

Such direct irradiation may cause some recombination of charges on the surface of electret causing additional electret discharges.

Such effect is also found to be dose rate dependent. Such effect may be of interest in characterizing alpha and beta sources. Further if one uses an EIC of very small volume for exposure to X rays, the possible measurement of radiation induced discharge of electret itself can be used for measuring high levels of X and gamma radiations.

Experimental Elevation Correction Factors for Commercially Available EIC

Radon monitors are usually calibrated in a standard radon test chamber at sea level. When such detectors are used at elevations other than the sea level, the results need a correction. The geometries of EIC radon monitors are complex and theoretical calculation of the correction factors is difficult, though attempted. The correction factors are determined using a radon source inside an experimental chamber that can be maintained at different pressures over extended time periods. The correction factors are determined [37] for three different models of commercially available EIC (E-PERM) for elevations up to 3000 meters. Theoretical results do agree in general trend, but not agree very well with the results in these experimental studies. This is because of limitations of several assumptions made in theoretical modeling. Manufacturer of the commercially available EIC recommend using the experimentally determined factors.

Techniques for Measuring Short Duration High Radon Concentration [85]

EICs are normally standardized for measuring relatively low radon concentrations usually found in homes and outdoor environment. However, there are situations when very high levels need to be measured. Examples of such situations are: unventilated uranium mines, radon therapy caves, waste storage silos and acute inhalation experiments. Concentrations of 50,000 and 100,000 pCi/L (2 to 4 MBq/m3) are not uncommon as found in radon in underground soil and in radon therapy caves. At such high levels exposure periods have to be small to avoid full discharge of electrets. Such situations do does not allow equilibrium to reach between radon and the associated decay products inside the chamber.

For this reason, the standard calibration factors will not give correct results. The suggested [85] correct procedure is to expose the detector to radon for desired exposure period, then move the detector to low radon area. Measure the final voltage of electret after a period of at least 3 hours. The deficiency in the ramp up time is compensated during the ramp down time. This procedure has been tested and found to give good results. This method is also suggested by Kunzamann [17].

Use of EIC As Personal Dosimeters for Radon Workers

Radon professionals who perform indoor radon measurement or radon mitigation require to keep a record of the personal radon exposure during their occupational work. For such a purpose it is desirable to use a personal radon dosimeter. Ideal such dosimeter should be an accurate integrator capable of rapidly responding to the varying radon concentrations, sensitive enough for a practical use and small to wear for the workers. A 50 ml EIC (LST) and a standard AT detector meet these specifications. Houle [84] devised an experiment to compare these two devices for their performances.

In this study triplicate EICs and ATs from five different manufacturers were exposed repeatedly to 60 minutes and then low radon air for 120 minutes in a specially designed radon chamber. Radon was provided by a Pylon radium source with a measured air flow through the source. Length of exposure was carefully recorded.

Grab samples of the test chamber were taken to confirm the radon concentration of the chamber. A second run was made exposing detectors to 15 minutes for radon and 120 minutes for low radon air. Results are expected to show the effectiveness of the devices as a personal dosimeter. Radon concentration of air used as radon air was about 230 pCi/L (8500 Bq/m3) and the room air was used as low radon air. Most of the AT detectors showed a higher response by about 50%. These can be used as personal dosimeters keeping in mind the possible positive bias. Author was unable to ascertain the possible reasons for this high positive bias. Only EIC detectors responded with in 3% of the expected integrated concentrations. These can be used as personal dosimeters with confidence.

Performance of EIC in Magnetic Field

EIC are used for measuring radon and other ionizing radiations caused by alpha, beta, gamma and neutrons. EIC is an ionization chamber that uses the electrostatic field provided by the electret. EICs do not have any electronic components and can be introduced into the area where high magnetic fields are present such as in accelerator areas. A systematic study [93] is conducted in the laboratories of National Institute of Standards and Technology. Relative responses are measured with and without the magnetic fields. Magnetic fields are varied from 100 to 9000 gauss. No significant effect is noticed while measuring alpha and gamma radiations, However a significant effect is found while measuring beta radiation from ^{90}Sr-Y. Depending upon the magnetic field orientation, the relative responses increased from1.0 to 2.7 in vertical orientation. Responses decreased from 1 to 0.6 in the magnetic fields in horizontal position. This is due to the setting up of a circular motion for the electron by the magnetic field, which may increase or decrease the path lengths of electrons. It is concluded that EIC can be used for measuring alpha (and hence radon) and x and gamma radiation in the range of magnetic field studied. However caution must be exercised while measuring beta radiation.

Inherent Discharge of Electrets

Stability of electrets used in EIC are well studied [10]. In spite of their high stability, electrets show some finite rate of discharge even when not exposed to ionizing radiations. This is called as inherent discharge of electrets. For example ST electrets show a discharge rate of 1 to 2 volts month. This is very small compared to the discharges that occur while measuring indoor radon. When encountered special cases such as measuring very low radon concentration using small volume (50 ml) EIC, the inherent discharge may contribute significantly for the final results. Manufacturer recommends a correction for such discharges. Analysis algorithms provided by the manufacturer such corrections are already built in.

Radon Metrology Using NIST ^{222}Rn Emission Standards

Development of radon emission standards has been an important contribution to the field of radon metrology.

For the first time these standards are usable for calibrating radon monitors in our own laboratory with traceabile to national standard. These are hermetically sealed polyethylene capsules filled with ^{226}Ra solution. These are available in four strengths: 5,50,500, and 5000 Bq of ^{226}Ra. The standards are specified both in terms of strength of radium and the emanation coefficients Volkovitsky [95]. These two parameters are sufficient to calculate the expected rate of radon emanation rate. These can be used in two modes: accumulator mode and steady state mode. Equation (7) gives the expected radon concentration at delay of T days when enclosed in an enclosed container, termed as accumulator.

Use of NIST Standard Used in an Accumulator Mode (Figure 11 in Part 1)

This is fully described by Kotrappa [56]

Equation (7) gives the radon concentration at any accumulation time T days.

$$RnC = \frac{f\ Ra\ (1 - \exp\ (-k \times T)}{(V)} \tag{7}$$

where RnC is the radon concentration in Bq m^3, in an accumulator of volume V (m^3)

Ra radium content in Bq

k is the decay constant of radon in day^{-1} units

T is the accumulation or elapsed time in days

f is the emanation coefficient characteristic of the source as provided by NIST.

If a CRM (continuous radon monitor) is enclosed in an accumulator, it should read the radon concentration as given by equation (7) for an accumulation time of T days

If a passive devices such as an EIC or AT device need to be calibrated, equation (7) need to be integrated from 0 to T days to provide the average concentration of radon after an accumulation time of T days. This is given by equation (8)

$$RnC = \frac{f\ Ra}{V}(1 - \frac{(1 - \exp(-k\ T)}{(k\ T)}) \tag{8}$$

Kotrappa [56] used this method to calibrate the EICs and a CRMs for radon.

Use of NIST Standard Used in a Steady Flow Mode

This is fully described by Kotrappa [57]

When air flows over the source at a known steady rate, it is possible to calculate the radon concentration in air. Equation (9) gives the radon emanation rate.

$$RnC = [f \cdot Ra \cdot k] / F \qquad --- \tag{9}$$

where RnC is the radon concentration in air Bq m^{-3} . f is the emanation fraction. F is the air flow rate in m^3 s^{-1}. Ra is the radium activity in Bq. k is the decay constant of radon in s^{-1}.

A practical one meter cube radon test chamber is designed and operated bases on the use of NIST standard used in a steady flow mode [94].

Another example is the calibration of radon flux monitor using NIST emanation standard [68].

EIC for Radon in and around Volcanoes

Because of their unique features, these have found use in and around active and passive volcanoes. The studies by Heiligmann [48, 49] published in Journal of Volcanology and Geothermal Research is of special interest.

Manufacturers Manuals

Standardized version of electret ion chambers, also called as E-PERM$^{(R)}$ are manufactured by Rad Elec Inc. Most of the applications are also developed by Rad Elec Inc. in association with several Institutions and Universities. Currently Rad Elec Inc is the only organization manufacturing and supplying the standardized EIC system. Rad Elec Inc. has the responsibility to supply operating manuals. The web site www.radelec.com provides the basic information needed including several basic applications and some downloadable publications.

PART 2. USEFUL APPLICATIONS OF RADON EIC

1. EIC for measuring American national ambient radon concentration
2. EIC for the measurement of thoron and the use of passive radon thoron discriminative monitors
3. EIC for measuring radon decay products in air using E-RPISU (electret -radon progeny integrating sampling unit)
4. EIC based radon flux monitor for *in situ* measurement of radon flux from ground, granite, concrete surfaces and for uranium exploration
5. EIC for uranium prospecting for accurate time-efficient surveys using EIC based radon gas monitors
6. EIC for uranium prospecting for accurate time-efficient surveys using EIC based radon flux monitors
7. EIC for characterizing soil and building materials
8. EIC for the measurement of dissolved radon in water

1. EIC for Measuring American National Ambient Radon Concentration

This is an important study [89, 41] carried out by US EPA (United States Environmental Protection Agency). This is of national as well as geological importance. This study covers the quarterly measurement of radon in all 50 US States for a two year period. The second phase covered 50 locations in one state (Nevada) covering various geologies. The detectors were located in well ventilated shelters located approximately two meters above the ground.

The gamma background needed for subtracting its ionization component was measured using co-located TLDs for identical sampling periods. To minimize the statistical error, a set of three detectors were used for each measurement. This study concluded that field measurements using SST E-PERM EICs have been made with acceptable errors and the devices have exhibited sufficient sensitivity for measuring ambient levels of radon concentrations. Coefficient of variations for the mean estimates were well below the 50% required for reproducibility, except for New Mexico (88%) because of its low mean value (0.16 pCi/L or about 6 Bqm^{-3}).

For those who need more details, please refer to the original publications.

The SST type of E-PERM EIC was chosen because of the following reasons:

1   The performance is not affected by varying temperatures and humidity found in all the locations in the US.

2   A study comparing the outdoor radon concentration measured by EIC and a calibrated RGM (radon gas monitor) agreed within 5 % of each other, over extended periods.

3   A study comparing the radon concentration measured by EIC and RGM in a radon test chamber agreed similarly.

4   USEPA confirmed that the Manufacturer is likely to be in business over the next decade, to assure supply of equipment, if the study needed to be repeated.

5   Limit of detection of radon, as determined independently by USEPA, is 2 Bq m^{-3}

The following observations are noted based on these studies:

Results ranged from 2.2 to 41.1 Bqm^{-3} for individual results with a median value of 14.4 Bq m^{-3}. This level is now officially considered to be the national ambient average radon concentration in the US. The annual mean ranged from a low of 5.9 to a high of 20.7 Bqm^{-3}.

The second study [41] has several interesting observations attempting to correlate outdoor radon concentration with local geology:

The mean outdoor radon in Nevada is comparable (15 Bqm^{-3}) to that observed in Nevada in a national study. However, the range is considerable (2.6 to 52 Bqm^{-3}). There appears to be some correlation with the radon soil gas data and the outdoor radon concentration. The towns with more than 20% of indoor radon concentration above the EPA action limit of 48 Bqm^{-3}, appear to correlate with higher outdoor radon concentrations.

The protocols used in this study have become a model for similar measurements.

2. EIC for the Measurement of Thoron and the Use of Passive Radon Thoron Discriminative Monitors

There are two major isotopes of radon gas, one is ^{222}Rn (usually called radon) that is released from uranium and the other is ^{220}Rn (usually called thoron) released from thorium. In view of the presence of both uranium and thorium in the earth crust, both radon and thoron are found in nature and in homes. Radon has a half life of 3.8 days, whereas thoron has a shorter half life of 55 seconds. In homes, radon is found in much higher concentrations than thoron because of the differences in half life. This may not be so in some regions, thorium is much more abundant than uranium in the ground [91].

There are some special situations where there can be high airborne thoron such as locations where thorium is processed or stored.

The radon measuring EIC is designed with a small diffusion inlet area leading to the diffusion time much longer (more than 5 minutes) than the half life of thoron. This minimizes the entry of thoron. Normally radon EIC has about 3% response to thoron and is considered insignificant, therefore the EIC is considered as a pure radon detector. This was confirmed in Japan [25], and was designated as R EIC. Thoron measuring EIC [45, 86] is designed with larger diffusion inlet areas (Figure 8) leading to the diffusion time much shorter than the half life of thoron to allow thoron to get into the sensitive volume. Such thoron EICs also register radon and some thoron. This is termed as R-T (radon-thoron). A pair of detectors (R and R-T) are needed to make a thoron measurement. The differential signal between R and RT units is used for calculating thoron concentration. Such a pair is called as "discriminative set used to discriminate and measure both radon and thoron".

Once calibrated properly, it is possible to calculate both radon and thoron concentration, when a mixture of these gases are present in the sampling atmosphere. Calibration is done using either a standard thoron test chamber [86] or a test chamber monitored by standard thoron measuring continuous monitors such as Rad 7 [45]. Different sets are now available in different volumes, 58 ml, 210 ml and 960 ml. Such availability extended the range and sensitivity for monitoring from a few days to one year. Vargas [51] found that the 210 ml set gave satisfactory results in one of the inter-comparison studies.

Equation (10) is used for the calculation.

$$Tn = \frac{(I-J)}{CF(I,J) \times D} - \frac{(K-L)}{CF(K,L) \times D} \tag{10}$$

where I and J are the initial and final electret voltages of RT monitor. K and L are the initial and final electret voltages of R monitor. D is the exposure period in days. Tn is the thoron concentration in units of $kBq\ m^{-3}$. CF is the calibration factor in units of volt drop per ($kBq\ m^{-3}$ days) for the corresponding pair of data.

Equations connecting the CF with MPV are available for all the three EIC sets of discriminative monitors. The detailed procedures of using these EIC sets to measure both radon and thoron concentrations are fully described [45, 86].

The EICs come in different volumes, providing different sensitivities. The thoron calibration factors for 58 ml, 210 ml and 960 ml volume R and RT pairs are respectively 2.8, 18.7 and 89 volts drop per ($kBq\ m^{-3}$-days) respectively.

These provide much wider sensitivities and ranges compared to alpha track based passive radon-thoron discriminative monitors.

Attenuation of Thoron in Membranes

In many cases both radon and thoron are present in the environment to be sampled. These two gases not only have different half lives, but also have different biological properties, with different action limits. While measuring indoor radon, thoron is an interference and should be stopped from entering the sensitive volume of true radon monitors. The methods used for stopping thoron usually take advantage of the differences in half lives of radon and thoron. Any such method should not stop radon, only thoron.

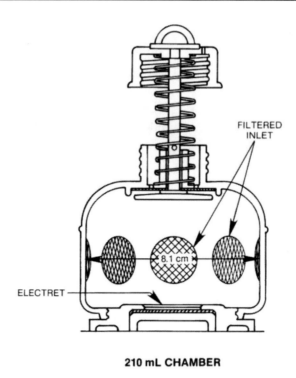

210 mL CHAMBER

Figure 8. Schematic of 210 ml thoron (R-T) measuring EIC.

Recently a study (Leung, 2007) used a thin layer (5 to 6 µm) of polyethylene (PE) in front of the passive entry of gas into the sensitive volume of the passive radon monitors. This stopped 92% of thoron, but allowed more than 98% of radon to go through to the sensitive volume. This is considered adequate for most passive radon monitors. It is easy to explain the performance of PE, based on the differences in the half lives between radon and thoron. The time taken to diffuse through PE is the same for both radon and thoron, but that diffusion time is very small relative to the half life of radon leading to insignificant decay of radon during the passage, whereas it is significant relative to the half life of thoron leading to the significant decay of thoron. Smaller or larger thicknesses of PE will not function satisfactorily. Leung optimized and demonstrated that 5 to 6 µm thick PE works satisfactorily. There are other methods of achieving the same results. In electret ion chambers (Kotrappa, [12] radon enters through a small opening. By controlling the ratio of the diffusion area to the sensitive volume, it is possible to control the diffusion entry time, thus stopping or minimizing thoron interference. The EIC for radon responds only to 3% of thoron while fully responding to radon. Another method used in flow-through radon monitors is to have a long loop of tube to allow the decay of thoron before entry into the sensitive volume. Even though PE works, it has some practical limitations. The membrane is very thin and electrostatic. It is difficult to position this on the inlet of the radon progeny filter in a stretched condition, and sealing the edges with an adhesive can be quite challenging. This may introduce uncertainty in the performance. In recent work [65] Tyvek membranes are studied for the attenuation of thoron. These have several advantages. Tyvek membranes of standard thicknesses, well defined properties and complete transparency to radon, are available commercially. These are antistatic and have relatively larger thicknesses for handling and sealing.

In this work [65], different layers of Tyvek membranes are introduced between the thoron source and the thoron detector, and thoron attenuation is measured, leading to attenuation of thoron for different thicknesses of Tyvek. The results can be used to control the thoron attenuation factors. 960 ml and RT pairs are used to measure thoron attenuation in different thicknesses of Tyvek membranes. A stack of 7 membranes (1 mm thick) reduces thoron by 50%, and a stack of 31 membranes (4 mm thick) reduces thoron by 95%. There is virtually no decay of radon for the Tyvek membranes even with 31 membranes (4 mm thick) whereas a stack of 31 membranes reduced the thoron concentration by 95%.

Applications Attenuators

A stack of Tyvek membranes can easily be built to be used as a thoron attenuator without attenuating radon. Such a stack can be inserted at the passive entry of any passive radon monitors such as AT monitors or other similar radon monitors. One of the important applications is in uranium exploration projects. 960 ml electret ion chambers are widely used in uranium exploration projects in Canada, (Charlton [96] and elsewhere. The procedure results in mapping of radon concentrations on the ground to indentify radon anomalies (ups and down) to locate where to look for uranium. It is also important to make sure that such anomalies are not caused by thoron. Using EIC radon monitors with a thoron attenuating stack of Tyvek in parallel with a regular radon monitor will prove whether the signal is due to thoron or not. Any uncertainty in uranium exploration work is solved.

3. EIC for Measuring Radon Decay Products in Air Using E-RPISU

Health effects of radon are attributed to the inhalation and deposition of the airborne decay products. Usually the concentration is referred to as potential; alpha energy concentration and is expressed in the units of working levels. Please see Figure 1 for the schematic views of the monitoring head. An air-sampling pump (0.5 to 2 liters per minute) is used to collect the radon progeny for a known sampling time on a 3.5 cm^2 filter sampler mounted on the side of an electret ion chamber. The flow rate can be adjusted for a desired flow rate. Recommended flow rate is 0.5 to 1 liter per minute. Sampling duration is usually between 1 to 7 days. The filter paper is mounted such that the progeny collected emit their radiation into the interior of the chamber. The alpha radiation emitted by the progeny collected on the filter ionizes air in the electret ion chamber. The ions are continuously collected by the electret, providing integrated alpha activity collected on the filter paper. The standard electrets and the readers used in the standard radon EIC System can be with this unit. The initial and final readings of the electret, the flow rates and the duration sampling (as recorded by the time totalizer) are used in a software to calculate progeny concentration in WL units. Software is provided for data analysis by the manufacturer.

These monitors have sufficient sensitivity to provide results with better than 10% precision at 0.01 WL for a 2 day measurement when used with LT electrets. These provide better than 10% precision at 0.001 WL, when used with ST Electrets. The sampling heads are designed to minimize the deposition losses during sampling.

First part of Figure 9 shows the schematic of the EIC radon progeny integrating sampling unit (E-RPISU) in off position for storage or shipping. The second part shows the schematic of the E-RPISU in the sampling mode.

E-RPISU™
(Electret-Radon Progeny Integrating Sampling Unit)
Schematic

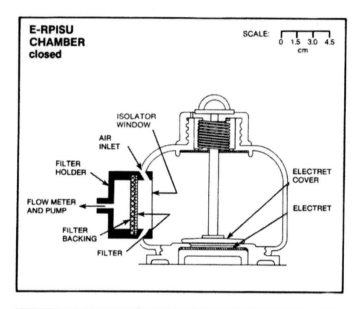

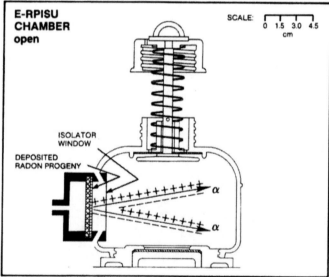

Figure 9. Schematic of EIC radon progeny monitor in air (E-RPISU).

A commercial unit that combines E-PERM and E-RPISUTeflon® is available (Rad Elec Inc.) for simultaneous measurement of radon, radon progeny, unattached radon progeny concentrations and equilibrium factors over extended periods. This also provides an accurate measurement of radon for determining equilibrium factors.

Development of this device was called "Radon Progeny Integrating-Sampling Unit" (E-RPISU®) was partially supported by US Department of Energy at Grand Junction, CO.

A scientific paper was published [97] The E-RPISU® Unit was evaluated by the USEPA, listed for use, and was used in several Radon Monitoring Proficiency Programs. Following is an extract from the USEPA Evaluation Report "During the evaluation, the E-RPISU® evolved as a viable and reliable instrument that may be used to measure radon decay products in units of WL with accuracy comparable with other decay product monitors when using the USEPA protocol. The unit did not show sensitivity to relative humidity or differing equilibrium conditions used in the Test Chamber. The E-RPISU® was well within 25% variation maximum allowed by USEPA when compared to calibrated WL Monitors, never more than 12%" Units were also entered into DOE-EML Inter-comparison project (EML-527, 1990) and showed results within 15 % of the target value.

The E-RPISU® is basically composed of an air sampling pump, a flow meter and an electret ion chamber with an appropriate filter holder and an electret. The protocols applicable to other progeny monitors are applicable to this unit also. Operating user manual is available from the manufacturer. The entire system needs to be operated simultaneously for the identical duration. This also provides an accurate measurement of radon for determining equilibrium factors. The software does such analysis. One of the methods of measuring unattached decay products of radon is to sample air through a wire mesh. Two E-RPISU units can be run together, one with the filter and the other with the mesh. Calculate radon progeny concentration on both using the same procedure. The ratio of the result with mesh to that of filter gives the unattached fraction.

4. EIC Based Radon Flux Monitor for *In Situ* Measurement of Ground, Granite, Concrete Surfaces and for Uranium Exploration

Recent interest in radon (^{222}Rn) emanation from building materials like granite and concrete has sparked the development of a measurement device that is suitable for field or home measurements. Based on tests with discrete component flux monitors, a large volume (960 ml) hemispherical electret ion chamber (EIC) was modified to work both as the accumulator and detector in a single device. Usually the flux monitors have two components, one component is an accumulator and the other component is a radon detector. The device entrance is covered by a carbon coated Tyvek sheet to allow radon from the surfaces to be characterized into the EIC chamber. An optional 4 mm thick Tyvek membrane can be introduced to minimize the response to thoron in the EIC. This device is calibrated with a NIST radon emanation standard whose radon emanation rate is precisely known. Side-by-side measurements with other emanation techniques on various granite surfaces in lab and field environments produce comparable emanation results. For low emitting building materials like concrete, a flux of 110 Bq m^{-2} d^{-1} (11 pCi ft^{-2} h^{-1}) can be measured with 10% precision using an ST electret for 8 hours. Sensitivities, ranges and applicable errors are discussed.

These can be used with a collar, if required as in the case of uranium exploration, or without a collar for measuring radon flux from ground or from concrete and granite surfaces. Manufacturer provides detailed user manual for different applications. A schematic figure illustrates different parts of the EIC flux monitors.

There has been an increased interest in the radon emanation from granite used for countertops and tiles in homes. Radon originates from naturally occurring uranium present in granites and other building materials made from stone.

The radon emanation rate of granite, along with the home's volume and ventilation rate determines its contribution to room radon. The radon emanation rate is also called the radon flux and is defined as radon activity released per unit area per unit time. While the appropriate scientific flux unit would be Bq m^{-2} s^{-1}, units like Bq m^{-2} d^{-1} and pCi ft^{-2} h^{-1} are easier to use in practice. Results are provided in these units with some conversion factors for other units in the text.

This monitor uses a well known "accumulation theory" of determining the radon flux from the objects (Kotrappa, 2009). The entire volume of the chamber serves as an accumulator and the electret ion chamber (EIC) part serves as a radon monitor. Equation (11) relates the average radon measured by the EIC and the radon flux F when the accumulator has no radon losses other than radioactive decay.

$$C(Rn)Av = \frac{(F \times A)}{V \times 0.1814}\left[1 - \left(\frac{1 - e^{-0.1814T}}{0.1814T}\right)\right]$$

(11)

Notation:

F is the radon flux in Bq m^{-2} d^{-1}

A is the area of the granite measured in m^2

A is also the area of the radon flux monitor window in m^2

(F × A) is the exhalation rate in Bq d^{-1}

0.1814 is the decay constant of radon in d^{-1}

C(Rn) Av measured by EIC, is the integrated average radon concentration in Bq m^{-3}

T is the accumulation time in days

V is the air volume of the accumulator in m^3

The radon flux F calculated using equation (11) leads to a flux in units of Bq m^{-2} d^{-1}.

Manufacturer provides spreadsheet for the analysis.

These have been used widely by radon measurement professionals and have shown acceptable results. These were used in an IAEA Inter-Comparison exercises by Vargas [46] published under the title: "Inter-comparison of different direct and indirect methods to determine radon flux from soil".

The study shows that the EIC worked properly and are in good agreement with other direct methods, both integrated and continuous. Due to condensation problems in certain environmental conditions, it is recommended that the EIC should not be used in such environments. The study carried out during solar hours provided better results. For this study, the EIC without collars were more appropriate. It can be concluded that EIC flux monitors are appropriate in field campaigns due to their performance, mobility and price. The flux monitors with collars are most appropriate during uranium exploration projects, which are done during 6 to 8 hours during day hours.

Inter-comparison study done by Kotrappa and Steck [67, 68] gave acceptable results when used on granites. These were the EIC flux monitors without collars and without thoron filters. Appropriate user manuals are available from the manufacture for different applications of the EIC radon flux monitors.

Several users have used these flux monitors for their studies [54, 55, 59, 60, 67, 68, 70, 71, 75, 78, 82].

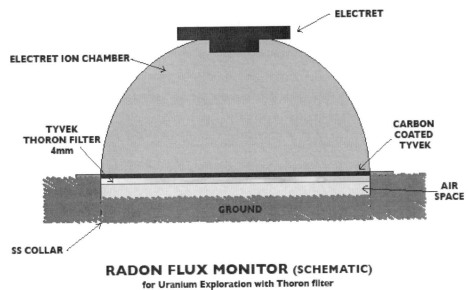

Figure 10. EIC based radon flux monitor.

Some investigators have used passive flux monitors in their work. These are similar to the currently used flux monitors (Figure 3), but are vented (four filtered holes). These were calibrated on uranium tailing beds [54] producing known radon flux. Note that accumulator type (figure 3) do not require calibration. Some workers found that the radon flux measured from ground does not agree with the measurements done with calibrated passive devices. Where relative values are needed the passive types have some advantages.

5. EIC for Uranium Prospecting for Accurate Time-Efficient Surveys Using EIC Based Radon Gas Monitors

Measurement of radon gas above the ground is one of the standard practice for uranium prospecting. Until now only passive detectors that can be used for such surveys is to use alpha track detectors which need several months for the results. When radon survey was done more than 20 years back, the techniques (AT techniques) were not that sophisticated and lots of data had to be rejected to draw some general conclusions [96]. With the availability of large volume EIC it is possible to measure radon concentrations in about three days and the results are arrived in the field, with good sensitivity, better than 0.5 pCi/L (about 20 Bq/m3). Recently a large scale survey was made in Canada [96] and concluded that it is possible to perform uranium prospecting for accurate time-efficient surveys of radon emissions in air is possible. This study was aimed at previously surveyed area to compare results with a comparison to earlier radon and He surveys using EIC radon monitors. The present efforts were aimed at getting large number of results in exploration setting in 3 to 4 days with an acceptable accuracy. Large volume (960 ml) Electret Ion Chambers (EIC) was chosen. Exposure duration was chosen to be about 3 days. Same staff members responsible for deployment and retrieval were able to do the analysis and complete the report within one week. This has sufficient sensitivity to measure 0.5 pCi/L (about 20 Bq/m3) in 3 days.

In addition, the EIC technology was also available for measuring dissolved radon in water in streams. Rad Elec's standard method is also usable for measuring radon in water samples collected from streams. Such measurements were also a part of prospecting.

Protocol

Large volume (960 ml) EIC (HST EIC) is enclosed in a Tyvek Bag. It is lowered into a small pit 2 feet x 2 feet x 6 inches. The pit is covered with another Tyvek sheet and edges were tied down with soil and pebbles in that area. The latter step prevents the cover to fly out. Tyvek, being transparent to radon, the EIC sees the radon in the surrounding areas. Analysis was done using standard spread sheet to calculate the results in any required format and units. Assuming the gamma level of 10 µR/h, the radon calculations were done. This is not exact because gamma background can be higher at certain locations due to the presence of uranium. Since gamma signal over and above normal background of 10 µR/h is a positive signal for the presence of uranium; combined signals are better suited for the purposes of uranium prospecting. Following is the summary of the report:

"Ur-Energy Inc. is focusing its exploration efforts on discovery of an Athabasca-style unconformity-associated uranium deposit in the Thelon Basin, NWT, and Canada. This work describes the exploration methodology and results to date of electret ion chamber (EIC) radon surveys employed in the summer of 2005 in the search for a deeply buried, high grade uranium deposit at Ur-Energy's Screech Lake Property. The survey has demonstrated that accurate radon gas measurements covering sizeable surface areas can be obtained within three days under variable summer field conditions. A total of 433 ground-EIC measurements were completed over a grid measuring 3 km by 1.5 km and centered on Screech Lake. In addition, 26 water samples from streams and lakes in the area were measured for radon content. Current work has confirmed, extended, and refined historic results obtained in past explorations that included radon and radiogenic helium surveys. The work was done using commercially available 960 ml HST EIC radon monitors. Additionally new radon anomalies have been discovered. This work has rekindled interest in using practical, short duration radon surveys for uranium exploration". (Ref) This study not only confirmed the earlier studies, but identified other locations due to better resolutions.

6. EIC for Uranium Prospecting for Accurate Time-Efficient Surveys Using EIC Based Radon Flux Monitors

Even though the methods described in previous section (5) was perfectly suitable for large scale surveys, manufacturers of EIC came up with another device for measuring radon flux from the ground. Radon flux is a better and faster index of radon emanation from the ground. Further such measurements can be done in 6 to 8 hours on the same day.

This method has all of the advantages of radon gas surveys, but is a more efficient, better quantifier of radon emanation from the ground. This has become the standard method used in recent uranium prospecting work.

Figure 10 illustrates the way a radon flux monitor with stainless steel collar can be deployed for uranium prospecting.

Protocols

Radon flux monitors with collars are fitted into the ground and sides are covered with the soil (Figure 3). Several hundreds of these can be anchored in 1 to 2 hours. At the end of the day these are taken out and assessed in the night. Such surveys saved several days compared to the 4A method. These are more sensitive because the detectors are right on the ground to be measured, where as in method 4A the radon measurement is the area.

Following extract is from the web site of RadonEX (Canadian uranium exploration company):

EIC based radon flux monitors in conjunction with radon in water provided revolutionary assessment of the uranium exploration results.

A program using EIC technologies, on the ice of Patterson Lake led directly to the drilling discovery of Patterson Lake South (PLS) uranium deposit - the biggest mineral discovery of 2013 in the world.

Conclusions: EIC radon based is a powerful technology that can revolutionize the uranium prospecting technology. The current technique has advanced the methodology for uranium prospecting from several months to 8 hours.

7. EIC for Characterizing Soil and Building Materials

A four liter glass jar with rubber seals has been successfully used as a convenient sealable accumulator for characterizing radon EICs using NIST radon emanation standards [56]. This is simply done by enclosing a NIST radon emanation source and the radon EIC unit inside the jar and sealing the jar. Using the characteristics of the NIST source (radioactivity of Ra in Bq and the emanation factor of the source) it is possible to calculate the expected radon concentration at the end of a known accumulation period. Comparison of the calculated radon concentration from calibrated radon EIC and the expected theoretical radon concentration, leads to the verification of calibration factors for the enclosed radon EIC.

Calibration constants of radon EIC can be corrected based on the results. Equation (12) gives the relationship between various parameters.

$$C(Rn)Av = \frac{(Ra \times f)}{V \times 0.1814}\left[1-\left(\frac{1-e^{-0.1814T}}{0.1814T}\right)\right]$$

(12)

Notation:

F is the radon flux in Bq m^{-2} d^{-1}

(Ra × f) is the exhalation rate in Bq d^{-1}

0.1814 is the decay constant of radon in d^{-1}

Ra is the radioactivity of radium in Bq

f is the radon emanation factor

(Rn) Av is the integrated average radon concentration in Bq m^{-3}

T is the accumulation time in days

V is the air volume of the accumulator in m³

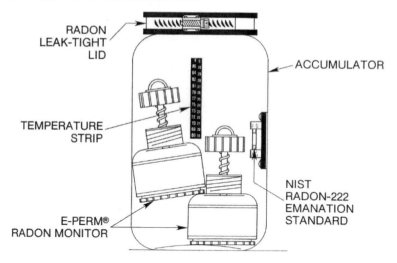

Figure 11. The accumulator system for calibrating EIC with NIST emanation standard.

Radon Emanating Radium Content for Soil Samples

In the place of NIST source, use the soil sample to be characterized. Measure the average radon concentration of radon after a known period of accumulation using a radon EIC. Knowing all the parameters in equation (12), solve for $(Ra \times f)$. This does not give radium content of the sample but it can be called as radon emanating radium content. If f is known then real radium content can be determined. In practice, 10 to 30 grams of soil is taken, then dried and ground to fine powder. The sample is then wetted with 20% water. Such a soil sample has an approximate f value of 0.2. Further divide $(Ra \times f)$ by f and the weight of the sample to calculate radium concentration of the soil sample in units of Bq/gram. This method is simple with no chemistry or radiochemical analysis involved. This was used by Heiligmann [48, 49] in his study on "distal degassing of radon and carbon dioxide on Galreras volcano, Columbia". Because of similarities of soil samples they just reported radon emanating radium concentration (Bq kg⁻¹) in soil samples of interest.

8. EIC for Characterizing Soil and Building Materials

Radon Exhalation Rates from Building Materials [40, 69]

Methods similar to methods used in section on Radon emanating radium content from soil, are usable for building materials such as samples of granite or granites. These are characterized in terms of radon flux from their surfaces. Equation (13) gives the relationship between different parameters when used for granites, enclosed in a 4 liter sealable glass jar.

$$C(Rn)Av = \frac{(F \times A)}{V \times 0.1814}\left[1-\left(\frac{1-e^{-0.1814T}}{0.1814T}\right)\right]$$

(13)

Notation:
F is the radon flux in Bq m^{-2} d^{-1}
A is the area of the granite measured in m^2
(F × A) is the exhalation rate in Bq d^{-1}
0.1814 is the decay constant of radon in d^{-1}
C(Rn) Av measured by EIC, is the integrated average radon concentration in Bqm^{-3}
T is the accumulation time in days
V is the air volume of the accumulator in m^3
The radon flux F is calculated using equation (13) that leads to a flux in units of Bq m^{-2} d^{-1}.

Enclose the sample and a radon EIC inside the sealable 4 liter jar. After the desired accumulation period analyze the radon RIC to determine the radon concentration. Calculate *(F × A)* from equation (13). Knowing the total surface area *A*, the radon flux *F* can be calculated in appropriate units. Kotrappa [40] has analyzed three samples for two different periods of accumulation (3 days and 5, 9 days). The results obtained were in the range of 20-30 Bq m^{-2} d^{-1} generally agreeing with the results reported in the literature. Methodology provides error analysis and lower methods of detection (LLD). For a two day accumulation period, LLD works out to be 7 Bq m^{-2} d^{-1}.

9. EIC for the Measurement of Radon in Water
EIC for the Measurement of Dissolved Radon in Water

Measurement of radon in water is important because of several reasons. Quite a large population use well water for drinking and other daily uses.

When water goes through geological rocks containing sources of radon such as uranium, it picks up radon that remains as dissolved radon in water in the wells.

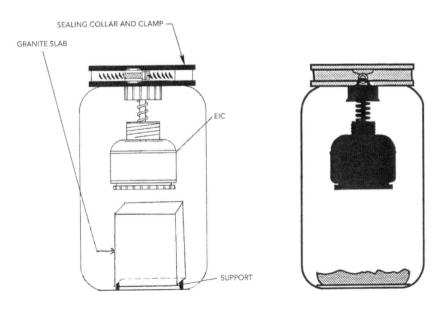

Figure 12. Left: EIC: for characterizing building materials Right: for charactering soil samples.

Just like radon in air in homes can vary from home to home, concentration of dissolved radon in water can vary home to home depending upon the geology of the location. Such water when used for showering, laundering, and other purposes, contributes to the radon in air. The USEPA has estimated that the dissolved radon concentration of 10,000 pCi/L leads about 1 pCi/l in air. Further, drinking such water can also have some biological effects on people. Normally, the concentration of dissolved radon in well water can vary from 1000 to several million pCi/L. While doing geophysical exploration for uranium, radon water steams have led to the discovery of uranium sources. Dissolved radon in water has been used as a tracer for many interesting research studies. The EIC method belongs to the general class of "de-emanation method" of the bottom of a glass measurement measuring radon in water. A small water sample of a known volume is placed in jar of a known volume. A radon EIC is suspended in the air phase above the water. The lid of the measurement jar is closed and sealed to make it radon-tight. Radon reaches equilibrium between the water and air phase. At the end of the desired exposure period, the measurement jar is opened and the EIC is removed. The average radon concentration in the air phase is calculated using the standard EIC procedure. A calculation using this air concentration in conjunction with the other parameters gives the radon concentration of the water. EIC methods are widely used as indoor radon monitors and are not affected by 100% relative humidity, and because of this property, it is possible to do in situ measurement of radon in water.

Kotrappa and Jester [43] have described the theory and practice of measuring the radon concentration in water using the EIC method.

Equation (14) is used for the calculation of dissolved radon concentration in water.

$$C(Rn)W = \frac{(Av\,C(Rn)A \times k\,T \times (\frac{VA}{VW} + 0.26)}{\exp(-k\,D) \times (1 - \exp(-k\,T))}$$

(14)

Notation:

$C(Rn)W$ Radon concentration of water sample, Bq L^{-1}

$AvC(Rn)\,A$ Average radon concentration in air as measured by radon EIC, Bq L^{-1}

D delay time between time of collection to start of analysis in days

T Analysis time between start of analysis to the end of analysis

k Decay constant of radon in day $^{-1}$

VA Volume of air in liters

VW Volume of water sample in liters

0.26 is the Oswald Coefficient (ratio of the radon concentration in the liquid phase to the radon concentration in air phase), normally neglected for small volume samples (50 to 200 ml samples).

The only parameter that needs to be measured is $AvC(Rn)\,A$ the radon concentration in air inside the measuring jar.

The method is schematically shown in Figure X. Water sample in glass bottle is held at the bottom of the jar in a clip. Radon EIC is inserted into the jar. Jar is sealed and analysis bottle is lifted up.

Water spills and releases radon into the jar. Radon EIC measures the average radon concentration in air inside the jar. Kotrappa and Jester [43] provides detailed analysis of errors and minimum methods of detection.

Additional Researches on This Topic

Dua and others [47, 87] adapted the method to make continuous measurement of dissolved radon in water. The USEPA [52] did a detailed evaluation of the EIC method in comparison with the standard liquid scintillation (LS) method for measurement of different concentrations. They found that the EIC method consistently gave 15% higher results compared to the LS method and recommended to apply this correction to be in agreement with the LS method. International users [61, 63, 64] have used the EIC method successfully in their projects.

The method was used for studying dissolved radon in water in different seasons [92] from the same well. Kitto [72] did an assessment of the EIC method in comparison with other methods. Kotrappa [80] also reviewed and compared different methods of measuring radon in water. References has listed several intersting uses for radon in water.

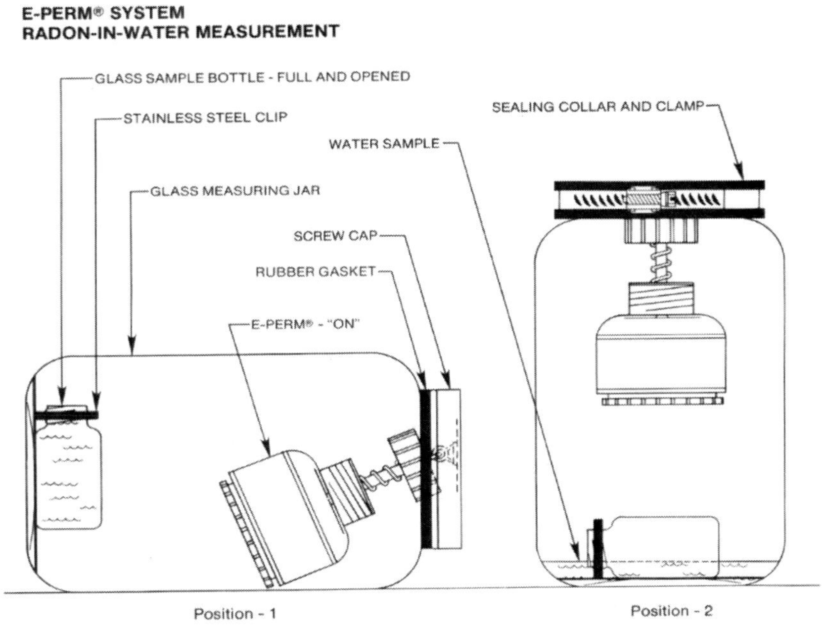

Figure 13. Schematic of EIC for measuring radon in water (change from position 1 to 2 to start the analysis).

ACKNOWLEDGMENTS

I am grateful to my associates at Rad Elec for their help in putting together this large document. In particular to Frederick Stieff, Alexandra Stieff, Prateek Paul (summer internee, now at Emery University) and John Davis. I also owe my gratitude to my past scientist associates, Lorin Stieff and John Dempsey, who contributed a great deal in the researches on EIC, as evidenced by being my associated authors in several cited publications. I am also thankful to my wife and associate Chandra Kotrappa not only in preparing this document but also for her contribution in researches leading to technologies needed for making electrets. My grateful thanks are due to Dr. Dan Steck and Andreas C. George, well known scientists in the field of radon and radon metrology, for reviewing this document and providing several useful comments to improve the document. Further they were kind enough to agree with me to include their summary report in the document.

Summary Comments from the Reviewers: Quotes

Comments by: Andreas C. George

Electret ion chamber technology has become an alternative and useful method for characterizing radon gas originating from various sources. The principle of detection of Electret Ion Chambers (EIC), has been investigated thoroughly in the past thirty years by several investigators and was found appropriate and useful for measuring environmental radon concentrations. The group headed by Dr. P Kotrappa researched and applied the EIC technology successfully to measure the concentration of radon indoors, outdoors and in the characterization of soil geology and building materials. The EIC technology has been proven to be very useful and practical in measuring the emanation of radon from soil surfaces. This particular use and those of other applications when used properly offer an alternate, less tedious and cost-effective method by providing results directly from reading the electret in the field. There is no need for sample transfer making the method direct and practical.

Here is my summary review: from Dan Steck Ph.D

This manuscript provides a thorough review of the electret ion chamber (EIC) technique and its practical applications. The description is detailed enough to be useful to radiation measurements specialists and basic enough to be useful to exploration geologists. Part 1 provides an excellent understanding of the underlying principle of radiation detection using EICs. Particular attention and detail are given for its use in airborne radon detection. Part 2 is a comprehensive description of the wide-ranging applications for radon and thoron measurements of concentration and flux from a variety of source material. Examples and documentation of performance are given for sophisticated laboratory uses as well as challenging field measurements. Part 3 provides almost 100 references for support and additional details. Overall, readers engaged in radon and thoron work will find this a useful work.

REFERENCES

[1] Edited by Sessler, G. M. Electrets, Third Edition in Two volumes, The Laplacian Press, *Series on Electrostatics*, Laplacian Press Morgan Hill, California 1998, 1, 41.

[2] Sessler,G. M. *ElectretsTopics in Applied Physics 33*, Berlin, Hedelberg, New York: Springfield-Verlag, Radiation Dosimeters 1980, 2, 25-26.

[3] Sessler, G. M., West, J. E. Production of high quasi-permanent charge densities in polymers foils of by application of breakdown fields, *Journal of Applied Physics* 1972, 43(3), 922-926.

[4] Bauser, H., Range, W. The electret ionization chamber: a dosimeter for long-term personnel monitoring. *Health Physics* 1997, 34, 97.

[5] Marvin, H. B. How to measure radiation with electrets, *Nucleonics* 1995, 13, 82.

[6] Kotrappa, P., Brubaker, T., Dempsey, J. C., Stieff, L. R. Electret Ion Chamber System for Measurement of Environmental Radon and Environmental Gamma Radiation, *Radiation Protection Dosimetry* 1992, 45, 107-110.

[7] Kotrappa, P., Dua, S. K., Pimpale, N. S., Gupta, P. C., Nambi, K. S. V., Bhagwat, A. M., Soman, S. D. Passive Measurement of Radon and Thoron Using TLD or SSNTD on Electrets, *Health Physics* 1982, 43(3), 399-404.

[8] Kotrappa, P., Dua, S. K., Gupta, P. C., Pimpale, N. S., Khan, A. H. Measurement of Potential Alpha Energy Concentration of Radon and Thoron Daughters Using and Electret Dosemeter, *Radiation Protection Dosimetry* 1983, 5(1), 49-56.

[9] Kotrappa, P., Dua, S. K., Gupta, P. C., Mayya, Y. S. Electret-A New Tool for Measuring Concentrations of Radon and Thoron in Air, *Health Physics* 1981, 41, 35-46.

[10] Kotrappa, P. Long Term Stability of Electrets Used in Electret Ion Chambers, *Journal of Electrostatics* 2008, 66, 407-409.

[11] Kotrappa, P., Dempsey, J. C., Hickey, J. R., Stieff, L. R. An Electret Passive Environmental ^{222}Rn Monitor Based on Ionization Measurement, *Health Physics* 1988, 54(1), 47-56.

[12] Kotrappa, P., Dempsey, J. C., Stieff, L. R., and Ramsey, R. W., A Practical Electret Passive Environmental Radon Monitor For Indoor Radon Measurements, *Health Physics* 1990, 58, 461-467.

[13] Kotrappa, P., Stieff, A., Stieff, F. Advanced Calibration Equation For E-Perm Electret Ion Chambers, *Proceedings of the 2013 International Radon Symposium*, 10-19 www. AARST.org listed under Radon Info, Radon research papers.

[14] Pretzsch, G., Dorschel, B. A Re-Usable Electret Dosemeter, *Radiation Protection Dosimetry* 1986, 12(4), 351-354.

[15] Kunzmann, S., Dorschel, B. Theoretical Model for Response Calculation of Electret Ionization Chambers for Radon Dosimetry, *Radiation Protection Dosimetry* 1996, 63 (4), 263-268.

[16] Kunzmann, S., Dorschel, B., Zeiske, U. Optimization of Electret Ionization Chambers for Radon Dosimetry on the Basis of a Theoretical Model, *Radiation Protection Dosimetry* 1996, 63(4), 269-274.

[17] Kunzmann, S., Dorschel, B., Havlik, F., Nikodemova, D. Studies on the Practical Application of Electret Ionisation Chambers for Radon Dosimetry Under Special Conditions, *Radiation Protection Dosimetry* 1996, 63(4), 275-279.

[18] Dorschel, B., Prokert, K., Seifert, H., Stoldt, C. Individual Dosimetry Using Electret Detectors, *Radiation Protection Dosimetry* 1990, 34, 141-144.

[19] Dorschel, B. Recent Developments in Detectors for Photon and Neutron Dosimetry, *Radiation Protection Dosimetry* 1990, 34, 103-106.

[20] Dorschel, B., Pretzsch, G. Properties of an Electret Ionisation Chamber for Individual Dosimetry in Photon Radiation Fields, *Radiation Protection Dosimetry* 1986, 12(4), 339-343.

[21] Dorschel, B., Seifert, H., Streubel, G. Some New Techniques for Neutron Radiation Protection Measurements, *Radiation Protection Dosimetry* 1992, 44, 355-361.

[22] Kunzmann, S., Dorschel, B., Hase, K., Zeiske, U. Studies on the Surface Charge Distribution of Radon Exposed Electrets within an Electret Ionization Chamber, *Radiation Protection Dosimetry* 1996, 66, 367-370.

[23] Kunzmann, S., Dorschel, B., Hase, K., Zeiske, U. Studies on the Surface Charge Distribution of Radon Exposed Electrets within an Electret Ionization Chamber, *Radiation Protection Dosimetry* 1996, 66, 367-370.

[24] Dorschel, B. Individual Dosimetry using electret detectors, *Radiation Protection Dosimetry* 1990, 34, 141-144.

[25] Sorimachi, A., Takahashi, H., Tokonami, S. Influence of the Presence of Humidity, Ambient Aerosols and Thoron on the Detection Responses of Electret Radon Monitors, *Radiation Measurements* 2009, 44, 111-115.

[26] Caresana, M., Campi, F., Ferrarini, M., Garlati, L., Porta, A. Uncertainties Evaluation for Electrets Based Devices Used in Radon Detection, *Radiation Protection Dosimetry* 2004, 1-6.

[27] Fallone, B. G., Podgorsak, E. B. Electrostatic fields in an ionization chamber electret, *Journal of Applied Physics* 1983, 54, 4739-4744.

[28] Fallone, B. G., Macdonald, B. A. Surface charge distribution during the charging and discharging of electrets produced in electret ionization chambers, *IEEE Transactions on electrical insulations* 1992, 27, 144-151.

[29] Macdonald, B. A., Fallone, B. G., Ryner, L. N. Feasibility study of an electret Dosimetry technique, *Physics in medicine and biology* 1992, 37, 1825-1836.

[30] Fallone, B. G., Macdonald, B. A., Ryner, L. R., Characteristics of a radiation charged electret dosimeter, *IEEE Transactions on electrical insulations* 1993, 28, 143-148.

[31] Fallone, B. G., Macdonald, B. A. Modeling of surface charge distributions in electret ionization chambers, *Review of Scientific Instruments* 1993, 64, 1627-1632.

[32] Macdonald, B. A., Fallone, B. G., Markovik, A. Radiation induced conductivity of Teflon in electret ionization chambers, *Journal of Physics D-Applied Physics* 1993, 26, 2015-2021.

[33] Macdonald, B. A., Fallone, B. G. Charge decay of electrets formed by ionizing radiation in air, *Journal of electrostatics* 1993, 31, 27-33.

[34] Macdonald, B. A., Fallone, B. G. Improved modeling of surface-charge distributions in electret ionization chambers, *Review of Scientific Instruments* 1994, 65, 730-735.

[35] Macdonald, B. A., Fallone, B. G. Effect of radiation rate on electret ionization chambers, *Physics in Medicine and Biology* 1995, 40, 1609-1618.

[36] Kotrappa, P. Performance of E-PERM™ electret ion chamber radon monitor at very high humidity, high temperature in the presence of microbial activity, *Health Physics* 2000, 228-228.

[37] Kotrappa, P. *Elevation correction factors for E-PERM radon monitors* 1992, 62:82-86.

[38] Stadtmann, H. Passive integrating radon monitors using electret ionization chamber, *Radiation Protection Dosimetry* 1990, 34, 179-182.

[39] Kitto, M. E., Haines, D. K., Arauzo, H. D. Emanation of Radon from Household Granite, *Health Physics* 2009, 96(4), 477-482.

[40] Kotrappa, P., Stieff, F. Radon Exhalation Rates from Building Materials Using Electret Ion Chamber Radon Monitors in Accumulators, *Health Physics* 2009, 97(2), 163-166.

[41] Price, J. G., Rigby, J. G., Christensen, L., Hess, R., LaPointe, D. D., Ramelli, A. R., Desilets, M., Hopper, R. D., Kluesner, T., Marshall, S. Radon in Outdoor Air in Nevada, *Health Physics* 1994, 66(4), 433-438.

[42] Kotrappa, P., Dempsey, J. C., Stieff, L. R. Recent Advances in Electret Ion Chamber Technology for Radiation Measurements, *Radiation Protection Dosimetry* 1993, 47, 461-464.

[43] Kotrappa, P., Jester, W. A. Electret Ion Chamber Radon Monitors Measure Dissolved ^{222}Rn in Water, *Health Physics* 1993, 64(4), 397-405.

[44] Kotrappa, P., Stieff, L. R. Elevation Correction Factors for E-Perm Radon Monitors, *Health Physics* 1992, 62(1), 82-86.

[45] Kotrappa, P., Steck, D. Electret Ion Chamber-Based Passive Radon-Thoron Discriminative Monitors, *Radiation Protection Dosimetry* 2010, 141(4), 386-389.

[46] Grossi, C., Vargas, A., Camacho, A., Lopez-Coto, I., Bolivar, J. P., Yu Xia, Conen, F. Inter-Comparison of Different Direct and Indirect Methods to Determine Radon Flux from Soil, *Radiation Measurements* 2011, 46, 112-118.

[47] Dua, S. K., Hopke, P. K., Kotrappa, P. Electret Method for Continuous Measurement of the Concentration of Radon in Water, *Health Physics* 1995, 68(1), 110-114.

[48] Heiligmann, M., Stix, J., McKnight, S., Williams, S. N., Dalva, M., Moore, T. Preliminary Soil Gas Measurements at Galeras Volcano, Columbia, *Santa Maria Decade Volcano Workshop* 1993.

[49] Heiligmann, M., Stix, J., William-Jones, G., Lollar, B. S., Garzon, G. Distal Degrassing of Radon and Carbon Doxide on Galeras Volcano, Colombia, *Journal of Volcanology and Geothermal Research* 1997, 77, 267-283.

[50] Kotrappa, P., Stieff, L. An advanced E-PERM system for simultaneous measurement of concentration of radon, radon progeny, equilibrium factors and unattached radon progeny" *Proceedings of the 2003 AARST International Symposium* 2003.

[51] Vargas, A., Ortega, W. Influence of Environmental Changes on Integrating Radon Detectors: Results of an Inter comparison Exercise, *Radiation Protection Dosimetry* 2007, 123, 529-536.

[52] Budd, G., Buntley, C. Operational Evaluation of the electret ion chamber method for determining radon in water concentrations, *International Radon Conference hosted by AARST (EPA reviewed paper)* 1993.

[53] Kotrappa, P. Radon Emanating Ra-226 Concentration from Soil, Radon from soil, *E-PERM System Manual* 2007.

[54] Kotrappa, P., Stieff, L. R. and Bigu, J. Passive E-PERM Radon Flux Monitors For Measuring Undisturbed Radon Flux from the Ground, *International Radon Symposium II-1.6* 1996.

[55] Rechcigl, J. A preliminary comparison of radon flux measurements using large area activated charcoal canister (LAACC) and Electret ion chambers (EIC), *International Radon Symposium, Florida Hosted by AARST* 1996.

[56] Kotrappa, P., Stieff, L. R. Application of NIST Radon-222 Emanation Standards for Calibrating Radon-222 Monitors, *Radiation Protection Dosimetry* 1994, 55, 211-218.

[57] Kotrappa, P., Stieff, L. R., Volkovitsky, P. Radon Monitor Calibration using NIST Radon Emanation Standards: Steady Flow Method, *Radiation Protection Dosimetry* 2004, 113, 70-74.

[58] Kotrappa, P. Electret Ion Chambers for Measuring Dissolved radon in Water: review of Methodology and Literature, *1998 International Symposium Sept. 14-16 1998 Sponsored by American association of Radon Scientists and Technologists.*

[59] Rechcigl, I. C., Stieff, L. R. Methods Of Measuring Ambient Atmospheric Radon And Radon Flux Associated With Phosphogypsum Treatment Of Florida Spodosol Soil, *Commun. Soil. Sci. Plant Anal.* 1992, 23, 2481-2594.

[60] Stieff, L. R., Kotrappa, P. Passive radon flux monitors for measuring undisturbed radon flux measurements, *International Radon Symposium, Florida Hosted by AARST II-1.1 to 1.6* 1996.

[61] Amrani, D. Groundwater radon measurements in Algeria, *Radiation Protection Dosimetry* 2000, 51, 173-180.

[62] Usman, S. Analysis of electret ion chamber radon detector response to Rn-222 and interference with background gamma radiation, *Health Physics* 1999, 76, 44-49.

[63] Sabol, J. Monitoring of Rn-222 in Taiwanese hot spring SPA waters using modified electret ion chambers, *Health Physics* 1995, 68, 100-104.

[64] Taipow, J. The determination of dissolved radon in water-supplies by the E-PERM system, *Applied Radiation and Isotopes*, 43: 95-101.

[65] Kotrappa, P., Stieff, L., Stieff, F., Attenuation of Thoron (Rn220) in Tyvek Membranes, *Proceedings of the 2014 International Radon Symposium Accepted for publication.* www.AARST.org listed under Radon Info, Radon research papers.

[66] Stieff, F., Kotrappa, P. Small Volume (53ML) EIC with On/Off Mechanism, *Proceedings of the 2012 International Radon Symposium*, 113-121 2 www.AARST.org listed under Radon Info, Radon research papers.

[67] Steck, D. Pre- and Post-Market Measurements of Gamma Radiation and Radon Emanation from a Large Sample of Decorative Granites, *Proceedings of the 2009 International Radon Symposium,* 28-51 2 www.AARST.org listed under Radon Info, Radon research papers.

[68] Kotrappa, P., Stieff, F., Steck, D. Radon Flux Monitor for In Situ Measurement of Granite and Concrete Surfaces, *Proceedings of the 2009 International Radon Symposium,* 77-89 2 www.AARST.org listed under Radon Info, Radon research papers.

[69] Kotrappa, P., Stieff, F. Electret Ion Chambers (EIC) to Measure Radon Exhalation Rates from Building Materials, *Proceedings of the 2008 International Radon Symposium,* 1-8 www.AARST.org listed under Radon Info, Radon research papers.

[70] Kitto, M., Haines, D., DiazArauzo, H. Emission of Radon from Decorative Stone, *Proceedings of the 2008 International Radon Symposium*, 1-4 2 www.AARST.org listed under Radon Info, Radon research papers.

[71] Brodhead, B. Measuring Radon and Thoron Emanation from Concrete and Granite with Continuous Radon Monitors and E-Perms, *Proceedings of the 2008 International Radon Symposium,* 1-26 2 www.AARST.org listed under Radon Info, Radon research papers.

[72] Kitto, M., Haines, D., Fielman, E., Menia, T., Bari, A. Assessment of the E-Perm Radon-In-Water Measurement Kit, *Proceedings of the 2007 International Radon Symposium,* 1-4 2 www.AARST.org listed under Radon Info, Radon research papers.

[73] Steck, D. Year to Year Indoor Radon Variation, *Proceedings of the 2007 International Radon Symposium,* 1-11 2 www.AARST.org listed under Radon Info, Radon research papers.

[74] George, A. Intercomparison of Sensitivity and Accuracy of Radon Measuring Instruments and Methods, *Proceedings of the 2005 International Radon Symposium*, 1-8 www.AARST.org listed under Radon Info, Radon research papers.

[75] Moorman, L. Radon Flux Measurements as Predictor for Indoor Radon Concentrations in New Home Residential Structures, *Proceedings of the 2005 International Radon Symposium*, 1-12 www.AARST.org listed under Radon Info, Radon research papers.

[76] Kotrappa, P., Stieff, R., Stieff, L. Advanced E-Perm System for Simultaneous Measurement of Radon Gas, Radon Progeny, Equilibrium Ratio and Unattached Radon Progeny, *Proceedings of the 2003 International Radon Symposium*, 25-34.

[77] www.AARST.org listed under Radon Info, Radon research papers.

[78] Lewis, R. Short-Term Electret Ion Chamber Blind Testing Program, *Proceedings of the 2003 International Radon Symposium*, 35-59 www.AARST.org listed under Radon Info, Radon research papers.

[79] Kitto, M. Relationship of Soil Radon Flux to Indoor Radon Entry Rates, *Proceedings of the 2002 International Radon Symposium*, 19-22 www.AARST.org listed under Radon Info, Radon research papers.

[80] Kotrappa, P. Practical Problems and Solutions for Transportation and Deployment of E-Perm Radon Monitors for Indoor Radon Measurements, *Proceedings of the 1999 International Radon Symposium*, 2.0-2.2 www.AARST.org listed under Radon Info, Radon research papers.

[81] Kotrappa, P. Electret Ion Chambers for Measurement of Dissolved Radon in Water-Review of Methodology and Literature, *Proceedings of the 1998 International Radon Symposium*, 1.1-1.6 2 www.AARST.org listed under Radon Info, Radon research papers.

[82] Jenkins, P. Results from the First Annual AARST Radon Measurements Intercomparison Exercise, *Proceedings of the 1998 International Radon Symposium*, 3.1-3.22 www.AARST.org listed under Radon Info, Radon research papers.

[83] Rechcigl, J., Alcordo, I., Roessler, C., Littell, R., Keaton, H., Williamson, J., Stieff, L., Kotrappa, P., Whitney, G. Radon Surface Flux Measurement Using Large Area Activated Charcoal Canisters (LAACC) and Electret Ion Chambers (EIC), *Proceedings of the 1997 International Radon Symposium*, 3.1-.3.11 www.AARST.org listed under Radon Info, Radon research papers.

[84] Sellars, B., Sabels, B. High Elevation Experiment with Electret Radon Monitors, *Proceedings of the 1996 International Radon Symposium*, 5.1-5.11 www.AARST.org listed under Radon Info, Radon research papers.

[85] Houle, P., Brodhead, B. A Chamber Exposure of EIC's and AT's t Simulate Their Use As Dosimeters for Radon Workers, *Proceedings of the 1995 International Radon Symposium*, 5.1-5.9 www.AARST.org listed under Radon Info, Radon research papers.

[86] Kotrappa, P., Stieff, L. The Measurement of Very High Levels of Radon Concentrations Using Electret Ion Chambers, *Proceedings of the 1994 International Radon Symposium*, 1.1-1.6 2 www.AARST.org listed under Radon Info, Radon research papers.

[87] Kotrappa, P., Stieff, L., Bigu, J. Measurement of 220 Radon (Thoron) Using Electret Ion Chambers, *Proceedings of the 1994 International Radon Symposium*, 2.1-2.7 www.AARST.org listed under Radon Info, Radon research papers.

[88] Kotrappa, P., Dua, S., Hopke, P. Electret Ion Chamber Method for Continuous Measurement of Concentration of Radon in Water, *Proceedings of the 1993 International Radon Symposium*, 27-30 www.AARST.org listed under Radon Info, Radon research paper.

[89] Kotrappa, P., Stieff, L. NIST Traceable Radon Calibration System—Calibrating True Integrating Radon Monitors-E-Perm, *Proceedings of the 1993 International Radon Symposium*, 28-36 www.AARST.org listed under Radon Info, Radon research papers.

[90] Hopper, R. D., Levy, R. A., Rankin, R. C., Boyd, M. A. National Ambient Radon Study In: *Proceedings of the 1991 International Symposium on Radon and Radon Reduction Technology, 2-5 April 1991, Philadelphia, PA, Research Triangle Park, NC: US Environmental Protection Agency, 1991, Proceedings of the 1991 International Radon Symposium*, 1-16 2 www.AARST.org listed under Radon Info, Radon research papers.

[91] Yates, J., Kotrappa, P. Electret Radon Sniffer as Diagnostic Tool for Optimizing Efficiency of Sub-Slab Ventilation Systems and −or Sealing, *Proceedings of the 1990 International Radon Symposium*, 1-33 www.AARST.org listed under Radon Info, Radon research papers.

[92] Kotrappa, P., Dempsey, J. The Significance of Measurement of Radon-220 (Thoron) in Homes, *Proceedings of the 1990 International Radon Symposium*, 1-9 www.AARST. org listed under Radon Info, Radon research papers.

[93] Kotrappa, P., Lowry, R. Measurement of Dissolved Radon in Water Over Different Seasons From a Typical Home in Maryland, *Proceedings of the 1990 International Radon Symposium*, 1-8 www.AARST.org listed under Radon Info, Radon research papers.

[94] Kotrappa, P., Stieff, L. R., Mengers, T. R., Shull, R. D. Performance of Electret Ion Chambers in Magnetic Field, *Health Physics* 2006, 90, 386-389.

[95] P. Kotrappa and F. Stieff. One meter cube NIST traceable radon test chamber, *Radiation protection Dosimetry* 2008, 128,500-502.

[96] P. Volkovitsky, NIST Rn emission standards, *Applied radiation and isotopes* 2006, 64 249-252.

[97] Charlton, J. D. and Kotrappa, P., Uranium prospecting for accurate time-efficient surveys of radon emissions in air and water with a comparison to earlier radon and he surveys, *Proceedings of the 20006 International Radon Symposium*, 1-8 www.AARST. org listed under Radon Info, Radon research papers.

[98] P. Kotrappa, J. C. Dempsey, L. R. Stieff, and R. W. Dempsey. An E-RPISU (Electret radon progeny integrating sampling unit) *The 1990 International Symposium on Radon and Radon Reduction Technology*" held in Atlanta GA and conducted by US EPA (Paper B-III-1) 1-9.

In: Radon
Editor: Audrey M. Stacks

ISBN: 978-1-63463-742-8
© 2015 Nova Science Publishers, Inc.

Chapter 2

SPACE-TIME DISTRIBUTION OF RADON-222 FROM GROUNDWATER-STREAMFLOW-ATMOSPHERE INTERACTIONS IN THE KARST SYSTEMS OF THE CAMPANIA REGION (SOUTHERN ITALY)

Michele Guida[1,], Domenico Guida[2] and Albina Cuomo[2]*
[1]Department of Physics "E. R. Caianiello",
University of Salerno, Fisciano (SA), Italy
[2]Department of Civil Engineering,
University of Salerno, Fisciano (SA), Italy

ABSTRACT

Karst systems provide 25% of the drinking water resources to the world's population and sustain aquatic life in most fluvial eco-systems. In contrast, the singular process of aquifer recharge, the particular mechanism of subterranean pathway and the complex interactions between surface and groundwater make these systems highly variable in space-time hydrological behaviour and vulnerable to contamination and pollution.

In order to provide a useful approach to integrate traditional approach at the above problem resolution, this chapter describes the findings from Radon-222 activity concentration monitoring data from stream-flow and in-stream springs measurement in typical Mediterranean karst landscapes.

The study areas which concern are located in the protected area of the Campania region (Southern Italy), primarily in the Cilento and Vallo di Diano National Park-European Geopark and Regional Park of Picentini Mnt and surrounding.

In these protected areas, the management of the relevant water resources requires adequate groundwater assessment by performing hydro-geomorphological and hydrological modelling supporting planning tasks in water protection for domestic drinking use, riverine wildlife preservation and water quality maintenance in application of the European Water Framework Directive (EWFD).

---

*Corresponding author. Department of Physics "E. R. Caianiello" and Faculty of Engineering, University of Salerno and National Institute of Nuclear Physics (I.N.F.N.), Giovanni Paolo II, 84084 Fisciano (SA), Italy. Tel. +39 089 965383, fax: +39 089 965 275. *E-mail address:* miguida@unisa.it.

In the framework of an interdisclinary research program carried out at regional scale, a Regional River Monitoring System has been realized, at basin, segment and reach scale to collect experimental data about ^{222}Rn activity concentration, in addition to physical-chemical and streamflow rate measurements.

Montly campaigns were performed from 2007 to 2010 by means field and laboratory Rn222 activity concentration measurements in the streamflow and inflow spring waters, using RAD7-H2O and Water Probe (Durridge, Inc., USA). Appropriate sampling procedures and measurement protocols have been tested, taking into account the different local hydrogeological and hydrological situations occurring along the karst systems. At segment scale, data elaborations from selected segment of the monitored rivers provide location and downstream influences of the groundwater inflows from river banks and bed, also in absence of valuable streamflow discharge increments. At river reach scale, more detailed data enable to improve a preliminary model in the ^{222}Rn degassing rate and provide a first contribution to surficial-groundwater seasonal hydrograph separation. The analysis of the seasonal monitoring data trends from karst springs confirms the the utility and reliability of the Radon222 as tracers in a general hydrogeological conceptual model, highlighting the complex behaviour of the multilevel groundwater circuits, the uppermost in caves, the middle in conduits and the lowermost in fracture network, corresponding to the differentiated recharge types in the fluvial-karst hydro-geomorphological system.

Ultimaly, these results are useful in the assessment of the high-level of Radon-in-air atmosphere diffusion enriched by Radon222 degassing from differen karst springs feeding surficial water bodies. Preliminary esperimental results from two study cases are provided in order to planning geosite fruition and irrigation water uses.

1. INTRODUCTION

Surface and groundwater resources assessment represents one of the main issues in socio-economic planning and management (FAO, 1986, 1998; Karanth K., 1987; Winter T., 1995) and requires more and more interdisciplinary-based scientific researches, particularly in hydrogeology, hydro-geomorphology and hydrology (Marine I. W., 1979; Tolstikhin I. N. and Kamensky I. L., 1969). that global fresh-water resources, stored in rivers, lakes, and aquifers, constitute less than 0.5 % of all the water on the Earth, and therefore, their uses have to be, necessarily, sustainable (Loucks D. and Gladwell J., 1999), especially, in the light of a global severe water scarcity scenario forecasted by 2025 (IAEA, 2002).

Karst aquifers provide 25% of the drinking water resources to the world's population and sustain aquatic life in most fluvial systems, inducing a lot of ecological services to human communities. Being characterized by complex links between surface and groundwater, they turn out to be very vulnerable to contamination and pollution (Smith, L., A., 2004).

In Mediterranean environments, karst aquifer groundwater represents more than 98% of the available fresh-water supply and, during summer times, feeds perennial streamflow through the aquifer-derived base flow, thus contributing to the total streamflow in a measure of 30% to 70% (Dassonville Luc and Fé d'Ostiani L., 2003; Tulipano L. et al., 2005; Longobardi and Villani, 2008). An understanding of a given aquifer flow characteristics and its interaction with adjacent surface water resources, is critical if the total water resource is to be managed sustainably (Loucks D. and Gladwell J., 1999; Simonovic S., 1998; Shah T. et al., 2001).

In order to assess and manage water resources, the European Water Framework Directive 2000/60/EC (EWFD, 2000) suggests an integrated approach, taking hydro-geological, hydro-geomorphological, hydrological, hydro-geochemical, physical and biological contributions into account (Menéndez M. and Pinero M. J., 2003), particularly for groundwater-streamflow interaction assessment and monitoring (Grath J., Ward R., Quevauviller P., 2007).

Especially in karst Mediterranean landscapes, the interdisciplinarity turns out be fundamental (Ford, D. C. and Williams P. W., 1989, 2007; Margat, J. and Vallée, D. 2000), considering their very complex recharge processes, surface and groundwater circulation and discharge outflow mechanisms (Vogel, R. M. and Kroll, C. N., 1992; Quevauviller, PH. 2005; Barbieri, M. et al., 2005).

Determination of the interaction between roundwater and surficial water in karst landscape is particularly difficult because of the complex hydraulic interconnections of fractures and solution openings in carbonate rocks with basin drainage network. In fact, J. V. Brahana and E. F. Hollyday (Brahana J. V. and Hollyday E. F., 1988) indicated that dry reaches of streams can be used as indicators of groundwater reservoirs.

In term of hydro-biological response to the hydrologic conditions in karst environments, a number of the organisms have also been found to be indicators of the acquifer-river interactions. Vervier and Gibert (1991) quantified the interactions between water, solutes, and organisms at the interface between a stream and a groundwater outlet from karst terrane. The location of the ecotone showed marked spatial fluctuations according to the prevailing hydrology and that interactions were strong during high flows and, on the contrary, negligible, during low flows. This turns out to be very important in the protected areas, with specific destinations for native fish life in the European, national and regional laws.

A substantial help in providing an answer to many problems of interest in karst hydro-geomorphology and hydrology has been given, in the last decades, by the use of isotopes (stable and unstable) like tracers, both in field investigations as in laboratory analysis (Levêque, P. S. et al., 1971; McDonnell, I. J., 2003; Solomon, D. K. et al. 1993, 1995, 1997; Goldscheider N. and Drew, D., 2007).

Besides the traditional and long time use of natural isotopes in hydrology and hydrogeology (Emblanch C. et al., 2003), one of the most interesting, promising and innovative approach to assess quantitatively the groundwater contributions to streamwaters and seawaters in natural environments, consists in measuring Radon222-in-water activity concentrations (Andrews, J. N. and Wood, D.F. 1972; Shapiro, M.H., 1985). Therefore, Radon-222 provided an useful natural tracer because its activity concentrations in groundwater turn out to be typically one order of magnitude or bigger than those ones occurring in surface waters (Rogers A., 1958).

Radon-222 (for sake of simplicity called simply "Radon" in the following) is a volatile gas with a half-life of 3.8 days, moderately soluble in water and atmosphere. It is released to groundwater from Radium-226 alpha decay, by means of permanent alpha recoil in micro-pore or fracture walls (Rama and Moore W. S., 1984) and progressive dissolution of the aquifer forming material that supplies more and more soluble Ra-226, subsequently decaying to Radon (Ellins K.K. et al. 1990). Due to its volatility, Radon gas quickly dissipates when exposed to the atmosphere producing a significant disequilibria between concentrations in ground water and surface water.

From the seminal work of Rogers (Rogers A., 1958), the assessment of space-time variations in Radon concentrations between surface and groundwater (Ellins K. K. et al, 1990; Lee, R. and Hollyday, E. F., 1987, 1991) provides insights in: i) testing infiltration-filtration models (Genereaux, D. P., and Hemond, H. F., 1990; Genereaux et al., 1993; Gudzenko, V.,

1992; Kraemer, T.F. and Genereux, D.P., 1998), ii) performing hydrograph separation (Hooper, R.P. and C.A. Shoemaker, 1986), iii) calculating residence times (Sultankhodzhaev, A. N. et al., 1971) iv) interpreting the role of "old water" in non-linear hydrological response of catchments, v) estimating shallow and deep water mixing (Hoehn E. and von Gunten H. R., 1989; Hamada H. 2000, Hakl J. et al., 1997; Semprini L. 1987; Gainon, F., et al., 2007), and vi) calculating flow velocities in homogeneous aquifers (Kafri U., 2001). Yoneda et al. (1991) used Radon-222 as a tracer to localize the single locations discrete points of groundwater inflow to a river in Japan and Ellins et al. (1990) to quantify groundwater inputs to a stream in Puerto Rico, and, then, Lee and Hollyday (1991) to assess that, in the period of minimal flow, groundwater contributed with 36% to the Carters Creek in Tennessee.

In addition, the use of Radon enables the researchers to trace groundwater migration pathways (Hoehener P. and Surbeck H., 2004), and to assess the time dependence of groundwater migration processes (Schubert M. et al., 2008). Infiltration of surface waters from a river to groundwater (Hoehn E. and von Gunten H. R., 1989) as well as flow dynamics in a karst system (Eisenlohr L. and Surbeck H., 1995) are just a few examples of applications where Radon-based methodology has been used successfully to gain additional information on environmental functioning.

This potential for using Radon , as a suitable aqueous tracer (Levêque, P. S., 1971), is due to its main characteristics: i) it occurs in the environment in an ubiquitous way; ii) it behaves like an inert substance; iii) its half-life of 3.8 days, differently from other aqueous environmental tracers, like stable isotopes; iv) it is easy to manage, fast to monitor and its measurements cheap to perform.

Usually, Radon-in-water activity concentrations is measured with respect to the typical expected or reference values in surficial, subsurficial and groundwaters, through either sampling (batch sampling) measurements, performed on fixed volume samples of collected waters from springs or along the riverbed and followed by laboratory analysis or through continuous monitoring measurements directly in the waters (Surbeck H., 2005). In the last case, in order to better implement such an approach, it is required the use of simple and not expensive, field-usable Radon-in-water monitors, with temporal resolutions of hours or less. Nowadays, are available radon-in-air devices fulfilling the above requirements, among which the most commonly used are the "AlphaGuard" (Genitron Instr.) and the "RAD7" (Durrige Co. Inc.). During several testing surveys a quick comparison between the Radon-in-air analyzer AlphaGuard (Genitron Instr.) have been also carried out, for one specific monitoring station, confirming the detailed inter-comparison performed by M. Schubert (Schubert M., 2008, Schubert M. et al., 2008).

Besides Radon activity concentrations, chemical and physical parameters (pH, water temperature, dissolved oxygen, TDS, water conductivity, water resistivity, etc.) have also been collected by means of multi-parametric instrument HI 9828 (Hanna Instruments S.r.l.).

Synthetically, the objectives of the Radon in-water monitoring program have been to (i) localize and quantify the contributions of groundwater along the main stream riverbed and banks; (ii) set up an adaptive methodology, based on monthly Radon activity concentration measurements in streamflow and springs, for the baseflow separation from the other streamflow components; (iii) verify the hydrodynamical behaviour of the karst circuits and their influence on streamflow.

In the following, the results of the investigations on the space-time variations in Radon activity concentration along the reference segments and reaches of the selected karst-conditioned river basin are illustrated and discussed (Figure 1).

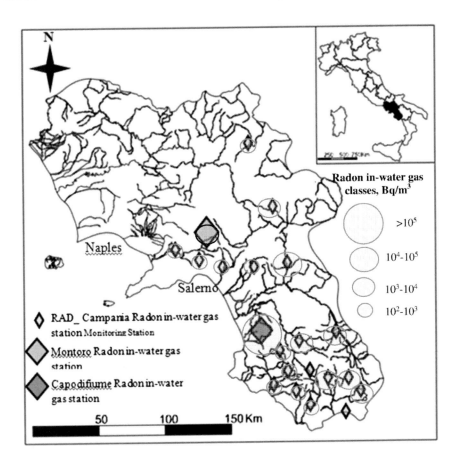

Figure 1. Location of the main monitoring stations and study areas and related radon in-water activity concentration classes in the middle and southern Campania region.

2. THE BUSSENTO RIVER BASIN STUDY AREA

2.1. Hydro-Geomorphological Setting

The Bussento river basin, located in the Cilento and Vallo di Diano National Park (Figure 2), is one of the largest and more complex fluvial systems of the southern Campania region (Southern Italy).

The "complexity" is due to the highly hydro-geomorphological conditioning induced by the karst landforms and processes, like summit karst highlands with dolines and poljes, lowlands with blind valleys, streams disappearing into sinkholes, cave systems, and karst-induced groundwater aquifers and gravitational karst-induced depressions.

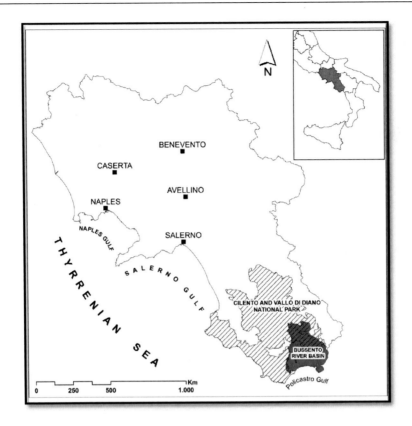

Figure 2. Location of the Bussento river basin.

In this sense, in the basin can be recognized the following sub-basins (colored areas in the Figure 3):

- Upper Bussento Sub-basin, in light blue, with two inter-connected endorheic basins (Vallivona and Sanza) and Karst Highlands, in light green transparent on light blue in the figure;
- Eastern Bussento Sub-basin, with terminal outlet to Capello Spring Oasis, near Casaletto Spartano village and small endorheic basins *("Affunnaturo")* and large Karst Highlands;
- Middle Bussento Sub-basin, heavily conditioned by the Middle Bussento Karst System, illustrated in the following paragraph;
- Western Bussento Sub-basin, in light brown on figure 3, greatly corresponding to the Sciarapotamo creek basin, conditioned by summit conglomerate aquifers and marly-clayey succession in lower sector;
- Lower Bussento Sub-basin, the blue area, on figure 3, includes the catchments from the Bulgheria and Roccagloriosa carbonate secondary aquifer and alluvial and coastal clastic valley fillings;
- Bussento-Policastro gulf system, a complex and still unknow groundwater system with interconnected aquifers and huge Submarine Groundwater Discharges.

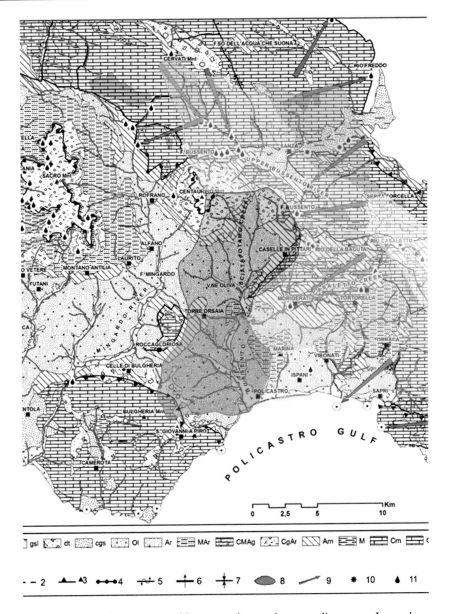

Figure 3. Hydro-geomorphological map of Bussento river and surrounding areas. Legend: Hydrogeological Complex: s)Sand and gravelly sand; gsl) Fluvial sandy gravel; dt) Slope debris; cgs) Sandy conglomerate; Ol) clayey Olistostrome; Ar) Sandstone and marls; MAr) Marl and Sandstone; CMAg) Marl and Conglomerate; CGAr) Conglomerate and Sandstone; Am) marly shale;M) Silty marl "fogliarina"; Cm) Marly limestone; C) Limestone; D) Dolostone. Symbology:1) Fault; 2) Hypotisized fault; 3) Overthrust; 4) Permeability limit; 5) Groundwater exchange; 6) Losing river; 7) Gaining river); 8)Karst summit; 9) Probable groundflow direction;10) Sinkhole; 11) Main spring; 12) Coastal and submarine spring. Sub-basins: light blue: Upper Bussento; green on light blue: Endorheic Upper Bussento basins; pink: Eastern Bussento; green: Middle Bussento River-Kars System; light brown: Western Bussento; blue: Lower Bussento.

The main stream originates from the upland springs of Mt. Cervati (1.888 m), one of the highest mountain ridges in Southern Apennines. Downstream, the river flows partly in wide alluvial valleys (i.e., Sanza Valley) and, partly, carving steep gorges and rapids, where a number of springs, delivering fresh water from karst aquifers into the streambed and banks, increase progressively the river discharge downstream.

The hydro-geomorphological setting of the river basin is strongly conditioned by a complex litho-structural arrangement derived from geological, tectonic and morphogenetic events occurred from Oligocene to Pleistocene along the Tyrrhenian Borderland of the southern Apennine chain (Bonardi 988, Bonardi et al. 1985). The chain is a NE-verging fold-thrust belt derived from an orogenic wedge, accreted by defomation and shortening by overthrusting of the sedimentary covers of several paleogeographic domains: Internal Sedimentary Domain in the Ligurian oceanic crust on the External Sedimentary Domain of Carbonate Platform-Continental Basin along the passive margin of the African plate. During the Plio-Pleistocene, the above cited fold-thrust belt is affected by polyphase uplift transtensive and trans-pressive movements, with general lowering toward the Tyrrhenian sea and juxtaposition of the clayey-marly successions in the erosional grabens, to the carbonate sequence, as karst summit horsts.

In the study area, the main structural features are the overthrust of the Internal Units on the Bifurto/Piaggine formations at the NE piedmont of the M.nt Centaurino and the Sanza trans-tensive line, along the southern piedmont of M.nt Cervati massif.

The main stream of the Bussento river originates from the upland springs of M.nt Cervati (1.888 m asl), one of the highest mountain ridge in the Southern Apennines, flowing downstream, partly in wide alluvial valleys (i.e., Sanza valley) and, partly, in steep gorges and canyons, where a number of springs delivering fresh water from karst aquifers into the streambed and banks, increasing progressively the river discharge. The upper right area is characterized by marly-arenaceous rocks outcrops (Marchese Hills), while the left upper area is characterized by limestone sequences (M.nt Rotondo and Serra Forcella highlands). More downstream, near Caselle in Pittari village, the Bussento river and adjacent neighbours minor creeks flow, respectivelys into "La Rupe " (Bussento Upper Cave), Orsivacca and Bacuta-Caravo stream sinks, channelling the entire fluvial surface streamflow drained from the upper Bussento basin into the a hypo-karst cave system and re-emerging about four kilometers downstream, in the neighbourhood of the Morigerati town, from the resurgence, called "Bussento Lower Cave". Downstream the resurgence, the Bussento river joints with Bussentino creek, originating from the eastern sectors of the drainage basin and, flows along deep canyons and deep gorges, carved into the meso-cainozoic litho-stratigraphic limestone sequences, prevalently constitute of limestone and marly limestone, referred to Alburno-Cervati Unit (D'Argenio et al., 1973). In the western and southern sectors of the basin (Sciarapotamo creek sub-basin), marly-clayey successions of the Liguride and "affinità Sicilide" Units (Cammarosano et al., 2000) dominate the hilly landscape, whereas they underlie the arenaceous-conglomerate sequences at the Centaurino M.nt. Downstream the confluence with Sciarapotamo creek, the Bussento river flows, as a meandering a terraced floodplain and, finally, in the Policastro coastal plain.

Surface and groundwater circulation in the basin results very complex. Groundwater inflows from outside of the hydrological watershed and groundwater outflows towards surrounding drainage systems, frequently, occur. This complexity is due to the occurrence of soils and rocks with highly different hydraulic permeabilities and to the highly hydrogeological conditioning induced by the karst features. Bussento river regime is also affected by a very complex hydropower and drinking water system, retaining and diverting the river discharge by dams, an artificial lake and weirs.

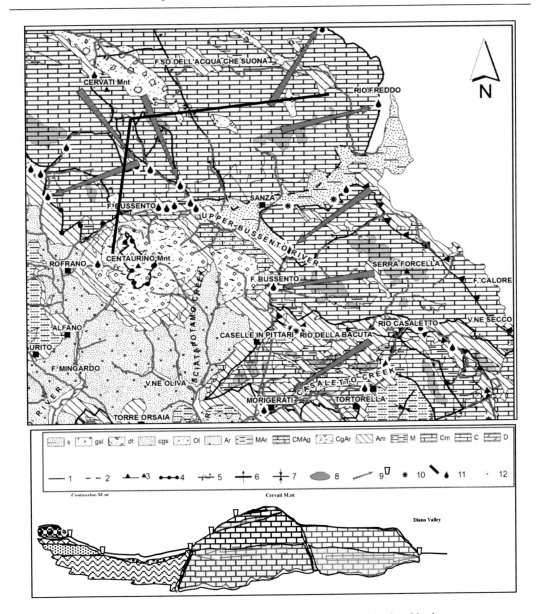

Figure 4. Hydro-geomorphological map of the Upper Bussento river and related hydro-geomorphological features. Legend: s. Sandy conglomerate complex; gsl. Gravelly sandy silty complex; dt. Debris complex; Ol. Blocky clayey olistostrome complex; Ar. Sandstone complex; MAr. Marly sandstone complex; CMAg. Marly conglomerate sandstone complex; Am. Silty Sandstone complex; M. Marly complex; Cm. Marly limestone complex; C. Limestone complex; D. Dolomite complex. 1. Permeability limit; 2. Buried permeability limit; 3. Overtrusth hydro-geological limit; 4. Syncline hydro-geological limit; 5. Overturned strata; 6. Horizontal strata; 7. Sincline; 8. Karst summit; 9. Hypotized groundwater flow direction; 10. Stream sink; 11. Main spring; 12. Section line.

The Bussento river regime is also affected by a very complex hydropower plant system, which retains and diverts the river discharge in the Sabetta reservoir and the Casaletto weirs, respectively, from the upper Bussento river and the Bussentino creek reaches segment to the Lower Bussento river segment.

The Upper Bussento Sub-basin and included catchments are located upstream the La Rupe Sinkhole and since 1960 upstream Sabetta Reservoir.

The mainstream originates from south-western upland valley of the Mount Cervati (Vallivona and Mezzana valleys), where many, low discharge springs from shallow aquifer in debris cover on marly-clayey bedrock originate ephemeral creek flowing into the Vallivona Affunnaturo sinkhole. From the Varco la Peta spring-resurgence, the Inferno creek flows southward, carving steep gorges in form of a typical bedrock stream, with cascade and rapids, where further springs (Montemezzano spring), along the streambed, increasing progressively the river discharge, as well as along the piedmont (Sanza Fistole spring groups). The true Bussento river begins downstream the junction of the above cited Inferno creek and the Persico creek. This last flows at the bottom of an asymmetric valley, characterized at the left side by the above cited steep, carbonate southern mountain front of M.nt Cervati and at the right by the gentle northern terrigenous mountain slope of the M.nt Centaurino (1551 m asl). The middle right side of the basin is characterize by marly-arenaceous rocks outcrop (M.nt Marchese hilly ridge), while the left middle side is characterized by limestone sequences at M.nt Rotondo and Serra Forcella (Figure 5).

Figure 5. The Bussento River Basin monitoring network. BS is the symbol for streamflow stations whereas PG is the symbol for marine stations, for discharge and radon measurements.

The geological constraints and the permeability variability in type and rate induce a very complex hydrogeological behaviour to coupled aquifer-river system (Figure 5).

In the northern sector of the basin, is located one of the main karst aquifer of the Southern Apennine: Cervati Aquifer. The hydrogeological boundaries of this aquifer are: at the northern side, a set of compressive tectonic lines (reverse faults, overthrusts), at the S and SW, is confined by the impermeable "bends" of the marly-clayey aquicludes, connected to carbonate aquifer by normal faults or stratigraphical limits. Inside aquifer, shear zones due to compressive and extensional tectonics originate intermediate aquitard, controlling a "segmentation" of multilayered aquifers, with karst spring at various elevations (see hydrogeological section below the Figure 5).

In general, in the Bussento river basin can be recognized main and secondary aquifers.

The M.nt Cervati karst carbonate aquifer, located in the northern side of the river basin, is one of the main aquifer of the southern Apennine; it is delimited at the North and N-E, by regional hydro-tectonic lines and at the SW and South by clayey aquicludes and higly fratured carbonate aquitards; minor hydro-structural lines induces multilayered and compartment aquifers (sub-aquifers), with centrifugal directions to the groundwater flows (Table 1). M.nt Forcella karst carbonate aquifer, located in the eastern sector of the Bussento river basin, having 75% in area outside of the Upper Bussento, feeds only the 13 Fistole Spring Group, emerging a few hundred meters upstream the end of the river segment, with a M.A.D 3 m³/s. M.nt Alta karst carbonate aquifer, located on the N-E sector of the basin feeds only the Farnetani Spring Group, with a M.A.D 1.5 m³/s, is interconnected with the Sanza Endorheic Basin and related sinkhole-cave system, feeding the Bonomo Mulino seasonal springs and resurgences. M.nt Centaurino multilayered terrigenous aquifer feeds several spring with a total M.A.D 0.1 m³/s.

Table 1. Hydrogeological characteristics of the springs from Cervati aquifer. M.A.D.: Mean Annual Discharge; GWFD: GroundWater Flow Direction

Spring name	Sub-aquifer name	Elevation (m a.s.l.)	M.A.D. (l/s)	GWFD	River Basin Receptor
Rio Freddo	M.nt Arsano	470	750	East	Tanagro
Fontanelle Soprane	M.nt Arsano	470	800	N-E	Tanagro
Fontanelle Sottane	M.nt Arsano	460	400	N-E	Tanagro
Varco la Peta	Vallivona	1200	40	Southern	Bussento
Montemezzano	Inferno creek	900	100	Southern	Bussento
Sanza Fistole Group	Basal Southern Cervati	550-470	300	Southern	Bussento
Faraone Fistole Group	Pedale Raia	450	400	S-W	Mingardo
Calore Group	Neviera	1150	100	North	Calore
Sant'Elena Group	Rotondo	420	400	N-W	Calore
Laurino Group	Scanno Tesoro	330-400	600	N-W	Calore

The main stream originates from south-western mountain slope of the Mount Cervati, where many, low discharge springs from shallow aquifer in debris cover on marly-clayey bedrock originate ephemeral creek inflowing into the Vallivona Affunnaturo sinkhole. From the Varco la Peta spring-resurgence, the Inferno creek flows southward, carving steep gorges in form of a typical bedrock stream, with cascade and rapids, where further springs

(Montemezzano spring), along the streambed, increase progressively the river discharge, as well as along the piedmont (Sanza Fistole spring groups).

2.2. The Streamflow and Geo-Chemical Monitoring Systeme and Dataset

Historical streamflow data consist of two short daily streamflow time series, recorded at the Caselle in Pittari and Sicilì gauging stations, from 1952-1968 and from 1952-1957, respectively.

The lack of up-to-date historical adequate streamflow time series, both on a temporal and on a spatial point of view, makes even more difficult a realistic calibration of a modelling approach. For this reason, a monitoring campaign, illustrated in the following paragraph, was planned to temporally and spatially extend the streamflow database.

On January 2003, the Sinistra Sele River Watershed Regional Agency, started a monitoring campaign with the aim to measure, in many different cross sections and on a monthly time base, the Bussento river discharge. Based on the above described geomorphological and hydrogeological settings, 25 gauge stations were indicated as significant to individuate the river and springs hydrological regime (Figure 6).

Table 2. Bussento Monitoring Stations

Code	Station Name	Basin Sector	Latitude	Longitude	Distance (m)
BS00	River Mouth	Lower	543605,3	4435295,3874	0
BS01*	SS18 Bridge	Lower	543365,5	4435974,715	740
BS02	Railway Bridge	Lower	542247,8	4438272,0435	3680
BS03	Vallonaro junction	Lower	541834,8402	4440099,9059	6200
BS04	SS517 Bridge	Western	543412,8693	4442664,5218	10930
BS12	Hydro-power	Middle Lower	543583,6	4442368,8007	10246
BS13*	Sicilì Bridge	Middle Lower	546446,5601	4442939,8484	14100
BS14	WWF Oasis	Middle Karst	553475,0	4445318,3639	22717
BS15	Capello Oasis	Eastern	546915,6	4444081,9437	15580
BS23	Morigerati bridge	Eastern	548065,3	4443510,8961	15995
BS24	Melette bridge	Eastern	557102,7175	4446756,1804	28734
BS25	Caravo Sinkhole	Middle Karst	548948,4602	4447695,8382	20500
BS17**	Sabetta Reservoir	Upper	547207,9202	4449424,2903	20900
BS18	Acquevive bridge	Upper	548000,0954	4451699,0969	23534
BS19	Farnetani bridge	Upper	546973,6582	4452744,0284	25550
BS20	Abate bridge	Upper	544406,0243	4453604,0162	28460
BS22	Varco Carro bridge	Upper	543049,7711	4454630,4533	30095
BS21	Inferno lower bridge	Upper	543083,6774	4454695,186	30300
BS16	Ciciniello upper bridge	Upper	545934,8916	4449803,428	22300

* Same location of the station having managed by the Campania Regional Civil Defense Sector.
**Same location of the Bussento at Caselle managed by National Hydrographic Service.

Since December 2009 a further monitoring campaign, focused on the upper and middle Bussento river basin (Figure 3 and Table 2), was started and managed by CUGRI on behalf of the Regional Agency for the Environmental Protection of Campania region (ARPAC), within a more comprehensive study on the radon-222 activity concentration in stream and spring waters. Besides radon concentration, more chemical and physical variables have been measured, such as pH, water temperature, dissolved oxygen, atmospheric pressure, electrical conductivity and water resistivity.

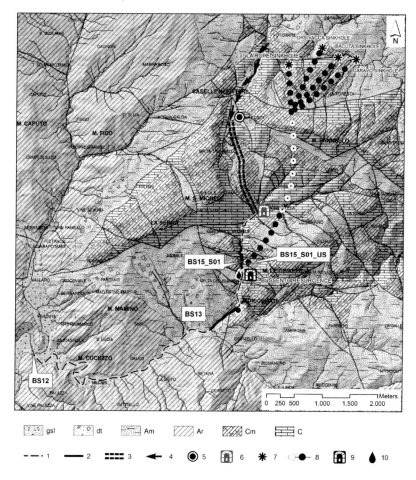

Figure 6. Hydro-geomorphological map of the MBKS. Legend: gsl) sandy gravel; dt) slope debris; Am) marly shale; Ar) sandstone and marls; Cm) Marly limestone; c) Limestone. Symbology: 1) River segment; 2) River reaches; 3) Abandoned subterranean flowpath; 4) Abandonned flowpath; 5) Abandoned stream sink; 6) Abandoned resurgence of the Palaeo-Bussento river ; 7) Active stream sink; 8) Explored (grey circle) and no-explored (withe circle) subterranean flowpath; 9) Active resurgence of the Bussento river; 10) Main karst spring.

Monitoring stations locations have been carefully identified to investigate the complex interaction between groundwater and streamflow, caused by the complex karst hydrogeological structure and system hydrodynamics. The monitoring timing of the river discharge was oriented to measure the delayed sub-surface flow and the baseflow component of the hydrograph. For this reason, several recession curves of historical data were analysed, deriving the more appropriate time from the flood peak discharge at which the delayed sub-surface and baseflow occur. Consequently, the monitoring campaigns were planned to measure the stream discharge at least seven days after the end of the rainfall event, while in dry periods the measures were conducted two times a months.

2.3. The Middle Bussento Karst System Case Study

In the Middle Bussento Karst System case study (MBKS, Figure 4), the Rn-222 activity concentration data, acquired as reported above, have been arranged in relation to the fluvial level hierarchy and scale analysis: firstly, at segment scale, managing the data collected only from the main stations; secondly, at reach scale, including also the data from the intermediate and complementary stations.

The middle-lower segment and the Sicilì bridge reach are located downstream the previous from the end of the Morigerati gorge to Sciarapotamo creek confluence, along which three reach can be recognized from downvalley: the downstream, in correspondence of the Bussento Hydropower Central, results a typical riffle-pool river (Montgomery and Buffington, 1998), as a entrenched meander in fluvial and strath terraces, the second upstream reach, Bottelli House reach, results a riffle-pool river along low order alluvial terraces and finally the third, the above cited Sicilì Bridge reach, a plane bed river slightly entrenched in alluvial terrace.

Segment River Scale Analysis

The first river segment (Middle Bussento Karst System Segment) begins at the Old Mill Spring to Bussentino creek junction gorge, comprises, from upstream, the following stations (Figure 5): BS25 (Bacuta-Caravo Station), at the one of the three Stream Sinks upstream the MBSKS; BS15_S01_US (Old Mill Spring Upstream Station), BS15_S01 (Old Mill Spring Station), BS15_S01_DS (Old Mill Spring Downstream Station) and BS15_S01_03 (Old Mill Spring Station just upstream the BS15_S02 reach junction.

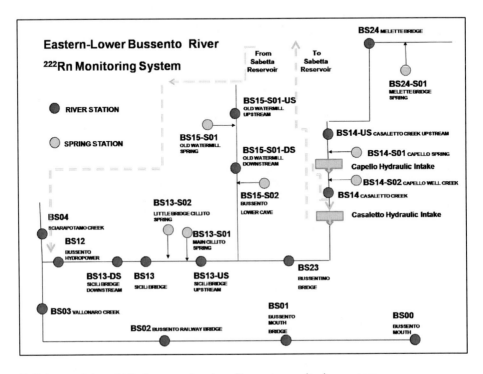

Figure 7. Scheme of the middle-lower and eastern Bussento monitoring system.

The second river segment, called Middle-lower Bussento, starts from the above cited junction gorge to Bussento Hydropower Central (Figure 3), comprising, from upstream, the following stations: BS13_US, BS13_01, BS13_, BS13_DS and BS12 (Figure 7), with the numerous groundwater inflows from the bank and bed fracture along the uppermost segment reach. The groundwater inflows are named Cillito spring group (main spring code BS13_01).

The correlation between annual mean values Rn-222 activity concentration data vs. topographic distance of the stations monitored along the two segments are shown on plots in figure 8 and 9, highlighting the location of the Radon high-content inflow from main spring stations.

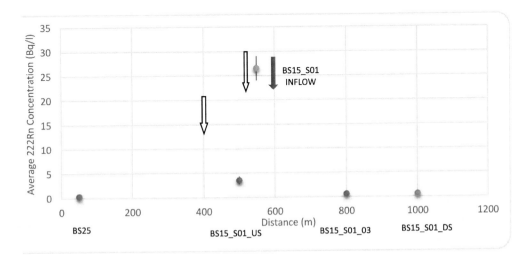

Figure 8. Average 222Rn concentration for 2008 data at the middle Bussento segment. 1) diffuse epikarst springs, not monitored; 2) monitored concentrated conduit karst spring (BS15_S01).

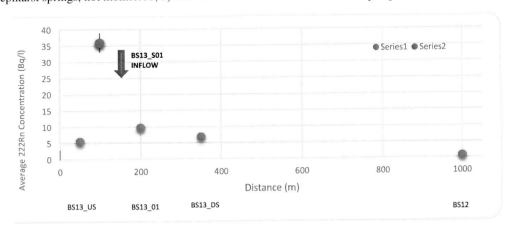

Figure 9. Average 222Rn concentration for 2008 data at the middle-lower Bussento segment.

Data analysis at segment scale highlights the spatial variations of Rn-222 activity concentration, detected along the medium and medium-lower Bussento river segments, suggests the following considerations: i) the in-water variations of Rn-222 activity concentration vs. the river long profile detect clearly the location of the surface-groundwater interactions, also where no discharge increments result from quantitative surveys (see

BS15_S01_US and BS13_DS values); ii) define roughly the linear extension downstream of the groundwater influx, strictly related to magnitude of groundwater inflow and hydraulic condition of each reach ; iii) prospect the approximate streamflow reference base value in Rn-222 activity concentration for the Bussento river, corresponding to the lower values detected from BS12 station (0,7 Bq/l, 2008 average data) and BS25 station (0,3 Bq/l, 2008 average data).

River Reach Scale Analysis

The rate of spatial in-stream groundwater influx results differentiate for the two segments of interest, in relation to groundwater hydro-chemical type, discharge magnitude, and hydraulic river constraints, related to hydro-geomorphological typology of stream. In order to understand this differentiation, due to a different degassing rate in Radon-222 activity concentration from free surface of streamflow, an analysis at reach level and more detailed scale has been performed along the *Sicilì Bridge reference reach* and *WWF Oasis reference reach*. In following, the results and data discussion for each reach is explained. The first reference reach is the so called *Sicilì Bridge* reference reach. It is located uppermost the middle-lower Bussento river segment (Figure 9), identified by the reference main station BS13 (*Sicilì Bridge*). The station BS13-US was chosen as upstream monitoring station. It is placed upstream the Cillito springs group, emerging along the right bank, from enlarged fractures into calcarenites, overlaid by the marly-clayey formation, functioning as regional aquiclude. Downstream the first spring outlet of the Cillito group, four monitoring secondary stations (BS13-03, BS13-02, BS13-01, BS1300) have been established in the river at a distance of 50 m from one to another (Figure 10).

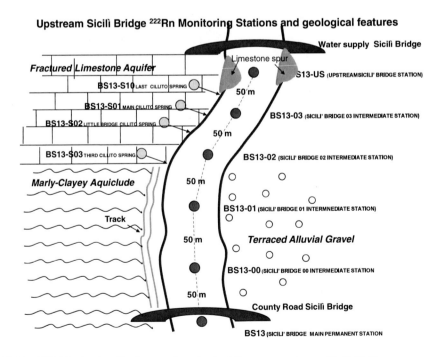

Figure 10. Rn-222 Monitoring Stations and geological features upstream Sicilì Bridge.

Downstream the main station BS13, another two monitoring stations have been established: BS13-DS, and BS13-DS-Jundra-DS, this one downstream the superficial inflow from Jundra creek. The results of the measurement campaigns carried out from September 2007 to December 2007, as testing procedure, and from January 2008 to December 2008, as experimental measurements, are plotted in figures 12-13.

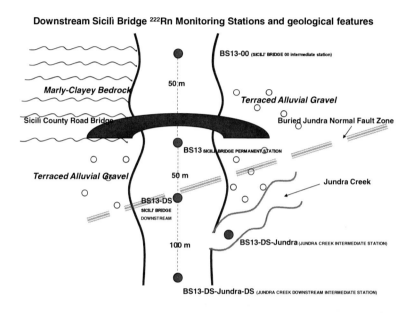

Figure 11. Rn-222 Monitoring Stations and geological features downstream the Sicilì Bridge.

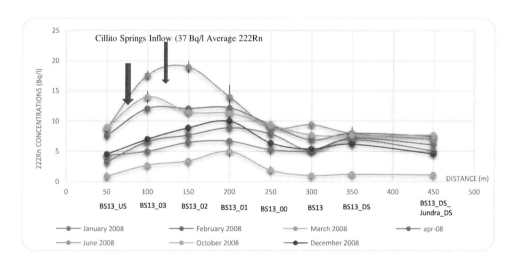

Figure 12. Plot of the 2008 measurement campaigns data at the Sicilì Bridge reference reach.

These results show, as expected, that concentration measured at the group of 4 stations from BS13_03 to BS13_00 increases because of the inflow from the lateral springs, whose water is richer in Rn-222. At the following stations there is a downstream decrease of Radon concentration due to Radon losses to the atmosphere (degassing). with the exception at the station BS13_DS, seems due to buried fault influence which showing a certain increase of concentration for almost all the measurement campaign.

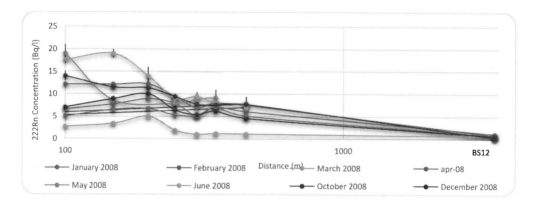

Figure 13. Plots of the 2008 measurement campaign data at the mid-lower Sicilì Bridge segment.

The plots from Figure 12 and Figure 13 also show:

1. Homogeneity in the general trend of the curves: there is, in fact, an increase in Radon activity concentration values starting from the station BS13_03 and then a decrease from the station BS13 which is not influenced by the springs.
2. Seasonality of radon relative concentrations, confirming in general that the measures made during the aquifer recharge period provide values of concentrations that are lower than the ones of the discharge period. There is also an intermediate stripe of values corresponding to the first part of the new recharge period with a decrease in the Radon activity concentration.
3. There is an anomalous increase in Radon concentration, for all the three periods considered,
4. between the stations BS13 and BS13_DS, that is at the moment subject of further investigations in order to determine whether it can be attributed just to statistical fluctuations or not.

An analysis of the Radon diffusion phenomenon from water to atmosphere has been made for the Sicilì segment. We hypothesize that Radon losses due to degassing can be explained by an exponential law, $e^{-\alpha L}$, according to the outcome of the application of the stagnant film model (Ellins et al. 1990, Wu et al. 2004), where L is the distance between two stations and α is a decay-like coefficient. So, the station BS13_02, can be considered as the higher point and the station BS12 as the lower one to calculate α for this segment.

The results reported in Table 3 show that the estimated value for α is higher in the discharge period (mean value: $8.3 \cdot 10^{-4}$ m^{-1}) than the one for the recharge period (mean value: $5.0 \cdot 10^{-4}$ m^{-1}).

The second case study concerns a reference reach of the Bussento river, within a WWF oasis Figure 14), in which there is a main spring (BS15_S01 – Old Watermill Spring) at which we measured an average 222Rn activity concentration of 36,5 Bq/l.

Table 3. Degassing coefficient

Measurement Campaign	Degassing Coefficient α (m^{-1})	R^2 Curve Fitting
January 08	6.4·10^{-4}	0.961
February 08	5.1·10^{-4}	0.945
March 08	4.3·10^{-4}	0.420
April 08	7.8·10^{-4}	0.968
May 08	6.4·10^{-4}	0.924
June 08	7.9·10^{-4}	0.884
October 08	10.0·10^{-4}	0.990
December 08	7.7·10^{-4}	0.952

As in the previous case a monitoring station (BS15_S01_US – Old Watermill Upstream) has been established above the inflow of the water coming from the spring, which, through a little cascade, falls into the river. Below the cascade and down the course of the river other 4 monitoring stations have been established. This part of the river is characterized by high turbulence which affects the Radon losses.

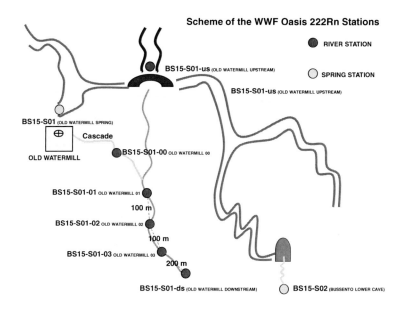

Figure 14. Rn-222 monitoring stations and geological features at WWF Oasis reference reach.

The results of the measurement campaigns are plotted in Figure 15. It can be noted a great increase of Radon concentration in correspondence of the stations below the spring inflow, and then a quick decrease.

Also for the WWF Oasis reach, a preliminary modeling has been made for the Radon degassing from water: in this case, because of the high turbulence of the river, we have a very sudden and sharp decrease of the Radon activity concentration values as shown in Figure 16. The highest point in the plot (corresponding to the monitoring station BS15_S01_01 below the inflow from the main spring BS15_S01) and the lowest one (corresponding to the last station BS15_S01_DS) have been considered, obtaining that the best curve fitting the plot is a power-law-like $y = K*x^{-\delta}$ with δ as Radon "degassing" coefficient in this case (Table 4).

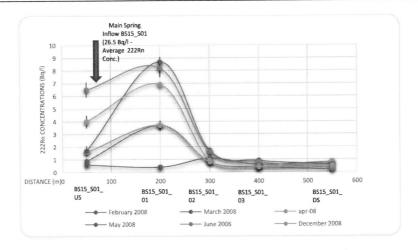

Figure 15. Results of the 2008 measurement campaigns at the WWF Oasis reference reach.

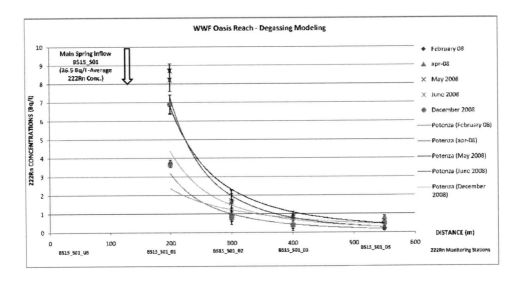

Figure 16. Radon degassing modeling for the data from the WWF Oasis reference reach (2008 measurement campaigns).

Table 4. Degassing coefficient at the WWF reach

Measurement Campaign	Coefficient K [Bq/l*m]	δ	R^2 Curve Fitting
February 08	10^7	2.87	0.979
April 08	$1.6*10^4$	1.66	0.578
May 08	$9*10^6$	2.65	0.938
June 08	$2*10^8$	3.22	0.989
December 08	$7*10^6$	2.69	0.795

Karst Spring Groundwater Analysis

During the above illustrated measurement campains, some karst springs along the Bussento river basin have been, also, monitored. Their importance is due to their content in

Radon, which is responsible of its activity concentration increase in the surface water. According to the results in Radon activity concentration, three "families", corresponding to the typologies of karst springs assumed in the conceptual model, have been identified (Table 5):

1. Fracture basal springs (i.e., B13_S01 and BS13_S02), with high values of Radon activity concentration (32.4 Bq/l (mean value) from the first one and 35.8 Bq/l from the second one) and with low standard deviation and variance values;
2. Conduit springs (i.e., BS15_S01) with very variable values (between 17.5 Bq/l (min) and 33.5 Bq/l (max)) and with low standard deviation and variance values;
3. Cave resurgence springs with highly variable values (between 0.5 Bq/l (min) and 6.5 Bq/l (max)).

Table 5. Radon activity concentration recorded at the Bussento karst springs

Spring Station	[^{222}Rn] Min (Bq/l)	[^{222}Rn] Max (Bq/l)	[^{222}Rn] Mean (Bq/l)	DEV ST	VAR
BS13_S01	28	45	32.4	0.8	0.68
BS13_S02	32	40	35.8	1.0	1.05
BS15_S01	17.5	33.5	25.6	1.1	1.20
BS15_S02	0.5	6.5	1.10	0.13	0.017

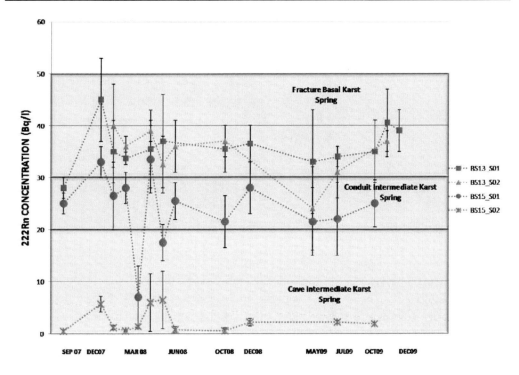

Figure 17. Seasonal variability in Radon concentration in the Bussento karst springs.

There is, therefore, a spatial variability in Radon activity concentration, which is shown in Figure 17. As for the seasonal variability, the two basal springs of the Cillito group do not show any relevant difference in Radon concentration during the year. At the conduit spring (BS15_S01), more varying values have been obtained: they are a little higher in the recharge period (average value: 26.4 Bq/l) than in the discharge one (mean value: 23 Bq/l). For the resurgence spring some higher values (6.5 Bq/l) have been obtained at the beginning of the discharge period, while in the other months there are data with little variability.

2.3. Upper Bussento Case Study

The Rn-222 concentration monitoring campaign of the Upper Bussento river was oriented to investigate the variability of gas in stream water and stream inflowing springs water and to separate the total streamflow in the subsurface and baseflow components.

In the Figure 18 as shown the monitoring stations map and the characteristics of each one.

Station Code	Drainage Area (DA) (Kmq)	Elevation (m.a.s.l.)	Pervious DA (Kmq)	Impervious DA (Kmq)
BSU17	85.15	912	64.08	21.07
BSU18	82.13	927	62.43	19.70
BSU19	66.84	927	49.49	17.35
BSU20	47.20	1079	38.74	8.46
BSU22	14.73	926	11.12	3.61

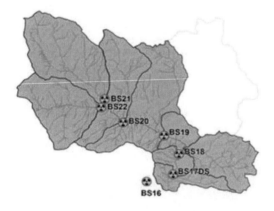

Figure 18. Monitoring stations in the upper Bussento river basin.

For the particular river reach under consideration there exist two significantly different temporal patterns: a first behavioural pattern, that records a temporal large fluctuation of radon concentration around a mean value and a second behavioural patter, that records a mainly constant radon concentration during the time.

The existence of these two patterns is strictly related to the presence or not of significant inflowing springs water contributions in particular gauging stations. Opposite to stations BSU17 and BSU18, where substantial stream inflowing springs water occur, stations BSU19, BSU20 and BSU22 are indeed featured by the absence of springs feeding the streams along the correspondent river reach. This physical characterization would explain the larger and constant concentration of radon for the first two stations and the lower and fluctuating radon concentration for the remaining stations. radon concentration fluctuation detected for the stations BSU19, BSU20 and BSU22 can be explained by the fact that, in these particular sections, surface flow component, which is the results of the fastest transformation of rainfall and is the poorest radon concentration stream water fraction, is significantly contributing to the total discharge.

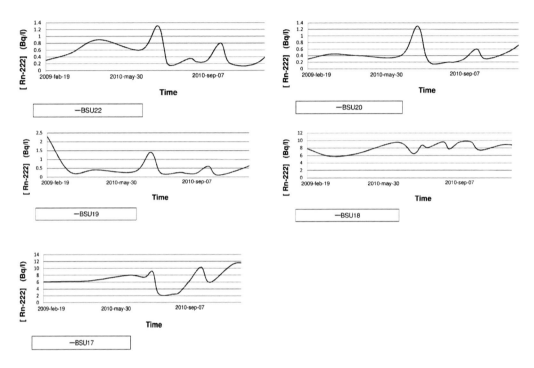

Figure 19. Temporal variation of radon concentration measured in a number of gauging stations along the river network.

The effect of the increase in the surface component of total discharge is also detectable for stations significantly affected by springs feeding, as a function of the proportion of groundwater versus surface water: stations BSU18, compared to station BSU17, receives a large fraction of groundwater contributing to total streamflow and is not thus affected from rainfall events, only increasing the surface component of total discharge.

As a proof, in figure 20, the temporal pattern of radon concentration measured in sections BSU17 and BSU18, is compared to the temporal pattern of radon concentration in spring water feeding the stream, in cross sections BS17S0N and BS18S0N, immediately up stream sections BSU17 and BSU18. It is evident, in particular for the station BSU18 for the same reasons previously indicated, that the temporal pattern of radon concentration in stream water (BSU18) strongly resembles the temporal pattern of radon in spring water (BS18S0N).

More insights about the radon concentration dynamics could be achieved by comparing, at the annual scale, the temporal pattern of precipitation, discharge and radon concentration (Figure 21).

During the winter season, the abundant precipitations recharge the deep water resources system and at the same time produce a significant surface component of total streamflow (flood conditions), with radon concentration approaching a rather constant and average value during the whole period. During the summer dry season, instead, when low flow conditions occur, the river discharge is mainly sustained by the baseflow, that is the outflow of deep water resources systems, characterized by the larger radon concentration because of the long residence times. Measured data confirm indeed that, in the period from May to September, the river discharge decrease and a consequent increment in radon concentration is instead detected.

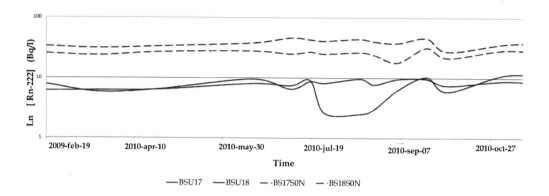

Figure 20. Temporal pattern of radon concentration in stream water (BSU17 and BSU18) and in spring water feeding the stream (BS17S0N and BS18S0N).

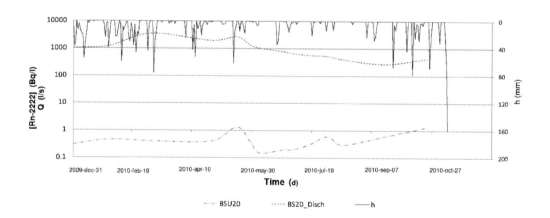

Figure 21. Precipitation, discharge and radon concentration temporal patterns.

Starting from the well-known assumption that water is composed of a set of well mixed end members, the collected data of radon concentration are used to illustrate an example of hydrograph separation into different flow components. To this aim, mass balance and mixing equations can be written, as described in Kendall and McDonnell (1998):

$$Q_T = Q_{SSF} + Q_{GW}$$

$$C_T Q_T = C_{SSF} Q_{SSF} + C_{GW} Q_{GW}$$

where:

Q_T is the total streamflow, Q_{SSF} is the sub-surface delayed flow, Q_{GW} is the groundwater flow, C_T is the Rn-222 value in total streamflow, C_{SSF} is the Rn-222 value in sub-surface delayed flow, C_{GW} is the Rn-222 value in groundwater flow.

As an example, the mixing equations are applied at cross section BSU18, which is one of the gauging sections where groundwater contributions are extremely large, to derive the Q_{SSF} and Q_{GW} components of total discharge. If Q_{SSF} and Q_{GW} are the unknown variables, application of equations requires observation and measures of all other variable. Q_T and C_T are indeed the only measured variables, whereas values for C_{SSF} and C_{GW} are inferred from measurements referred to different cross sections.

The Rn-222 content of river water is strongly affected by volatilization to the atmosphere, and this must be accounted for in using radon data to estimate a possible groundwater influx from subsurface water sources (Kies A., 2005). If C_{DS} and C_{US} are the radon concentration measured downstream and upstream cross sections, and L is the length of the river segment between the mentioned cross sections, the relationship between radon concentrations is described by the following equation (Wu Y. et al., 2004):

$$C_{DS} = C_{US} \times e^{-\alpha L}$$

Model equation is applied between sections BS18_S0N and BSU18, assuming C_{US} as the radon concentration in section BS18_S0N, which is the river inflow spring water section, and determining the C_{DS} as the radon concentration in section BS18, at a distance of about 1 Km, also representing the C_{GW} concentration in section BSU18. Application of the volatilization model requires a value for the parameter α, previously calibrated on a specific river reach of the Bussento network (Guadagnuolo D., 2009), whose hydro-geomorphological settings are similar to the one of the river reach investigated in this report and resulting in an α coefficient equal to 0.9. C_{SSF} radon concentration at cross section BSU18, that is the concentration of sub-surface flow, is computed as the mean value of radon concentration measure in sections BSU22, BSU20 and BSU19, where deep water resources contribution are negligible and representative of the sub-surface flow. Results are illustrated in Figure 22.

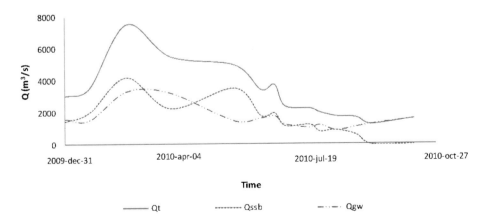

Figure 22. Hydrograph separation based on mixing equation solution and environmental tracers measurements.

3. THE BUSSENTO RIVER BASIN PHYSICAL MODEL

The implementation of the Radon measurement techniques along the surface and groundwater bodies in the Bussento river basin has given us the prospective of the concrete possibility of using these methodologies in a karst mediterranean environment. It has been established the possibility of localizing groundwater influx in riverbed, also in absence of discharge variations (due to riverbed deposits).

The aim of this chapter is to provide a physical scheme of the complex recharge, storage and routing system of MBKS monthly measurement campaigns of Rn222 activity concentration have been performed in the Bussento river basin (§ 2.2) in such river stations chosen according to their relevance for the study of the interactions between groundwater and surface waters.

The analysis of the seasonal data trends from karst springs (§2.3) confirms a hydrogeological conceptual model, highlighting the complex behaviour of a multilevel groundwater circuits, the uppermost in caves, the mid in conduits and the lowermost in fracture network, corresponding to the differentiated recharge types in the fluvial-karst hydro-geomorphological system.

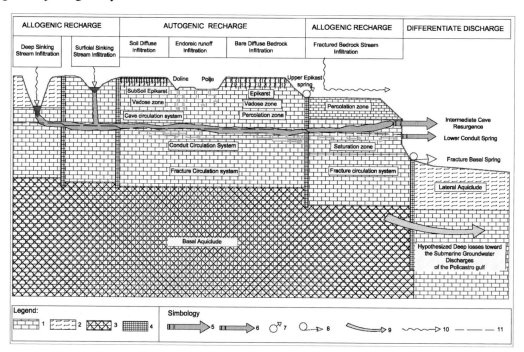

Figure 23. Conceptual hydro-geomorphological model of the Middle Bussento River karst system (MBKS).

In order to provide a physical scheme of the Middle Bussento karst area, a preliminary, physical-based conceptual model has been built-up, accounting for an interconnected sequence of geologic substrates, structural discontinuities, type and rate in permeability distribution, recharge areas and discharge points, that collectively attempt a conceptualization of the karst aquifers-river interactions (White W. B., 1969, 1977, 1988, 2003), focusing on the variety of hydro-geomorphologic settings and their influences on the streamflow regime.

With reference to the work done by G. Iaccarino et al. (Iaccarino G. et al. 1988), and by D. Guida et al. (Guida D. et al., 2006) , the conceptual hydro-geomorphological model of the MBKS, contains three nested hydrological domains (Figure 23): i) a hydrogeologic domain; ii) a hydro-geomorphological domain and iii) an aquifer-river domain.

The hydrogeologic domain represents the 3-D structure of aquifer, aquitard and aquiclude, conditioning the groundwater circulation and storage, vertically differentiated in the classic subdivision of karst hydro-structure (Bakalowicz M., 1995): epikarst, vadose, percolation and saturated or phreatic zones (Ford D., C. and Williams P., W., 1989). The last one is hydro-dynamically subdivided in cave, conduit and fracture routing system (White W. B., 1969). The hydro-geomorphological domain comprises karst and fluvial landforms and processes, conditioning groundwater recharge ("karst input control", *sensu* Ford D.C. and Williams P., W., 2007), by means of the infiltration and runoff processes, including: a) allogenic recharge from surrounding impervious drainage basins into deep and shallow sinking stream infiltration points, and fractured bedrock stream infiltration; b) autogenic recharge, including sub-soil and bare diffuse epikarst infiltration, endorheic runoff infiltration in dolines and poljes; c) groundwater discharge ("karst ouput control", *sensu* Ford D.C. and Williams P., W., 2007), differentiated in the groundwater-river interactions within the aquifer-river domain. This last comprises the complex interactions between the streambed-springs system, which generally results in a downstream river discharge increasing, occurring generally in typical bedrock streams, flowing in gorge and canyons carved in enlarged fractured limestone sequences. Following the routing karst system, the springs inflowing into streamlow can be characterized in: i) upper epikarst springs, ii) intermediate cave resurgence spring, iii) lower conduit springs and iv) basal fracture springs. Figure 4 highlights, also, the hypothesized deep losses towards the Submarine Groundwater Discharges (SGD), emerging in the Policastro gulf (Guida et al., 1980). Each of the mentioned components correspond, in the modelling conceptualization of the scheme, to a linear storage, which releases streamflow as a function of the water storage and of a characteristic delay time. The characteristic time indicates that there is a delay between the recharge to the system and the output from the system itself, and this delay is greater for deeper acquifers. The number of storages, each representing then a different process, contribute to the total streamflow through a recharge coefficient, that is a measure of the magnitude of the single storage.

4. UPPER TUSCIANO RIVER CASE STUDY

In this chapter are compared the data of Electrical conductividy (EC) and Rn-222 as tracers, in detecting groundwater contributions, also along the reaches where only the discharge measurement result in stream flow decrease.

The drainage river basin, named Tusciano river basin (Figure 24), is located in the Picentini Regional Park (province of Salerno, Campania region, Southern Italy, Figure 24).

The Tusciano river (41 km long) originates from Mt. Polveracchio (1.790m asl) that is the highest peaks of the Picentini Mountain System (Cuomo et al. 2011) and his river basin (Figure 22) extends to the sea for an area of about 260 km^2 .

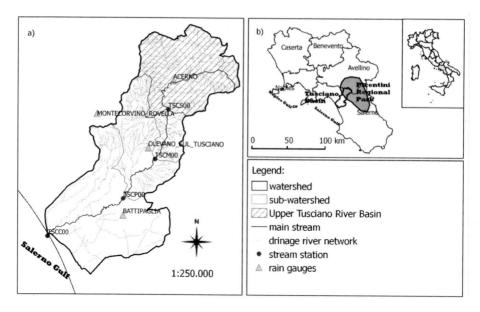

Figure 24. Tusciano River Basin.

The investigated drainage area study concerns the Upper Tusciano river basin (about 80 km^2) whose main stream is about 15 km long (Figure 24). The drainage pattern is weakly trellised and ramified. Table 6 contains the main catchment morphometric parameters.

Table 6. Main catchment morphometric parameters

	Rc	Ru	Rf	Ra	D	n1	F 1	ga
Isc	0,59	1,30	0,32	0,64	4,47	147	6,46E-06	2,90
_04	0,46	1,47	0,33	0,65	3,60	231	5,40E-06	10,33
_03	0,53	1,37	0,39	0,70	3,93	392	5,92E-06	8,26
_02	0,50	1,41	0,34	0,66	3,96	416	6,06E-06	8,13
_01	0,45	1,48	0,33	0,65	3,93	434	6,05E-06	8,17
_00	0,49	1,43	0,35	0,67	3,95	488	6,16E-06	7,65

Code: Rc: circularity ratio; Ru: uniformity coefficient; Rf: form factor; Ra: length ratio; D: drainage density; n1: number of first order fluvial branches; F1: frequency of first order fluvial branches; ga: density of hierarchical anomaly.

The Upper Tusciano river stream flow originates by junction of two headwater torrents. On the left, it receivs a lot of waters through the Bardiglia torrent feeding by the Bardiglia and Savuco springs, both coming from the NE hillside of the M.nt Polveracchio (Figure 25) . On the right valley side, the river segment receives the stream waters from the Pinzarrino catchment and the Ausino spring group.

This latter is located upper stream the Isca della Serra valley, while on the left side it circumvents the Acerno village until arriving at Acqua Buona segment, where it swells for the water coming from the Molari valley (Figure 25).

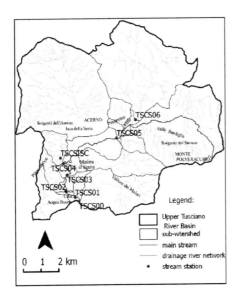 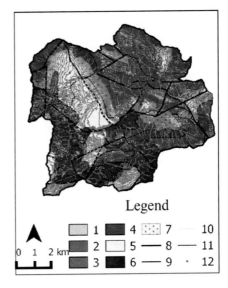

Figure 25. Upper Tusciano river basin and monitoring system stations (on the left); . Hydrogeological Map (on the right). Legend:1. Pelitic-limestone Complex; 2. M.te Marzano and M.ti della Maddalena Unit Limestone Complex; 3. Calcareous complex of Picentino-Taburno Unit; 4. Picentini-Taburno Unit Limestone –marly complex; 5. Lacustrine complex; 6. Lagonegresi I e II Unit Cherty-marly complex; 7. Continental, epiclastic deposit complex; 8.Faults; 9.Main stream; 10.Drainage river network; 11. Contour; 12.Springs.

The Picentini Mountains hydrogeological unit includes calcareous-dolomitic structures that represent one of the most important aquifers in southern Italy, thanks to the abundant orographic rainfalls produced by very wet frontal systems from Tyrrhenian sea.

The groundwater flow is very high (85-90%) compared to surface runoff, due to the intense infiltration phenomena and to the presence of widespread carbonate bedrock discontinuities. The basin is mainly composed of limestone and dolomite, secondarily of lacustrian and epiclastic deposits complex, as shown in Figure 25.It has been chosen as the most significant reaches the one included between Casa Isca location and Acqua Buona because it is representative of the geomorphical stream behavior for its mountain section and in a situation not overly influenced by anthropogenic interventions.

In order to gain useful and effective insights derived from Radon activity concentration measurements and elaborations, monthly measurement campaigns have been performed in the Tusciano river basin. Simultaneously with the Rn 222 measurements were performed electrical conductivity and discharge measurements. Discharge measurements (2010-2014) have been gauged about a week after rain events, in order to estimate as much as possible base flow component of the total hydrograph.

Besides Radon activity concentrations chemical and physical parameters (pH, water temperature, dissolved oxygen, TDS, water conductivity (EC), water resistivity, etc..) have been collected using the multi-parametric HI 9828 (HANNA Instruments S.r.l.).

The stations, whose coordinate locations have been measured by means of GPS GS20 Professional Data Mapper Leica Geosystems, have been chosen according to their relevance for the study of the interactions between groundwater and surface waters. Each monitoring station has been labeled with an alphanumeric code, beginning with the three letters TSC

(TSC stands for Tusciano) followed by a string of bits, containing the station ID number plus a code for river stations (Figure 26).

Stations	Mean [m^3/s]
ISC + 04	1,4
TSC_03	1,6
TSC_02	1,59
TSC_01	1,93
TSC_00	2,27

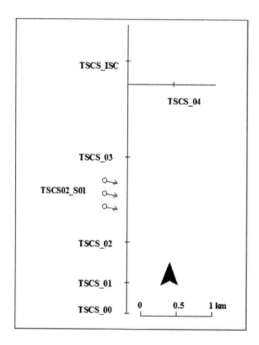

Figure 26. Tusciano stylized main stream and the average of discharge at the main stream stations and average monthly discharge data.

The results of the measurement campaigns highlight, as shown in Figure 27, clear and gradual increase in discharge along the reach between the closure section TSCS_00 and the ones upstream (junction between TSC_ISC and TSCS_04).

Nevertheless was recorded an anomalous decrease in streamflow discharge in the months of March, June, July, November 2011, March, May, June 2012, January, July 2013 and finally in March 2014 from TSC_03 to TSC_02, identified as the most representative reach. During the dry season this anomalous behavior did not occur.

Analyzing the average monthly discharge data, reported in Figure 27, was confirmed the anomaly of the decrease in streamflow discharge between the two above mentioned stations; conversely the number of spring should increase progressively and significantly the river discharge.

Simultaneously with the flow measurements, in order to establish the role of the groundwater on the streamflow, were operated radon-discharge-EC measurements.

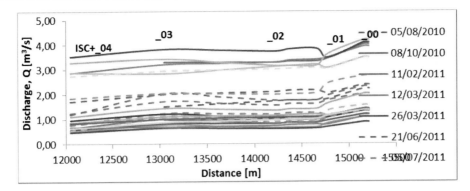

Figure 27. Discharge at the Tusciano main stream stations [m³/s]. The dashed lines refer to the campaigns carried out during the recharging period (from October to May) while the continuous lines to streamflow recession period (from May to October).

These were carried out from August 2010 to June 2012. Figures 28 and 29 contain all the Radon activity concentration and EC data recorded at all the river stations. The data pattern shows, as expected, that most of all Radon concentrations and EC values measured at the group of the stations between TSCS_03 to TSCS_02 increase because of the inflow of the lateral resurgence springs, whose water is richer in Radon e contribute to the increase in conductivity. In the months following the campaign of June 15, 2012 was continued to perform only discharge and EC measurements that confirm as referred above.

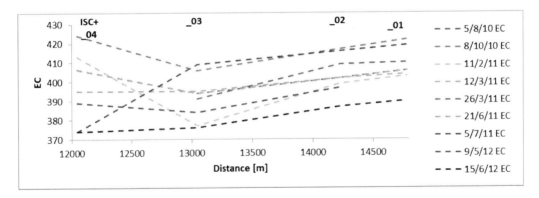

Figure 28. EC values at the main stream stations.

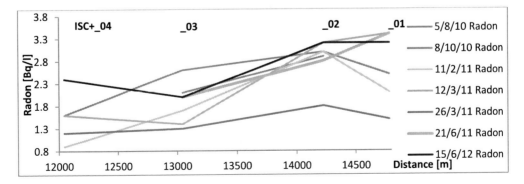

Figure 29. Radon concentration [Bq/l] at the main stream stations.

5. THE LABSO-LAURA KARST SYSTEM CASE STUDY

This chapter deals with the use of radon as tracer in hydrogeological studies. The study areas are located in the Montoro alluvial plain (Campania region, southern Italy). The Labso and Laura springs, located in Montoro plain, close to the Borgo and Preturo villages, are part of a hydro-geomorphological district characterized by the hydrogeological features depicted in the hydrogeological map sketch, modified from hydrogeological map of southern Italy (Allocca et al. 2007), Figure 30.

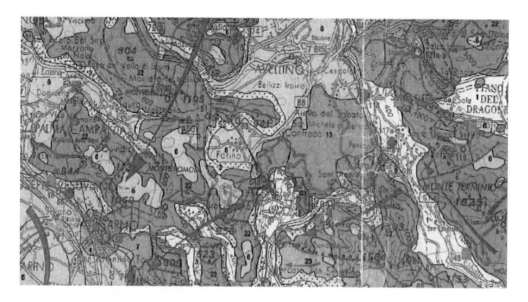

Figure 30. Sketch of the hydrogeological map of the study area, modified from Allocca et al. 2007.

As shown in Figure 30, the blue arrows on the hydrogeological map, demonstrate most of the groundwater flow towards Sarno and Solofra, only a small part flows towards Laura and Labso springs coming from the Forino plains, indistinctly.

According to the previous hydrological setting and to the aims of the study, since 2012 winter, a monitoring activity was begun, building-up an appropriate monitoring system and identifying and characterizing the stations to be monitored (Tables 7-8 and Fig. 31-32).

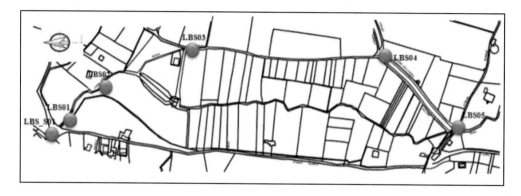

Figure 31. Location of the Labso reach stations.

Table 7. Labso spring and downstream reach stations

ID	CODE	KIND	GEOGRAPHICAL COORDINATES AND ALTITUDE				ABSOLUTE DIST.(m)	RELATIVE DIST.(m)
			X	Y	Z(m s.l.m)		
0	LBS_S01	Spring	40°50'24.31"	14°46'07.29"	209	0	0
1	LBS01	Stream Bed	40°50'37.13"	14°46'04.50"	205	11	11
2	LBS02	Stream Bed	40°50'33.98"	14°46'08.59"	192	197	186
3	LBS03	Stream Bed	40°50'33.78"	14°46'08.71"	189	375	178
4	LBS04	Stream Bed	40°50'29.14"	14°46'10.72"	185	760	385
5	LBS05	Stream Bed	40°50'16.48"	14°46'10.02"	187	965	205
6	LBS06	Stream Bed	40°50'11.99"	14°46'04.17"	186	987	22

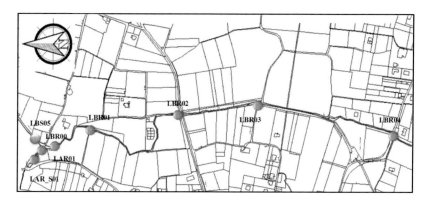

Figure 32. Location of the Laura reach stations.

Table 8. Laura spring-Viara channel reach stations

ID	CODE	KIND	GEOGRAPHICAL COORDINATES AND ALTITUDE				RELATIVE DIST.	ABSOLUTE DIST.
			X	Y	Z (m slm)		
0	LAR_S01	Laura Spring	40°50'12.47"N	14°46'2.70"E	189	0	0
1	LAR01	Stream Bed	40°50'11.85"N	14°46'3.75"E	188	53,2	53,2
2	LBR00	Stream Bed	40°50'10.88"N	14°46'3.61"E	187	31,7	84,9
3	LBR01	Stream Bed	40°50'6.99"N	14°46'5.90"E	183	147	231,9
4	LBR02	Stream Bed	40°49'58.12"N	14°46'7.84"E	182	415	646,9
5	LBR03	Stream Bed	40°49'50.38"N	14°46'9.14"E	181	254	900,9
6	LBR04	Stream Bed	40°49'36.88"N	14°46'5.48"E	178	457,73	1358,63
7	LBR05	Stream Bed	40°49'20.14"N	14°45'56.18"E	175	617,07	1975,7
8	LBR06	Stream Bed	40°49'13.20"N	14°45'50.74"E	174	262,9	2238,6
9	LBR07	Stream Bed	40°49'6.69"N	14°45'42.82"E	173	299,3	2537,9
10	LBR08	Stream Bed	40°48'56.34"N	14°45'38.87"E	172	364	2901,9
11	LBR09	Stream Bed	40°48'46.88"N	14°45'39.81"E	171	295	3196,9
12	LBR10	Stream Bed	40°48'39.76"N	14°45'40.97"E	170	234	3430,9
13	LBR11	Stream Bed	40°48'33.03"N	14°45'38.71"E	168	221,2	3652,1
14	LBR12	Stream Bed	40°48'22.46"N	14°45'41.54"E	162	361,8	4013,9

The monitoring program was addressed to collect hydro-chemical spring and stream flow data during the discharge period of the aquifer feeding the above springs. From May to October 2013, in the stations above listed, four monitoring campaigns were carried out and data on water flow (Q), average velocity (V_{mean}), specific electrical conductivity (EC), temperature (T) were collected. At same time, water samples were taken for the laboratory evaluation of activity concentrations in Radon-in-water using RAD7-H2O monitor by Durridge Inc. (USA).

Tables 9 and 10 show the dataset and relative plots referred to the 30/10/2013 campaign on both Labso and Laura springs and related downstream reaches.

Table 9. Example of monitoring dataset from Labso spring to Labso-Laura junction reach

ID	Distance (m)	Sampling Time 30/10/2013	Measurement Time 19/07/2013	Radon (Bq/l)	Temperature °C	Electric Conductivity (mS/cm)	TSD/ppm	Discharge (l/s)
LBS_S01	0	09:25	19:21	5,4	13	500	207	57
LBS01	11	10:54	09:50	5,3	13-13	500-556	250-278	57
LBS02	197	11:15	09:50	3,6	13.5-13.5	490-464	241-232	33
LBS03	375	11:40	11:05	1,5	14-14	510-497	280-261	25
LBS04	760	12:40	11:07	0,8	15	498	249	25
LBS05	965	12:00	12:28	0,4	17.5-17.5	502-560	251-280	25
LBR00	987	13:15	16:42	4,8	13		0	98

Table 10. Monitoring dataset from Laura spring to Labso-Laura reach – Viara channel

ID	Distance (m)	Sampling Time 30/10/2013	Measurement Time 19/07/2013	Radon (Bq/l)	Temperature °C	Electric Conductivity (mS/cm)	TSD/ppm	Discharge (l/s)
LAR_S01	0	12:03	18:36	10,0	13/12,5	450/449	225\250	89
LAR01	33,5	12:50	12:45	8,6	12,5/13	440/499	220\249	67
LBR00	73,2	13:15	16:42	4,8	13/12,7	455/489	228\245	98
LBR01	252,2	12:22	16:57	3,9	13,5	455	228	
LBR02	659,3	14:57	14:26	3,0	14	462	231	
LBR03	913,3	14:50	14:24	1,6	15,6	463	232	
LBR04	1371,03	15:05	15:46	0,9	15	465	233	
LBR05	1988,1	14:20	15:47	0,5	16.5	467	234	
LBR06	2251	15:20	17:47	0,2	17	444	222	
LBR07	2550,3	15:30	17:34	0,5	16,7	463	232	
LBR08	2914,3	15:40	15:50	0,2	17	465	233	
LBR09	3209,3	15:45	14:30	0,6	17,6	463	232	
LBR10	3443,3	15:50	13:06	0,2	17	460	230	
LBR11	3664,5	16:00	10:09	1,2	17,0	465	233	
LBR12	4026,3	14:10	11:23	0,2	17,4/17,2	444/510	222\255	95

Once collected the data and defined the mean values of the basic parameters, the Labso and Laura spring data have been compared and interpreted. In spite of the space proximity, about 1000 m, and short monitoring time, some significant differences in the following hydrogeological parameters were found (Table 11):

Table 11. Comparative analysis between the Labso and the Spring Laura hydrogeological and hydro-chemical parameters. a: recession coefficient; W: storage dynamic volume; T: residence time

	Period:May-October			Period:June-October							
	a_1	$W_1(m^3)$	$T_1(gg)$	a_2	$W_2(m^3)$	$T_2(gg)$	$Q_m(l/s)$	s_Q (l/s)	EC(mS/cm)	s_{EC} (mS/cm)	Radon(Bq/l)
Spring Labso	0,008	2,662	121	0,005	1,98	203	130	86	500	18	6
Spring Laura	0,007	3,960	135	0,006	2,96	169	202	104	450	22	13,5

An initial attention should be paid on analysing the first three parameters (a1, W1, T1), which were initially defined for the period from May to October and later, when it was found that, in the month of May, the aquifer was still in recharging phase, also for the period from June to October. In particular, in the first case, the depletion coefficient a1 of Labso spring is higher than the one for the Laura spring. It follows that the storage W1 is lower and thus the residence time is 121 days. This, initially, was a proper explanation for the average value of the activity concentration of 222Rn found in Labso spring, 6 Bq/l about half value as compared to 13 Bq/l of Laura one.

The preliminary analysis was subsequently varied when it was evaluated in the second case a2 in which the values were reversed, in particular of the residence times, 203 days for Labso spring and 169 days for Laura, respectively.

Furthermore, the EC as compared to the Radon values should decrease rather than increase compared to that of Laura Source, considering the well-known Rn-EC positive correlation in the specific literature (Guadagnuolo et al. 2013).

The data analysis above shortly illustrated can suggest new insights about the different hydrogeological constraints controlling the variable hydro-chemical response of the two springs. Figure 33 shows a more detailed hydrogeological interpretation of the aquifers feeding the Labso and Laura springs, supported by the data collected and analyzed in the above illustrated data analysis.

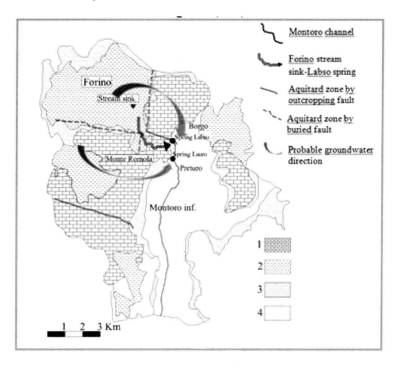

Figure 33. Reinterpretation of the hydro-geological setting of the Montoro study area. Legend: 1) Stratified limestone sequence; 2) Pyroclastic deposits and soils; 3) Piedmont loose debris, 4) Floodplain deposits.

The lower values in radon from the Labso than Laura spring can be explained considering a mixing mechanism between groundwater and surficial water entering the aquifer by the stream sink located in the eastern part of the Forino intermontane plain. The above cited apparent discrepancy between Labso-Laura EC values compared with radon values can be explained as due to anthropogenic causes. Ultimately, from hydrogeological point-of-view, we can conclude that the recharge zone of Labso spring is both autogenic and allogenic, while Laura spring have only autogenic recharge from M.t Romolo aquifer (Figg. 34 and 35).

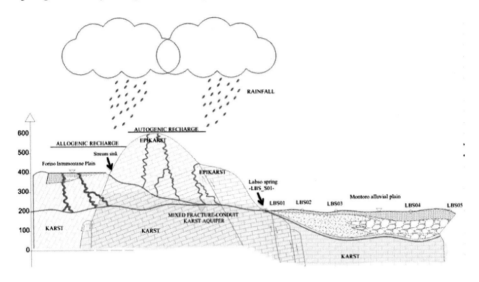

Figure 34. Hydrogeological interpretative cross-section from Forino stream sink to Labso spring.

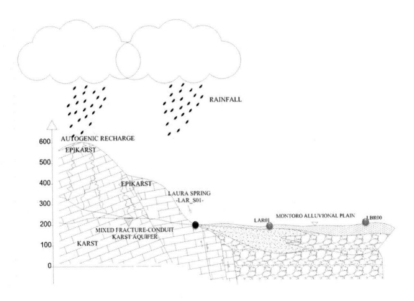

Figure 35. Hydrogeological model –Spring Laura.

Concerning the interaction between the ground and surface waters of the canal, reference is made to the data obtained during the various campaigns, particularly comparing the various 222Rn activity concentrations representing them in a single plot in which there is an immediate perception of its spatial and temporal variation.

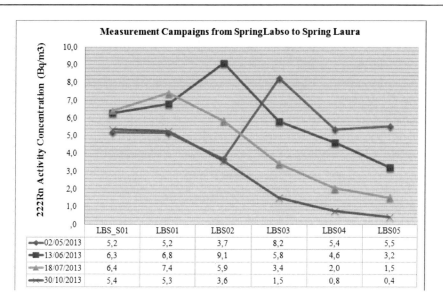

Figure 36. Variation of the 222Rn activity concentration as a function of the distance.

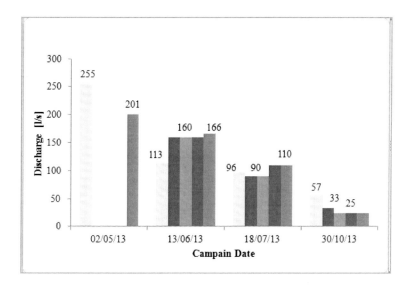

Figure 37. Discharge variation at monitored stations.

Particularly, considering firstly the case of Labso spring, as shown in the above figure, it is found that in the campaign dated 2/05/2013 there is a spike at the LBS03 station. In the campaign dated 13/06/2013, instead, at the LBS02 and on 18/07/2013 there is a spike in LBS01 and to conclude with the last campaign dated 30/10/2013, its spike is at the source. Actually, this fact can be explained by considering that the Radon activity concentration in the waters is significantly influenced by groundwater and geological nature of the site. Since the materials which characterize the soil in examination are not listed as potential active sources of Radon gas and of direct gamma radiations, the only explanation is linked to the flow of the river. In fact, considering the graph drawn in Figure 37, it is observed that: in the first campaign, at the station LBS03 in which there is a peak, no flow rate was detected, as shown in the histogram by the blank space between the peaks. Subsequently, in the 13/06/2013 campaign, in order to evaluate this anomaly, was found radon subjected to the phenomenon of degassing.

Therefore, a decrease of Radon should be noticed rather than an increase and further discharge measurements carried out also in correspondence of LBS03, highlighting some increase in stream flow, compared to the station LBS_S01 and LBS01. In fact, the values ranging from 113 l/s to 160 l/s, correspond to an increase that probably had already occurred in station LBS02, corresponding to the point in which the maximum value of Radon is present. This is revealed also for the third and final campaigns in which the discharge peak values 96 l/s in LBS01 and 57 l/s at station LBS01 can be noticed. Therefore, it is confirmed that Radon increases as the flow increases; this does not seem true if the above plot. In fact what must be pinpointed is that, although in the 13/06/2013 campaign the 166 l/s value is higher, but the 6 l/s have to be considered, as already determined in the previous section. In particular, this behavior of the discharge can be explained due to the fact that when the aquifer is in the phase of recharging has a higher water-table and in this case it is very likely that it rises up to the point of a spring release in farther points than the spring, a phenomenon occurred in the 02/05/2013 campaign at the LBS03 located at 375 m from Labso spring (Figure 38).

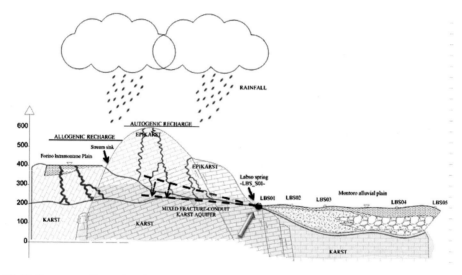

Figure 38. Reference to Campaign 2/05/2013-Rappresentation of the elevation of the water-table with spring release at the LBS03.

On the contrary, when the aquifer is discharge period, a water-table lowering exists, causing an upstream outleting closer to the Labso source point (Figure 39).

A further analysis was made at the merging station and, as the graph drawn in Figure 40 shows, there is a varying trend of Radon in time at the same station.

The analysis represents one of the first works concerning the monitoring of radon in Labso and Laura Sources, situated in the plain of Montoro Inferiore. In particular, through the use of gas Radon-222 as a natural tracer, an attempt was made to characterize and study the interactions between groundwater and surface water flows (river-groundwater interactions) in accordance with regulatory guidance; from this analysis the values of Radon recorded were quite low, and since radioactive mineral waters are classified as "light, medium and strong", it is carried out within the following limits:

1) Light up to 1110 Bq per liter (weakly radioactive);
2) Medium from 1110 Bq to 5550 Bq per liter (radioactive);
3) Strong from 5550 Bq per liter onwards (highly radioactive).

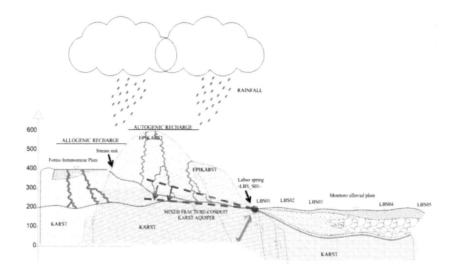

Figure 39. Reference to Campaign 2/05/2013- Rappresentation of lowering of the water-table with a spring release at the LBS_S01.

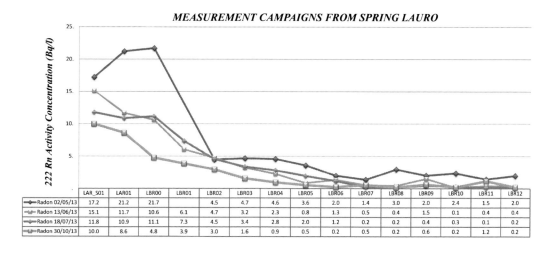

Figure 40. Variation of the 222Rn activity concentration function of the distance.

Therefore, the waters of Labso-Laura Source, can be defined as "weakly radioactive", as they possess a smaller amount of radioactivity of 1110 Bq/l. In addition to a purely scientific aspect, another factor took our attention, that is the protection of human health, certainly not negligible. Although radon concentrations in the water of the spring are not very high, it was however decided to conduct this study because, over the last decade, some studies have demonstrated that inhalation of high concentrations of radon can significantly increase the risk of lung cancer. In this regard, a protocol of measurements in air has been developed in order to understand how and in what quantity the gas present in the water evaporates into the air. The data obtained from the measurements of radon concentration reveal rather low values, which are considered below the range of typical values reported in international literature. Therefore, in this specific case, the Radon does not represent a risk to take into account with regard to planning.

6. Assessing Radon-in-Air from Streamflow: Labso-Laura and Capodifiume Springs

During the last decade, many studies have demonstrated that inhalation of high concentrations of radon can significantly increase the risk of lung cancer. In this regard, a protocol of measurements in air have to be developed in order to understand how and in what quantity the gas present in the water can diffuse in the air. In particular, through the use of gas Radon-222 as a natural tracer, an attempt was made to characterize and study the interactions between groundwater, surface water flows (river-groundwater interactions) and degassing in the atmosphere. In addition to a purely scientific aspect, another factor took our attention, that is the protection of human health, certainly not negligible. The field measurements campaigns have been carried out during the 2013 spring-autumn time on monitoring stations located along springs and downstream irrigation channels. The experiments were performed in two study areas, the first is the Capodifiume spring group located in the Cilento Geopark Geosite and the second located in the Montoro alluvial plain (southern Italy).

6.1. The Study Area of Capodifiume

The Capodifiume spring group, located downhill the town of Capaccio (SA), are one of the most important geosites of European Geopark of Cilento, Vallo di Diano and Alburni. The water of these springs, have a relevant discharge (annual average: 3 m^3/s) originating the homonymous river. They are unusable both for irrigation and drinking water because of high mineralization. In Holocene times, the high salt content deposited the travertine plastron (several km^2 wide and several tens of meters thick), where was founded the greek-roman city of Poseidonia-Paestum.

The springs of Capodifiume show higher values of ^{222}Rn concentrations (average of 230 Bq/l) and salinity compared to other karst springs in the Cilento, Vallo di Diano and Alburni European Geopark and Campania region (Figure 2). Following the previous hydrogeological studies and based on monthly monitoring Radon activity, an experiment was programmed in order to define the influence of the Radon degassing on in-air Radon concentration and diffusion. Assuming that the volatilization of the gas from the water surface is the main contribution to the degassing, in 06/08/2013 a preliminary test measurement was performed at the Lateral Spring LS01. Based on the results of this test, in 28/09/2013, a complete experiment was carried out, performing three set of measurements, positioning for each three RAD7 instruments along a transect at a distance of 0 m, 5 m and 10 meters from the spring stations to find a trend of the ^{222}Rn activity concentration in the air along the wind direction at the moment. Velocity and direction of the wind were also monitored using a Ventus weather station, detecting an average wind speed of 4,9 km/h in a constant NW wind direction. Once were identified the wind constant direction, three sets of measures were performed in that direction at 14:00, 15:10 and 16:30, respectively (Figure 42).

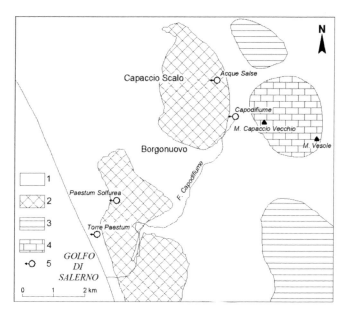

Figure 41. Hydrogeological sketch: 1) Fluvial sediments; 2) Travertines; 3) Clays, sandstones, marls, calcareous marls; 4) Limestones; 5) Springs.

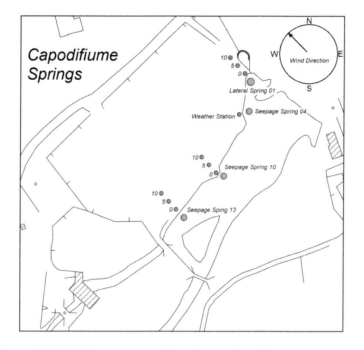

Figure 42. Location of the RAD7 instruments and weather station during the campaign of 28/09/2013.

The resulting concentrations shown evaluable decreasing trend, as reported in Table 12.

Table 12. Capodifiume Radon Experiment dataset

Station ^{222}Rn	Distance	Measurement Date	^{222}Rn Activity Concentration (Bq/m³)	RAD7 Instrument
Lateral Spring 01	0 m	28/09/2013 14:00	442 ± 41.5	2449
	5 m	28/09/2013 14:00	404 ± 34.2	2361
	10 m	28/09/2013 14:00	363 ± 64.7	1972
Seepage Spring 10	0 m	28/09/2013 15:10	278 ± 58.9	2449
	5 m	28/09/2013 15:10	250 ± 53.7	2361
	10 m	28/09/2013 15:10	198 ± 37.2	1972
Seepage Spring 13	0 m	28/09/2013 16:30	257 ± 72.3	2449
	5 m	28/09/2013 16:30	226 ± 91	2361
	10 m	28/09/2013 16:30	164 ± 28.5	1972

The reduction of the measured concentrations was quite gradual (about 8-9% every 5 m), so that at a distance of 10 m from the spring LS01 (Lateral Spring 01), the ^{222}Rn in-air activity concentration remained still about 363 Bq/m³. The particular emission of this spring that is weakly turbulent suggests a greater volatilization of radon in the air compared to the slower diffuse emissions of the Seepage Springs 10 and 13.

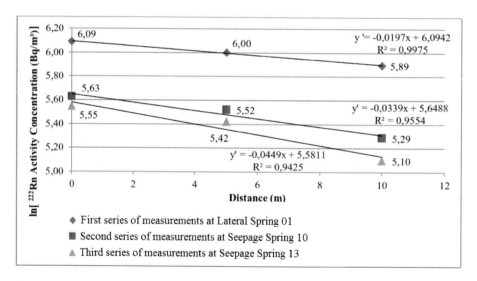

Figure 43. Decrease of 222Rn concentrations in air as a function of distance (campaign 28/09/2013).

Assuming that the springs are point source, the transfer coefficient of radon from the water body to the air was calculated with the following equation (Vinson et al. 2008):

$$C_T = \frac{\text{Increment of airborne radon added by water}}{\text{Radon activity in water}}$$

The Table 13 contains the average concentration in-water versus the corresponding average concentration Radon in-air at the selected stations.

Table 13. Radon in-water and in in-air concentration

^{222}Rn Station	Average concentrations in water (Bq/m^3)	Measurement date	^{222}Rn activity concentration in air (Bq/m^3)
		Transfer Coefficient C_T	
LS01	264000 ± 12000	06/08/2013	$402 \pm 31,2$
		28/09/2013	$442 \pm 41,5$
SS10	206000 ± 8580	28/09/2013	$278 \pm 58,9$
SS13	194000 ± 9240	28/09/2013	$257 \pm 72,3$

Comparing the average values of ^{222}Rn activity concentration in water with the corresponding values measured in air at 0 m distance from the springs, the resulting C_T values are listed in Table 14.

Table 14. Transfer Coefficient CT with its error ΔCT

Station	Measurement Date	C_T	ΔC_T
LS01	06/08/2013	0,0015	0,0002
	28/09/2013	0,0017	0,0002
SS10	28/09/2013	0,0013	0,0003
SS13	28/09/2013	0,0013	0,0003

ΔC_T was calculated in the following way:

$$\Delta C_T = C_T * \left(\frac{|\Delta A|}{A} + \frac{|\Delta B|}{B} \right),$$

where:

A = ^{222}Rn activity concentration in air;
ΔA = ^{222}Rn activity concentration in air measurement error;
B = ^{222}Rn activity concentration in water;
ΔB = ^{222}Rn activity concentration in water measurement error.

The results obtained are comparable to each other in order of magnitude, and for the two diffuse sources SS10 and SS13, taking into account the measurement errors, the transfer coefficient C_T is very similar (Figure 44).

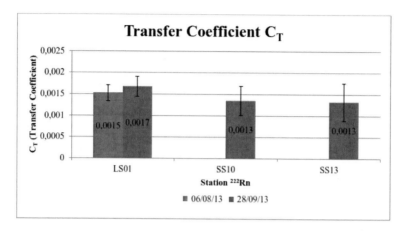

Figure 44. Transfer Coefficient CT.

Compared to the value proposed in the scientific literature, about the transfer of radon into homes from domestic services ($Ct=10^{-4}$, from Hess et al., 1982, 1990 and Nazaroff et al., 1987), the transfer coefficient estimated was approximately higher of an order of magnitude ($1.3-1.7 \times 10^{-3}$).

6.2. The Montoro Study Area

The Labso and Lauro springs, located in Montoro valley closed to the Borgo and Preturo villages, are part of a geological and hydro-geomorphological district characterized by the hydrogeological features recognized and depicted in the Figure 28.

Based on the results of the Capodifiume experiment, the same activity was repeated at the Labso and Laura spring, where previous radon in-water dataset was available. Consequently, in date 30/10/2013 two Radon in-air experiments were carried out at and surrounding the Labso and Laura springs. The Figure 45 depicts the location of measurement transects and Table 15 lists the resulting dataset.

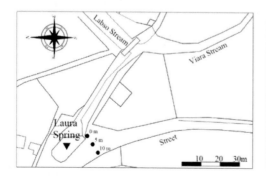 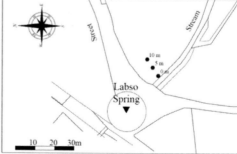

Figure 45. Location of the measurement transect.

Table 15. Data resulting from Labso-Lauro Radon in-air experiments

Station ^{222}Rn	Distance (m)	Measurement Date	^{222}Rn Activity Concentration (Bq/m^3)	RAD7 Instrument
Spring LBS_S01	0	30/10/2013 ore 10:00	37,3±18,7	2449
	5		36,8±18,4	2361
	10		39,8±23	1972
Spring LAR_S01	0	30/10/2013 ore 12:00	37±20,6	2449
	5		0	2361
	10		0	1972

Comparing the Radon data recorded at the two spring again, despite Laura spring has a higher concentration of ^{222}Rn activity in the water, no concentration of Radon is present in the air at a 5m-distance; differently for Labso spring, although its lower , there is a concentration of radon at a distance of 10 m from the channel bank. This difference is due to the fact that during the measurement in the air, near the two sources, there were different weather conditions. In fact, in the case of Laura there was wind with a speed of about 6 km/h, while in the case of Labso source it was approximately equal to 0 km/h.

6.3. Radon in-Air Assessment from Spring Water Radon Degassing

In consideration of the lacking in legislation regulating the threshold values of exposure to Radon in outdoor areas, in this chapter is considered the regulations provided by the Legislative Decree April 9, 2008, n ° 81 regarding workers and by the Recommendation of the EURATOM n.143/90 February 21, 1990, regarding the population, both designed for confined spaces (indoor). The above LD April 9, 2008 provides that the following levels of action must not be exceeded:

- 500 Bq/m³ annual average (corresponding to 3 mSv/year for 2,000 working hours) for workers while in underground sites or on the surface in those identified areas;
- 1 mSv/year effective dose for workers in the workplace involving the use or storage of materials containing natural radionuclides and in spas;
- 0.3 mSv/year for people involved in flying activities.

Once rated an average annual exposure time in the workplace according to which the solar year is 2000 hours (about 170 hours per month), it is possible to calculate the dose to which workers are exposed in the surveyed areas, assuming that the measured concentration corresponds to the annual average value. An example of calculation referred to the Labso and Laura springs is listed in Tables 16, 17.

Referring to the case study of Capodifiume, in the following is examined the case of a part-time worker with a work schedule of no more than 20 hours per week (4 hours per day for 5 days), trying to verify the compliance with the limits prescribed by the regulations (1 μSv / year).

Table 16. LBS_S01 dose values estimated for workers at a distance 0, 5, 10 m from the edge spring at the station

Station LBS_S01			
Distance (m)	^{222}Rn Activity Concentrazion (Bq/m^3)	Dose	
		mSv/anno	μSv/h
0	37,3	0,224 ± 0,1122	0,1119 ± 0,0561
5	36,8	0,221 ± 0,1104	0,1104 ± 0,0552
10	39,8	0,239 ± 0,138	0,1194 ± 0,069

Table 17. LAR_S01 dose values estimated for workers at a distance 0, 5, 10 m from the edge spring at the station

Station LAR_S01			
Distance (m)	^{222}Rn Activity Concentrazion (Bq/m^3)	Dose	
		mSv/anno	μSv/h
0	37,3	0,224 ± 0,124	0,1119 ± 0,0618
5	0	0,00 ± 0,00	0,00 ± 0,00
10	0	0,00 ± 0,00	0,00 ± 0,00

Table 18. Dose values (mSv / h) estimated, in the case of workers, at distance 0, 5 and 10 m from shore of the lake at stations LS01, SS10 and SS13

Station LS01 (Lateral Spring 01)			
Distance (m)	^{222}Rn Activity Concentration (Bq/m^3)	Dose	
		mSv/year	μSv/h
0	442 ± 41,5	2,652 ± 0,249	**1,326 ± 0,124**
5	404 ± 34,02	2,424 ± 0,204	**1,212 ± 0,102**
10	363 ± 64,7	2,178 ± 0,388	**1,089 ± 0,194**
Station SS10 (Seepage Spring 10)			
Distance (m)	^{222}Rn Activity Concentration (Bq/m^3)	Dose	
		mSv/year	μSv/h
0	278 ± 58,9	1,668 ± 0,353	**0,834 ± 0,176**
5	250 ± 53,7	1,5 ± 0,322	**0,75 ± 0,161**
10	198 ± 37,2	1,188 ± 0,223	**0,594 ± 0,111**
Station SS13 (Seepage Spring 13)			
Distance (m)	^{222}Rn Activity Concentration (Bq/m^3)	Dose	
		mSv/year	μSv/h
0	257 ± 72,3	1,542 ± 0,433	**0,771 ± 0,216**
5	226 ± 91	1,356 ± 0,546	**0,678 ± 0,273**
10	164 ± 28,5	0,984 ± 0,171	**0,492 ± 0,085**

Table 19. Values of dose (μSv / year) received from a part-time worker for 960 h/years

4 h/d (960 h/year)	0 m mSv/year	5 m mSv/year	10 m mSv/year
LS01	1,27 ± 0,12	1,16 ± 0,10	1,05 ± 0,19
SS10	0,80 ± 0,17	0,72 ± 0,15	0,57 ± 0,11
SS13	0,74 ± 0,21	0,65 ± 0,26	0,47 ± 0,08

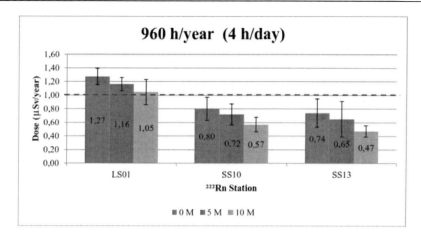

Figure 46. Annual dose received by a part-time worker in proximity of the various stations at different distances.

Table 20. Dose values estimated, in the case of the population at a distance 0, 5, 10 m from the edge spring at the station LBS_S01

Station LBS_S01			
Distance (m)	^{222}Rn Activity Concentrazion (Bq/m^3)	Dose	
		mSv/anno	µSv/h
0	37,3	1,865 ± 0,935	0,213 ± 0,107
5	36,8	1,84 ± 0,92	0,21 ± 0,11
10	39,8	1,99 ± 1,15	0,227 ± 0,131

As can be observed near the Lateral Spring 01 at a distance of 10 m from the shore the worker is exposed to an annual dose exceeding 1 mSv / year.

The study does not fall into any of the three working conditions provided by law, but wanting to make a working assumption, the only more likely condition is the working activities involving the use or storage of materials containing natural radionuclides as, regardless of the distinction between indoor or outdoor work, it has to do with ionizing radiations above those to which people are normally exposed. In this case, in fact, the greatest contribution to ionizing radiation entirely attributable to radon that evaporates from water and disperses into the surrounding area. The recommendation of the EURATOM n.143/90 February 21, 1990 on the protection of the public against exposure to radon in indoor provides for the establishment of an adequate system for reducing any exposure to indoor radon concentrations:

1) With respect to existing buildings, the reference level must not exceed an effective dose equivalent to 20 mSv per year, which, for practical purposes, can be considered equivalent to an annual average concentration of radon of 400 Bq/m³;
2) As for the buildings to be constructed, the reference level must not exceed an effective dose equivalent to 10 mSv per year, which, for practical purposes, can be considered equivalent to an annual average concentration of radon gas of 200 Bq/m³ . Bearing this recommendation, it is therefore possible to calculate, through a simple

proportion, the exposure values corresponding to the results obtained from the measurements (Table 20, 21, 22), considering the limit of 200 Bq/m³, corresponding to an effective dose equivalent to 10 mSv per year.

Table 21. Dose values estimated, in the case of the population at a distance 0, 5, 10 m from the edge spring at the station LAR_S01

Station LAR_S01			
Distance (m)	^{222}Rn Activity Concentrazion (Bq/m^3)	Dose	
		mSv/anno	µSv/h
0	37,3	1,865 ± 1,03	0,213 ± 0,118
5	0	0,00 ± 0,00	0,00 ± 0,00
10	0	0,00 ± 0,00	0,00 ± 0,00

Table 22. Dose values (mSv / h) estimated, in the case of population, at distance 0, 5 and 10 m from shore of the lake at stations LS01, SS10 and SS13

Station LS01 (Lateral Spring 01)			
Distance (m)	^{222}Rn Activity Concentration (Bq/m^3)	Dose	
		mSv/year	µSv/h
0	442 ± 41,5	22,1 ± 2,075	2,523 ± 0,236
5	404 ± 34,02	20,2 ± 1,701	2,306 ± 0,194
10	363 ± 64,7	18,15 ± 3,235	2,072 ± 0,369
Station SS10 (Seepage Spring 10)			
Distance (m)	^{222}Rn Activity Concentration (Bq/m^3)	Dose	
		mSv/year	µSv/h
0	278 ± 58,9	13,9 ± 2,945	1,587 ± 0,336
5	250 ± 53,7	12,5 ± 2,685	1,427 ± 0,306
10	198 ± 37,2	9,9 ± 1,86	1,13 ± 0,212
Station SS13 (Seepage Spring 13)			
Distance (m)	^{222}Rn Activity Concentration (Bq/m^3)	Dose	
		mSv/year	µSv/h
0	257 ± 72,3	12,85 ± 3,615	1,467 ± 0,412
5	226 ± 91	11,3 ± 4,55	1,289 ± 0,519
10	164 ± 28,5	8,2 ± 1,425	0,936 ± 0,162

The value of estimated dose received per hour associated with exposure to radon is, in all stations and at all distances examined, significantly higher (Figure 47) than the dose associated with the natural background (from 0.10 to 0.12 µSv/h).

Unlike the case of the springs of Labso and Laura, near the springs of Capodifiume at all stations and at all distances examined, the calculated values demonstrate that the dose received per hour associated with exposure to radon, is significantly higher than the natural background γ (approximately an order of magnitude). This condition could pose a serious risk to human health and therefore requires further study and, in the meanwhile, regulation and mitigation measures.

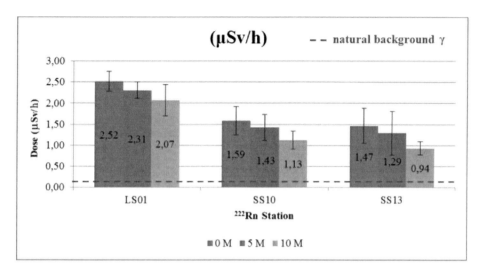

Figure 47. Hourly dose received by population in proximity of the various sections at different distances.

The analysis represents one of the first attempt concerning the monitoring of radon in selected study areas: Labso and Laura Sources, situated in the plain of Montoro Inferiore and Capodifiume, springs.. The results obtained from the analysis performed on the base of the measurements of radon concentration available for the two case study suggest a protocol and method to assess this scenario not considered in national and European legislation, providing an experimental procedure in the Radon risk planning.

CONCLUSION

The implementation of the Radon measurement techniques have confirmed the prospective of using these methodologies in a karst Mediterranean environment to investigate the complex interactions and exchanges between streamflow and groundwater. In particular the Rn-222 concentration analysis at the Bussento river basin performed for different space-temporal scale of analysis will be considered an appropriate method for the identification of springs inflowing the river, for their classification with respect to the origins of the karst springs (fracture basal karst spring, conduit and cave intermediate karst spring) and to made the Hydro-geological scheme of the Bussento river basin. The same procedure, applied to the Labso and Laura Springs in Montoro municipality, was usefull in distinguishing an allogenic-autogenic recharge for the Labso spring and an exclusively autogenic recharge for the Laura spring and then in the reinterpretation of the hydro-geological setting of Montoro study area. In addition, experimental Rn-222 data concentrations and physical-chemical data and streamflow rate used as constraint for the Rn222 analysis was usefull in the interpretation of some anomalous in the discharge pattern. In fact, in the Tusciano river basin was possible identifying losses of river discharge in the aquifer preliminary identified with the decrease of discharge downstream the river and then conrfimed by the Rn222 and Ec pattern.

The future aim of this research program is to continue and improve these studies using Radon as a naturally occurring tracer in order to investigate the complex interactions and exchanges between streamflow and groundwater confirming radon as an effective environmental tracer in hydro-geomorphological analysis.

REFERENCES

Allocca V., Celico F., Celico P., De Vita P., Fabbrocino S., Mattia S., Monacelli G., Musilli I., Piscopo V., Scalise A.R., Summa G. & Tranfaglia G. (2007) - Note illustrative della Carta idrogeologica dell'Italia meridionale (Responsabili Scientifici: Celico P., De Vita P., Monacelli G., Tranfaglia G.) - Istituto Poligrafico e Zecca dello Stato, ISBN 88-448-0215-5, p. 211, con carte allegate, ISBN 88-448-0223-6 (3 tavole fuori testo).

Andrews, J. N. and D. F. Woods, Mechanism of radon release in rock matrices and into groundwaters, *Trans. Inst. Min. Metall. Sec.,* B81, 1972, 198-209.

Bakalowicz M., The infiltration zone of karst aquifers, *Hydrogeologie,* 4, 1995, 3-21.

Barbieri, M., Boschetti, T., Petitta, M., Tallini M., Stable isotopes (2H, 18O and 87Sr/86Sr) hydrochemistry monitoring for groundwater hydrodynamics analysis in a karst aquifer (Gran Sasso, Central Italy), *Applied Geochemistry,* 20(11), 2005, 2063-2081.

Bisci, C., et al., in: Landslide Recognition, 1996, pp. 150-160.

Bonardi, G., Ciampo, G. & Perrone, V. (1985). La formazione di Albidona nell'Appennino calabro-lucano: ulteriori dati stratigrafici e relazioni con le unità esterne appenniniche, *Bollettino della Società Geologica Italiana,* vol. 104, Roma.

BonardiG., La Carta Geologica dell'Appennino Meridionale, Mem. Soc. Geol. It., 41. Presented at the Congress of Italian Geological Society, Sorrento, Italy, 1988.

Brahana J.V., and Hollyday E.F., Dry stream reaches in carbonate terranes − Surface indicators of ground-water reservoirs, *American Water Resources Association,* 24(3), 1988, 577-580.

Cammarosano A. et al., Il substrato del Gruppo del Cilento tra il M. Vesalo e il M. Sacro (Cilento, Appennino Meridionale), *Bollettino della Società geologica italiana,* 119 (2), 2000, 395 - 405. (in Italian).

Cuomo, A., Guida, D., & Palmieri, V. (2011). Digital orographic map of peninsular and insular Italy. *Journal of Maps,* 7(1), 447-463.

D'Argenio B., Pescatore T. & Scandone P., Schema geologico dell'Appennino meridionale (Campania-Lucania). Atti del Convegno: Moderne vedute sulla geologia dell'Appennino. *Ac. Naz. Lincei, Quad.,* 1973, 183.

Dassonville L. and Fé d'Ostiani L., Mediterranean watershed management: overcoming water crisis in the Mediterranean, Watershed Management: Water Resources for the Future - Working Paper 9, Chapter 6, Watershed Management & Sustainable Mountain Development Conference Proceedings (Porto Cervo, Sassari, Sardinia, Italy, 22-24 October 2003).

DURRIDGE, RAD7 RADH20 Radon in Water Accessory. Owner's Manual (Bedford, MA.), 2012, USA.

DURRIDGE, RAD7 RADON DETECTOR. Owner's Manual (Bedford, MA.), 2012, USA.

Eisenlohr L. and Surbeck H. , Radon as a natural tracer to study transport processes in a karst system. An example in the Swiss Jura, C.R. *Acad. Sci.*, Paris, 321, série IIa, 1995, 761-767.

Ellins, K. K., Roman-Mas A., Lee R., Using 222Rn to examine Groundwater/surface discharge interaction in the Rio Grande de Manati, Puerto Rico, *Journal of Hydrology*, 115, 1990, 319-341.

Emblanch C., Zuppi G.M., Mudry J., Blavoux B., Batiot C., Carbon-13 of TDIC to quantify the role of the unsaturated zone: the example of the Vaucluse karst systems Southeastern France, *Journal of Hydrology*, 279, 2003, 262-274.

EWFD, 2000. Directive 2000/60/EC of the European Parliament and of the Council of 23 October 2000 establishing a framework for Community action in the field of water policy (OJ L 327, 22.12.2000, p. 1). (The EU Water Framework Directive (EWFD) - integrated river basin management for Europe, (http://ec.europa.eu/environment/water/ water-framework/index_en.html).

FAO, Developing participatory and integrated watershed management, *Community Forestry Case Study Series,* 13, Rome, 1998.

FAO, Strategies, approaches and systems in integrated watershed management, *Conservation Guide,* 14, Rome, 1986.

Ford, D. C. and Williams P. W., Karst geomorphology and hydrology, London, Unwin, Hyman, 1989.

Ford, D. C. and Williams, P. W., Karst Hydrogeology and Geomorphology. (Reprint with corrections of the 1989 title "Karst geomorphology and hydrology"). J. Wiley and Sons Ltd, Chichester, 2007.

Gainon F., Goldscheider N., Surbeck H., Conceptual model for the origin of high radon levels in spring waters –the example of the St. Placidus spring, Grisons, Swiss AlpsSwiss, *J. Geosci.*, 100, 2007, 251–262.

Genereaux, D. P., and Hemond, H. F., Naturally occurring radon-222 as a tracer for streamflow generation: steady state methodology and field example, *Water Resources and Resources*, 26 (12), 1990, 3065 - 3075.

Genereaux, D. P., Hemond, H. F., and Mulholland, P. J., Use of radon-222 and calcium as tracers in a three-end-member mixing model for streamflow generation on the wet fork of Walker Branch watershed, *J. Hydrol.*, 142, 1993, 167- 211.

Goldscheider N. and Drew D., Methods in Karst Hydrogeology. Taylor & Francis Group, London, 2007, UK.

Grath J., Ward R., Quevauviller P., Common Implementation Strategy for the Water Framework Directive (2000/60/EC), Guidance Document No. 15 – Guidance on Groundwater Monitoring. Publications of the European Community, Luxembourg ISBN 92-79-04558-X - ISSN 1725-1087.

Guadagnuolo, D. (2009). Investigation of the groundwater-river interaction, using Radon-222 as a natural tracer, in a karst Mediterranean environment like in the case study of the Bussento river basin, PhDThesis. University of Salerno.

Guadagnuolo D., Guida D., Guida M., Cuomo A., Siervo V., Schubert M., Knoller K., Aloia A. (2013). "Origin of high radon levels in karst spring mixed waters – the case-study of the Capodifiume spring group, National Park of the Cilento and Vallo di Diano - European Geopark - Southern Italy", 12th European Geoparks Conference, Ascea (SA), 3-4 Settembre 2013.

Gudzenko V., Radon in subsurface water studies. In: Isotopes of Noble Gases as Tracers in Environmental Studies. Proceedings of a Consultants Meeting, *International Atomic Energy Agency*, Vienna, 1992, 249-261.

Guida, D., Guida, M., Luise, D., Salzano, G., Vallario, A., Idrogeologia del Cilento (Campania), *Geologica Romana* – Vol. XIX, 1980, 349-369.

Guida, D., Longobardi, A., Ragone G., Villani P., Hydrogeological and hydrological modeling for water resources management in karstic landscape at the basin scale. A case study: the Bussento river basin. 3rd EGU General Assembly, Vienna, Austria, 02-07 April 2006.

Hakl J., Hunyadi I., Csige I., Geczy G., Lenart L., Varhegyi A., Radon transport phenomena studied in karst caves: International experiences on radon levels and exposures. *International Conference on Nuclear Tracks in Solids* No. 18, (Cairo, Egypte, 01/09/1996) vol. 28, no. 1-6, 1997, 675 - 684.

Hamada H., Estimation of groundwater flow rate using the decay of 222Rn in a well, *Journal of Environmental Radioactivity*, 47, 2000, 1-13.

Hess, C.T., Vietti, M.A., Lachapelle, E.B., Guillemette, J.F., 1990. Radon. transferred from drinking water into house air. In: Cothern, C.R., Rebers, P.A. (Eds.), Radon, Radium, and Uranium in Drinking Water. Lewis Publishers, Chelsea, Michigan, pp. 51–67 (Chapter 5).

Hess, C.T., Weiffenbach, C.V., Norton, S.A., 1982. Variations of airborne and waterborne Rn-222 in houses in Maine. *Environ. Int.* 8, 59–66.

Hoehener P., Surbeck H., 222Rn as a tracer for nonaqueous phase liquid in the vadose zone: experiments and analytical method, *Vadose Zone Journal*, 3, 2004, 1276-1285.

Hoehn E. and von Gunten H. R., Radon in Groundwater: a tool to assess infiltration from surface waters to aquifers, *Water Resources and Research*, 25, 1989, 1795 - 1803.

Hooper, R.P. and C.A. Shoemaker A comparison of chemical and isotopic streamflow separation, *Water Resour. Res.*, 22 (10), 1986, 1444-1454.

Iaccarino G., Guida D., Basso C., Caratteristiche idrogeologiche della struttura carbonatica di Morigerati (Cilento Meridionale), *Memorie Società Geologica Italiana*, 41, 1988, 1065-1077, 2ff., 5 tabb, 1 tav. (in Italian).

IAEA. Rickwood P., 2002. *IAEA Bulletin*, 44/1/2002, pp. 21-24.

KafriU., Radon in Groundwater as a tracer to assess flow velocities: two test cases from Israel, *Environmental Geology*, vol. 40, Issue 3, 2001, 392-398.

Karanth K., Groundwater assessment. *Development and management*, Tata McGraw-Hill, 1987.

Kendale, C. & McDonnell, I.J. (1998). Isotope tracers in catchment hydrology, Elsevier, New York, pp. 40 – 41, ISBN 0-444-50155-X.

Kies, A., Hofmann, H., Tosheva, Z., Hofmann, L. & Pfister, L. (2005). Using Radon-222 for hydrograph separation in a micro basin (Luxembourg), *Annals of geophysics*, Vol.48, No.1, pp. 101-107.

Kraemer T.F. and Genereaux D.P., Applications of Uranium- and Thorium-Series Radionuclides in Catchment Hydrology Studies. in: Isotope Tracers in Catchment Hydrology (Kendall C. and McDonnell J.J. (Eds.)). Elsevier, Amsterdam, 1998, 679-722.

Lee R. and Hollyday E. F., Radon measurement in streams to determine location and magnitude of ground-water seepage. in: Radon, radium, and other radioactivity in groundwater. (Graves B. (Ed.)), 1987, 241-249. Lewis Publishers, Inc. (Chelsea, Mich.).

Lee R. and Hollyday E. F., Use of radon measurements in Carters Creek, Maury County, Tennessee, to determine location and magnitude of groundwater seepage. in: Field studies of radon in rocks, soils and water. (Gundersen, L.C. and Wanty, R.B. (Eds.)), 1991, 237-242. U. S. Geological Survey Bulletin.

Levêque P. S., Maurin C., Severac I., 1971. Le 222Rn traceur naturel complementaire en hydrologie souterranie. C. R. Hebd. *Seances Acad. Sci.*, 272, 18, 2290.

Levêque P. S., Maurin C., Severac I., Le 222Rn traceur naturel complementaire en hydrologie souterranie, C. R. Hebd. *Seances Acad. Sci.*, 272, 18, 1971, 2290.

Longobardi A. and Villani P., Baseflow index regionalization analysis in a Mediterranean environment and data scarcity context: role of the catchment permeability index, *Journal of Hydrology*, 355, 2008, 63-75.

Loucks D. and Gladwell J., *Sustainability criteria for water resource systems*, Cambridge University Press, 1999.

Margat J. and Vallée D., Water Resources and Uses in the Mediterranean Countries. Figures and Facts. The Mediterranean in Figures. Blue Plan for the Mediterranean. Regional Activity Centre, Sophia-Antipolis, France, 224 pp., 2000.

Marine I. W., The use of naturally occurring helium to estimate groundwater velocities for studies of geologic storage of radioactive waste, *Water Resources and Research*, 15, 1979, 1130-1136.

McDonnell I. J., Where does water go when it rains? Moving beyond the variable source area concept of rainfall-runoff response, *Hydrological Processes*, 17, 2003, 1869- 1875.

Menéndez M. and Pinero M. J., Common Implementation Strategy for the Water Framework Directive (2000/60/EC) - Guidance Document No 11, Planning Processes Produced by Working Group 2.9 – Planning Processes, Office for Official Publications of the European Communities, Luxembourg: 2003 - ISBN 92-894-5614-0- ISSN 1725-1087.

Montgomery D. R. and Buffington J. M., Channel processes, classification, and response. In: River Ecology and Management (Naiman R.J. and Bilby R.E., Eds.), 1998, Springer Verlag.

Nazaroff, W.W., Doyle, S.M., Nero, A.V., Sextro, R.G., 1987. Potable water as a source of airborne 222Rn in US dwellings: a review and assessment. *Health Phys.* 32, 281–295.

Quevauviller PH., 2005. Groundwater monitoring in the context of EU legislation: reality and integration needs, The Royal Society of Chemistry, *J. Environ. Monit.*, 7, 2005, 89–102.

Rama and Moore, W. S., Mechanism of transport of U–Th series radioisotopes from solids into ground water, *Geochim. Cosmochim. Acta*, 48, 1984, 395–399.

Rogers A., Physical behaviour and geologic control of radon in mountain streams, U.S. *Geological Survey Bulletin*, 1052 – E, 1958.

Schubert M., *Personal Communication*, 2008.

Schubert M., Schmidt A., Paschke A., Lopez A., Balcazar M., In situ determination of radon in surface water bodies by means of hydrophobic membrane tubing, *Radiation Measurements*, 43, 2008, 111 – 120.

Semprini L., Radon-222 concentration in groundwater from a test zone of a shallow alluvial aquifer in the Santa Clara Valley, California. in: Radon in Groundwater, (Graves B., (Ed.)). Lewis Publishers, Clelsea, MI, 1987, 205-218.

Shah T., Molden D., Sahthiradelvel, Seckler D., The global situation of groundwater: overview of opportunities and challenges, *International Water Management Institute*, 2001.

Shapiro M.H., Rice A., Mendenhall M.H., Melvin D. and Tombrello T.A., Recognition of environmentally caused variations in radon time series, *Pure and Appl. Geophys.*, 122, 1984, 309-326.

Simonovic S., Water resources engineering and sustainable development, Proceedings of the XXVI Congress of Hydraulics (Catania, Italy, 1998).

Smith L., Using Radon-222 as a tracer of mixing between surface and ground water in the Santa Fe river sink/rise system. *Thesis presented to the graduate school of the University of Florida,* 2004.

Solomon D. K., Cook P. G., and Sanford W. E.,. Dissolved Gases in Subsurface Hydrology. in: Isotope Tracers in Catchment Hydrology (Kendall C. and McDonnell J.J. (Eds.)). Elsevier, 1997, pp. 291-318.

Solomon D. K., Poreda R. J., Cook P. G. and Hunt A., Site characterization using 3H/3He groundwater ages (Cape Cod, MA.), *Ground Water,* 33, 1995, 988-996.

Solomon D. K., Schiff S. L., Poreda R. J. and Clarke W. B., A validation of the 3H/3He method for determining groundwater recharge, *Water Resources and Research*, 29, 9, 1993, 2851-2962.

Sultankhodzhaev A. N., Spiridonov A. I., Tyminsij V. G. , Underground water's radiogenic and radioactive gas ratios (He/Rn and Xe/Rn) in groundwaters and their utilization for groundwater age estimation, *Uzbek Geol. J.* 5, 1971,p. 41.

Surbeck H., Dissolved gases as natural tracers in karst hydrology; radon and beyond. UNESCO Chair "ErdélyiMihály" School of Advanced Hydrogeology, Budapest, Hungary, 22-27 August 2005.

Tolstikhin I. N. and Kamensky I. L., Determination of groundwater age by the T-3He method, Geochem. Int., 6, 1969, 810-811.

Tulipano L., Fidelibus D., Panagopoulos A., COST Action 621 Groundwater management of coastal karstic aquifers, Final report, 2005.

Vervier P. and Gibert J., Dynamics of surface water/groundwater ecotones in a karstic aquifer, *Freshwater Biol.*, 26(2), 1991, 241-250.

Vinson D.S., Campbell Ted R. & Avner Vengosha. Radon transfer from groundwater used in showers to indoor air. *Applied Geochemistry* 23 (2008) 2676–2685.

Vogel R. M. and Kroll C.N., Regional geohydrologic-geomorphic relationships for the estimation of low-flow statistics, *Water Resources and Research*, 28 (9), 1992, 2451-2458.

White W. B., Conceptual models for karstic aquifers. re-published from: Karst Modeling: Special Publication 5 (Palmer A.N., Palmer M.V., and Sasowsky I.D. (Eds.)). The Karst Waters Institute, Charles Town, West Virginia (USA), 2003, 11-16.

White W. B., Conceptual models for limestone acquifers, Groundwater, 7 (3), 1969, 15-21.

White W. B., . Conceptual models for carbonate acquifers: revised. In: Hydrologic Problems in Karst Terrain (Dilamarter, R. R. and Casallany, S. C. (Eds)). WesternKentuckyUniversity, *Bowling Green*, KY, 1977, 176-187.

White, W. B., Geomorphology and hydrology of karst terrains, OxfordUniversity Press, 1988, Oxford.

Winter T., Management of groundwater and surface water, *Review of Geophysics,* Supplement 33, 1995.

Wu Y., Wen X. and Zhang Y., Analysis of the exchange of groundwater and river water by using Radon-222 in the middle HeiheBasin of northwesternChina, *Environmental Geology,* 45(5), 2004, 647–653.

Yoneda, M., Inoue, Y., and Takine, N., Location of groundwater seepage points into a river by measurement of 222Rn concentration in water using activated charcoal passive collectors, *Journal of Hydrology*, 124, 1991, 307–316.

In: Radon
Editor: Audrey M. Stacks

ISBN: 978-1-63463-742-8
© 2015 Nova Science Publishers, Inc.

Chapter 3

HISTORY-RADON-GEOLOGY CONNECTIONS

I. Burian, J. Merta and P. Otahal

National Institute for Nuclear, Chemical and Biological
Protection (SUJCHBO), Czech Republic

ABSTRACT

The history of Central Europe (Bohemia) is connected with natural radioactivity. In the first part of the 20th century a first maximum allowed concentration of radon for uranium miners was implemented. Recently we measured a very high radon concentration in soil air at depths 0.8 m (3 $MBq \cdot m^{-3}$) in the center of the town Jachymov (Joachimsthal) – where Mme Curie discovered the element radium.

In the second half of the 20th century uranium became a strategic raw material and prospecting for uranium deposits became very intense. One of the prospecting methods was placing a deposition foil at the depth of about one meter and then measuring the surface activity caused mostly by ^{218}Po (first radon decay product) deposited on it.

Other sources of information about radium concentration can be obtained using gamma dose rate measurements, accomplished either by walking, using cars, or flying in airplanes. A map of gamma dose rates of the whole country was published in 1977 ([5]).

There is a rough correlation between uranium and radium concentrations, and of radon, in soil. This can be demonstrated based on large number of measurement.

In the 1980s the uranium mining boom was decreasing and all over the world the radon industry was born. In Bohemia (Czech Republic) it is very justified, due to high radon concentration in soil, and to bad insulation of building foundations.

At present the evaluation of radon risk index in building sites is compulsory. This evaluation is based on at least 15 measurements of radon-in-soil at the depth 0.8 m. There are two more necessary parameters utilized: the permeability of soil and estimation of geologic structure (possibility of change of parameters in vertical profile).

The reason of this detailed local evaluation is that, due to complicated geologic structure, within a distance of the order of meters the concentration of radon in soil air could be dramatically different.

For quality control of results the measuring companies are certified by SONS (State Office for Nuclear Safety). One of the necessary requirements is the accuracy of a measuring device here the checking is delegated to our Authorized Metrologic Center, which is traceable to PTB Braunschweig, Germany).

There exists a correlation between geologic characteristics – uranium (radium) content in soil (gamma dose) - and radon in soil (influenced by permeability and the presence or absence of geological faults). There is also a correlation between global estimation of radon risk emission (map of risk 1:200 000) and radon in underground water. These relationships will be shown here using results of one of our research tasks.

INTRODUCTION

The Bohemian Massif in Central Europe was affected by several mountain building movements in the past (such as Moldanubian folding, Proterozoic folding, Caledonian folding, Variscean folding, Alpine folding), which occurred from Proterozoic era up to Terciary. These movements were followed by marine ingression and regressions associated with the origin of the sedimentary cover.

Volcanic activity occurred frequently, rocks underwent regional and contact metamorphoses, which created tectonic and structural settings of the landscape.

All orogenetic processes gradually converted the Bohemian Massif into a stable shield (kratogen) resisting to following orogenetic deformations.

Afterwards, Quaternary era was an epoch of relative dynamic calmness. It can be supposed that regional movement was a phenomenon generating transport of rock material including radioactive elements.

Magmatic rocks exhibit a relationship between occurrence of radioactive minerals and acidity of rocks. Concentrations of uranium and radium in acidic magmatic rocks are up to seven times higher in comparison with basic rocks.

Sedimentary rocks have contents of radioactive elements generally slightly lower. Clays belong to the most radioactive sediments, while quartzites, limestones and dolomites are the least radioactive.

Radioactivity of metamorphic rocks is somewhere between the above genetic types of rocks, and its great variability is typical.

We are dealing with natural radioactivity, of course. Natural radioactivity has been occurring in the Bohemian basin (Bohemia – western part of the Czech Republic) for centuries. Around WWI many institutions were anxious to get ^{226}Ra found in the Ore Mountains, in the mining town of Joachimsthal (Jachymov).

In all decay chain series there is a gaseous isotope, of which most followed is ^{222}Rn. The reasons for studying radon are at least three:

1　General thirst of people for knowledge
2　Search for the possibility of earthquake prediction
3　Trying to understand the details, which could help to predict the concentration in dwellings, and so lower the cases of lung cancers.

Main source of radon is the geological basement, where radon is released from grains of radioactive minerals into integranular spaces, in dependence on the coefficient of emanation and other physical factors.

Radon risk of a site increases with the radioactivity of basement forming rocks (e.g., acid magmatic rocks), and with their increased permeability (e.g., conglomerates).

Higher radon flows can be expected in tectonically disturbed areas where faults are communication channels for radon movement.

Miners' lung cancer occurrence has increased due to inhalation of radon progeny - [1], [2]. Incidentally, there were about 200000 people who worked in uranium mines in (former) Czechoslovakia.

International Commission on Radiation Protection (ICRP) assumes Linear No-Threshold (LNT) hypothesis, therefore even small doses cause an increase in morbidity. In consequence, "radon dwellings" may not be harmless [3]. In view of the mentioned subsoil full of micro cracks, also in the floors, the average Rn concentration in Czech dwellings is 100 Bq/m^3. The highest found level was above 100 000 Bq/m^3, in a village Petrovice located on syenites (durbachites), which are part of the central Bohemian pluton occurring between Moldanubian and Proterozoic blocks.

Based on the above, the Czech Republic has devoted, for an extended period of time, a significant attention to limiting the infiltration of radon into dwellings, as well as with questions trying to understand these matters. Part of this effort is measuring Rn or radon progeny in older buildings, using passive solid state nuclear track detectors (SSNTDs), and measuring using electrets etc., in newly built houses. Before the building starts, a company (whose qualifications are certified by the State Office for Nuclear Safety - SONS) performs evaluation of the site.

The recommended procedure is described in [4]. Importantly, the instruments used for measurements have to be calibrated in our Authorized Metrologic Center (AMC), traceable to Physikalisch-Technische Bundesanstalt in Braunschweig, Germany.

Determination of radon index of a building site is based on measurement of radon activity concentration in soil gas sampled from the depth of 0.8 m. Minimal number of measured field stations is fifteen, and the third quartile is applied as representative value of group of observed data. The highest value of radon in soil gas observed in the Czech Republic reached 3000 kBq.m^{-3}.

The second important parameter for radon index estimate is the permeability of soils. Soil permeability can be either determined by an instrument measuring velocity of soil gas movement caused by pumping under pressure generated by means of a weight, or estimated by expert assessment of soil samples from shallow hand drills.

Further, a pedological character is determined by macroscopic description a core obtained by means of handheld drill.

Finally a gamma field is measured at 1m above surface.

Part of our institute is performing such measurements; the results of such measurements are given below.

RESEARCH RESULTS

Basic Correlations Radon – Gamma

The measurements were performed on 325 sites. The dose is expressed in µGy/h, the Rn soil concentration in kBq/m^3. The third quartile of the Rn concentrations is plotted.

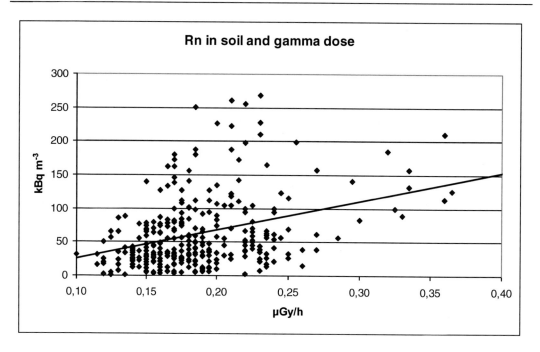

Figure 1. Correlation between soil Rn concentration and dose/hour.

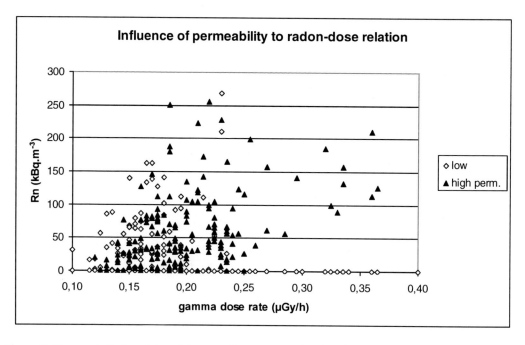

Figure 2. Gamma (µGy/hour) (x-axis) versus Rn concentration (y-axis).

We can assume that the Rn concentration in the layer at the surface depends on radium concentration in soil, as well as on soil permeability for Rn.

The hypothesis has not been confirmed. The surface layer of soil plays a negligible role in the relation between gamma dose rate (^{226}Ra content in subsoil) and Rn concentration.

Table 1. Permeability classes

permeability	k (10^{-14} m^2)
low	k < 5.2
medium	5.2 < k < 1800
high	k > 1800

Table 2. Classification of rocks

Magmatic rocks:	
1	granites, granodioris
2	syenites, durbachites
3	diorites
4	porphyries, porphyrites, aplites, pegmatites
Effusive rocks	
5	spilites, basalts, diabases, phonolites, melaphyres, tuffs, tuffites
Sedimentary rocks	
6	sandstones, arcoses, greywackes, conglomerates, quartzites, siltstones
7	clays, claystones, clay shales
8	limestones, dolomites
9	Loess's, secondary loess's, wind-blown sands, gravels, sandy gravels, terraces, alluvium
Metamorphic rocks	
10	silicates, shists
11	phyllites, mica shists, orthogneisses, paragneisses
12	marbles, quartzites

There are even some absurd tendencies – lower permeability leads to somewhat higher Rn concentrations!

Permeability was divided according to Table 1.

Correlations with Geological Estimates

The results of measurements were compared with estimates of the character of rock. The rocks were divided into following categories.

In some categories, due to scarcity of data, we are not including them into plots or tables.

The scatter of the points is considerable, still certain difference between g.c.1 (granites) and g.c.6 (sandstones) exists (see also Table). On the x-axis is uninteresting serial number of measurement.

On the x-axis is uninteresting serial number of measurement.

It is clear here that higher Rn concentration is present only in granites, and not in sandstones and similar rocks. Permeability probably plays an insignificant role.

It is further investigated the permeability.

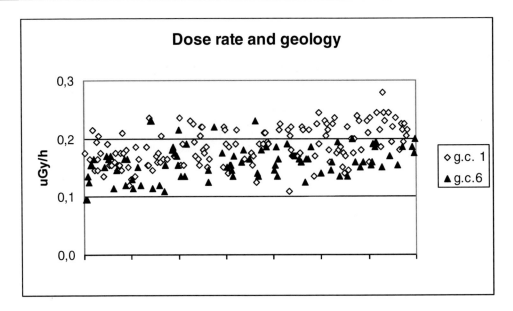

Figure 3. Dose rate for various rocks.

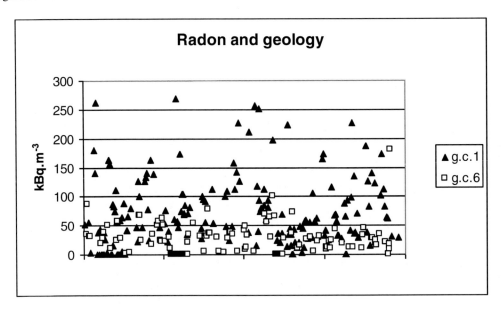

Figure 4. Radon concentration in soil for various rocks (1 - granites; 6 - sandstones).

On the x-axis is uninteresting serial number of measurement.

None of the values are quite uniquely different for various geological categories.

In the case of dose rate there appear some trends, as found by measurements across the whole country [5]. In case of Rn there is lower concentration for sands and clays.

No trends can be discerned for permeability.

It should be noted that this quantity was measured only for shallow region of the soil. So the correlation with rock type is very vague, in the rock itself (esp. granites) the gas permeability will be orders of magnitude smaller.

Table 3. Effect of rocks characteristic to dose rate and Rn soil concentration

	Dose rate (μGy/h)	Rn (low permeability)	Rn (high permeability)
g.c.1	0.20±0.05 n=167	AM 79 kBq·m^{-3} n=38	AM 72 kBq·m^{-3} n=66
g.s.6	0.17±0.05 n=99	AM 38 kBq·m^{-3} n=16	AM 25 kBq·m^{-3} n=43

Table 4. Influence of rock type to dose rate and radon in soil

geol. category	dose rate (μGy/h)	radon concentration (kBq·m^{-3})	frequency of permeability
1 (granites)	0.20±0.05 n=167	AM 78; 87^x/2,3	0.26 low; 0.45 high
4 (porhyries)	0.21±0.06 n=22	AM 84; 84^x/2,2	~0.20 low; ~0.50 high
6 (sandstones)	0.17±0.05 n=99	AM 30; 34^x/2,2	0.22 low; 0.52 high
7 (clays)	0.16±0.02 n=28	AM 35; 38^x/2,4	~0.44 low; ~0.37 high
11 (phyllites)	0.18±0.03 n=27	AM 70; 68^x/2,0	~0.14 low; ~0.45 high

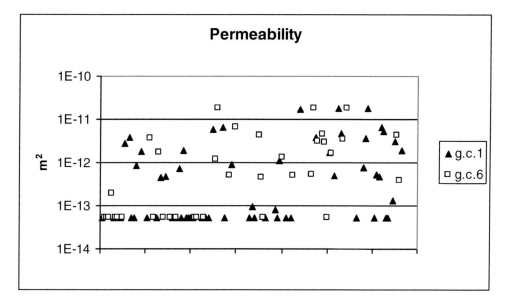

Figure 5. Permeability over different rocks (1 = granites, 6 = sandstone).

Radon Flux

In some cases radon flux from subsoil was measured. Basically, a rise in Rn concentration is followed in a 13 l vessel put on the surface. There were two such measurements performed on each site.

Results can be correlated with the estimates of geological categories.

g.c.1 (granites) – flux is 520 Bq·m^{-2}·h^{-1} (GSD 2.7)

g.c.6 (sandstones) - flux is less: 260 Bq·m^{-2}·h^{-1} (GSD 2.1)

Flux has to be correlated to the Rn content in the subsoil. The relationship of the flux to Rn soil concentration is basically the velocity of radon movement. For all determined permeabilities it is about 1 cm/hour.

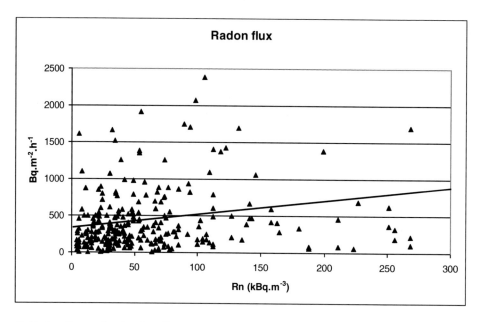

Figure 6. Radon flux and radon in soil.

The Possibilities of Using Maps

Based on years of measurements of many sites, a map was created of Rn risk, scale 1:200 000.. The regions are divided into three categories; however, the boundaries delineating these regions have only probabilistic character. It is clear from the following table 5. Based on our measurements, in regions with low risk, there are sites with high volume Rn activity in the subsoil. It agrees with the findings of other institutions measuring building sites. A detailed evaluation of a radon risk for individual sites requires then measurements of the whole site area.

Table 5. Correlation between site measurements and risk estimates

	$a_V < 20$ kBq·m^{-3}	20 kBq·m^{-3} < a_V < 70 kBq·m^{-3}	$a_V > 70$ kBq·m^{-3}
RI=1 (lowest)	40%	56%	4%
RI=2	18%	57%	25%
RI=3	9%	43%	47%

Clearly, there exist correlation between Rn in subsoil (characterized here by radon index (RI) taken from the map scale 1:200 000) and the Rn concentration in water.

Radon in Water

Many measurements have been made of various radionuclides in private wells. There appears certain correlation between U (and ^{226}Ra) – correlation coefficient about 0.3.

Table 6. Radon concentration in water

	RI:	1 (lowest)	2	3
n		99	241	82
Rn concentration	AM	45	56	195
in water (Bq/l)	GM $^x/$ GSD	38 $^x/$ 4.3	56 $^x/$ 4.1	281 $^x/$ 5.3

CONCLUSION

The results presented here should be understood as probabilistic, concerning averages of groups of measurements. Even for low Rn contents in subsoil there are cases of high Rn flux etc. Our data have been collected from an area 7000 km^2 (the whole country is about 79 000 km^2), however, we expect our results apply to the whole country.

It is clear it is very problematic trying to predict the Rn flux from the measurements of other quantities, or even from the estimate of geological character of the subsoil.

What we found was expected: in subsoils with higher Ra concentration (higher dose rate), there is higher radon concentration in soil. Higher Rn concentrations are above granites, not over sands or similar rocks. Permeability apparently does not play a significant role. Rn flux is higher for granites than for sandstones. Rn flux goes up with higher Rn in subsoil. Permeability probably does not play a significant role. Clearly, locations with high estimates of higher Rn content in subsoils (based on the risk map), there are higher concentrations in wells. According to our measurements, at locations with low radon risk estimates based on the risk map, we can find some sites with high volume Rn activities in the subsoils.

APPENDIX

Table 7. Ways of acquirements

aspect	"measurement"
geologic category	estimation
↓	
U, Ra concentration in rock	gamma dose rate
↓	
radon in soil	concentration in soil air
transport of Rn in surface layer	permeability
radon flow from surface	radon flux
↓	
Rn concentration in dwellings	SSNTD
detriment	epidemiologic study

This chapter is not dealing with aspects in the lower part of the table.

ACKNOWLEDGMENTS

Part of research was supported by the project by the Czech Ministry of the Interior VG20102014035.

REFERENCES

[1]   Sevc, J., Placek, V. *Pracov. Lek.* 1966, 18, 438-442 (in Czech).
[2]   Tomasek, L. *J. Radiol. Prot.* 2012, 32, 301-314.
[3]   Tomasek, L. *Neoplasma* 2012, 59 5, 559-565.
[4]   www.sujb.cz/fileadmin/sujb/docs/radiacni-ochrana/121031_Doporuceni_RIP.pdf (Czech).
[5]   Matolin, M. *Geophysik und Geologie*, Akademie-Verlag: Berlin GE, 1977, Band I., Heft 3, 75-86.

In: Radon
Editor: Audrey M. Stacks

ISBN: 978-1-63463-742-8
© 2015 Nova Science Publishers, Inc.

Chapter 4

EFFICIENCY OF FOUR DIFFERENT METHODS OF VENTILATION FOR RADON MITIGATION IN HOUSES

Lydia Leleyter[*], Benoît Riffault, Benoît Basset, Mélanie Lemoine, Hakim Hamdoun and Fabienne Baraud
UR-ABTE- Bd du Mal Juin Campus 2 Bat Sciences 2, Caen Cedex

ABSTRACT

Calvados (France) is a region that is naturally rich in radon. The radon contents in a nursery school and an individual house, located in a same town in Calvados, are measured. These two buildings do not have either basement or underfloor space and are directly on the elevation. It turned out that important radon concentration, with volume activity that could even exceeds 1000 Bq/m^3, were observed in both cases. In France, the public authorities distinguish three levels of exposure (below 400 Bq/m^3, between 400 and 1000 Bq/m^3 and over 1000 Bq/m^3) with respect to the management of the risk associated with radon in places accessible to the public (according to the "Arrêté du 22 juillet 2004, J.O. 11 août 2004"). In case of volume activities exceeding 1000 Bq/m^3 the French public authorities recommend that "important corrective actions must necessarily and rapidly be undertaken".

To decrease the radon concentration in confined atmosphere, two natural ventilation methods and two forced ventilation techniques were tested. The four studied techniques are:

- Punctual Natural Ventilation (*PNV*), which consists in renewing the internal air by simple aeration twice a day for 15 minutes.
- Continue Natural Ventilation (*CNV*) which consists in creating additional aeration in the first floor (10 cm diameter) and/or aeration of the basement thanks to some cavities linked to the outside by a pipe.
- Insufflating Mechanical Ventilation (*IMV*) which consists in setting off the air through the house roof and injecting fresh air into the house (at a temperature over or equal to 15°C).

---

[*] Corresponding author: lydia.reinert@unicaen.fr.

- Mechanical Ventilation with Double flow (*MVD*) which assures the air renewal by insufflating fresh air in to the living areas and extraction of the used air from the wet rooms.

The PVN efficiency is evidenced by an important reduction in radon levels (closed to 0 Bq/m^3), when the windows are opened. However, as soon as the windows are closed, at dusk, when the outside temperatures decrease, as a consequence of the stack effect, the convective transfer of radon, hence the radon concentration increases quickly. Thus this technique was not satisfactory.

The other 3 tested techniques (CNV, IMV and MVD) were perfectly adapted and completely satisfactory. Indeed they allowed a very important decrease (from 80 to 95%) of the radon average concentrations. Moreover, the radon concentrations peaks are consistently below 300 Bq/m^3, even during the cold periods which are characterized by strong convective transfer of radon in the buildings as a consequence of the stack effect.

I. INTRODUCTION

Radon 222 is a natural, odourless, colourless and insipid radioactive gas and a son of the radioactive uranium 238 family. The radon disintegration (alpha emission) produces some radionuclides and finally stable lead. Radon 222 can migrate through the subsurface and vent into the atmosphere. The quantities of radon, which escape from a ground, depend on the geochemistry of the ground (volcanic or granitic regions being more concerned by these emanations than sedimentary regions), and weather conditions. The atmospheric distribution leads, generally, to a fast dilution of the radon emanating from the ground. However, if radon exhaled on surface finds itself trapped in enclosed spaces such as buildings, it can then be accumulated and may potentially reach levels which are detrimental to the health of the inhabitants (lung cancer; Tirmarche, 1998; Baysson and Tirmarche, 2008; Darby et al., 2005 ; Catelinois et al., 2006; Collignan and Sullerot, 2008; Baysson et al., 2005; Melloni et al., 2000).

To reduce the radon concentrations in a building, two types of techniques can be applied:

1) To increase the internal air renewal thanks to a better ventilation, which could be natural (such as PNV: Punctual Natural Ventilation or CNV: Continue Natural Ventilation) or forced (such as IMV: Insufflating Mechanical Ventilation or MVD: Mechanical Ventilation Double flow).
2) To decrease the quantity of incoming radon by waterproofness improvement of the building.

In France, public authorities distinguish three levels of exposure (according to the "Arrêté du 22 juillet 2004, J.O. du 11 août 2004") with respect to the management of the risk associated with radon in public areas:

- Below 400 Bq/m^3 the situation does not justify particular corrective actions.
- Between 400 and 1000 Bq/m^3 it is recommended to undertake simple corrective actions.

- Over 1000 Bq/m^3 important corrective actions must necessarily and rapidly be undertaken.

It is also important to note that the recommended radon values to trigger corrective actions depend on national policy. For example: the Agency of American Environmental Protection recommends corrective action if radon concentrations in houses exceed 148 Bq/m^3 (Melloni et al. 2000), whereas the United Kingdom domestic Action Level is 200 Bq/m^3 (Green et al. 2002, Groves-Kirkby et al. 2008).

In France, a department is classified "as priority" when the average radon concentration is over or equal to 400 Bq/m^3, which corresponds to 31 departments. The national second plan of actions (2011-2015) plans in particular the extension of the statutory standards in housing environments, pursuit of measures in public areas, and new cartography of high-risk areas (ASN, 2013).

The objective of this chapter is to test the efficiency of natural or forced ventilations in order to decrease the content in radon, in two buildings (one private house and an nursery school) in the same village located in one of these 31 priority departments (according to Arrêté du 22 juillet 2004, J.O. du 11 août 2004).

II. MATERIALS AND METHODS

The radon measures are realized by means of a dosimeter (Measure Man from Sarad GmbH company), placed 1 m below the ground floor, which allows a continuous semi-quantitative measure with enough precision. External temperatures are taken from the MétéoFrance Website (http://france.meteofrance.com/). Both studied buildings (nursery school and individual house) do not have either basement or underfloor space and are directly on the elevation. Several techniques of reduction of the content in radon are studied here:

- Punctual Natural Ventilation (PNV): consists in renewing the internal air by aeration twice a day for 15 minutes. The internal air renewal allows to dilute the radon initially concentrated in the building. This method is very simple and a free reduction technique of the radon concentrations.
- Continue Natural Ventilation (CNV) which consists in creating additional aeration in the ground floor (diameter 10 cm) and/or aeration of the basement thanks to some cavities linked to the outside by a pipe.
- Insufflating Mechanical Ventilation (IMV) which consists in setting off the air though the house roof and injecting fresh air into the house (at a temperature over or equal to 15°C).
- Mechanical Ventilation with Double flow (MVD) which assures the air renewal by insufflating fresh air into the living areas and extraction of the used air from the wet rooms.

III. Results and Discussion

A. Ponctual Natural Ventilation (PNV)

The technique of natural ventilation (opened windows) for 15 minutes, was applied, each morning and each evening, to the tested individual house between June 27th and July 8th, 2008. During this period the radon concentrations were measured in a ground floor bedroom during 5 days. Then the detector was placed in the dining room (with a fireplace) located on the ground floor too.

Figure 1 shows the results over the whole tested period. The mean radon concentration measured was 194 Bq/m^3 in the bedroom and 1123 Bq/m^3 in the dining room (table 1). We can note that periods of day are characterized by weak radon concentrations whereas nights are characterized by elevated radon concentrations. This result involves that the presence of radon, in the house, has very probably for origin a direct exhalation of the underground radon, via a convective transfer. Indeed the higher values at nights underline the importance of the stack effect in which convective transfer of radon in the house was accentuated when the outside temperature was cold (Rydock et al., 2001, Zmazek and Vaupotič, 2007, Singh et al. 2008; Leleyter *and al.*, 2010 and 2012). The presence of the fireplace in the room seems to accentuate this phenomenon, by facilitating the thermal drawing (and thus the radon exhalation of the basement) (Leleyter et al. 2010 and 2012).

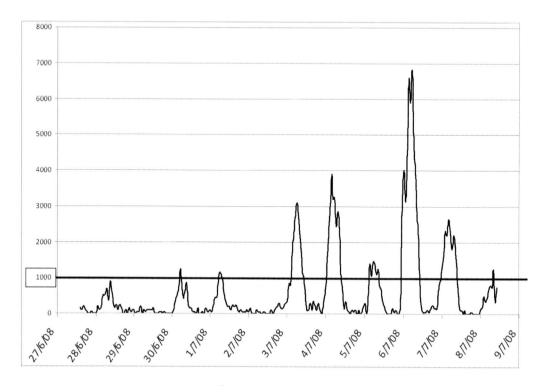

Figure 1. Radon concentrations in Bq/m^3 in the individual tested house, in the bedroom (from the beginning of the measures until 02/07/08 at 2 pm), then in the dining room till the end of the measures.

The efficiency of PNV is here clearly demonstrated, because with the opening of windows, the concentration in radon became close to 0 Bq/m^3. We can notice however that with the lock of windows, at the beginning of the night period, as soon as the outside temperature decreased, the radon concentrations in the house quickly increased. These important radon concentrations at nights (in particular in the dining room), could be unsatisfactory : PNV is not sufficient enough for this house.

B. Continue Natural Ventilation (CNV) and Insufflating Mechanical Ventilation (IMV)

To improve indoor air quality in this tested house, architectural modifications, allowing to have a continuous ventilation, are made to decrease the convective transfer and limit the observed night-peaks. Thus four additional low aerations (10 cm in diameter, 20 cm from floor level) were created in the ground floor, in order to have at least one low aeration in each room. Furthermore, to limit the radon accumulation in the underlying rubble, six cavities (50 cm^3) linked outside by a pipe (10 cm diameter) were installed. Thus preferential pathways for the venting of radon from the subsurface in the immediate vicinity of the house were created (Leleyter et al. 2010 and 2012). Moreover, IMV, which injects outside air into the house (82 m^3 per hour by a minimal 15°C temperature), was installed to produce a light overpressure within this house.

Following these modifications, new measures were made in the bedroom and in the dinning room in diverse periods, with or without IMV. The average values are shown on Table 1.

Table 1. Average values in radon [Rn] in Bq / m^3, measured on 5 days, in the bedroom (B) or the dining room (D) of the house with PNV or CNV or CNV combined with IMV. Average temperatures (T°C) according to the site of Meteo France, at the Caen-Carpiquet station

methods	PNV		CNV			CNV+IMV			
places	B	D	B		D	B		D	
[Rn]	194	1123	212	37	214	197	25	139	56
T(°C)	18	16	10	19	9	6	20	5	16
Month / Year	06-07/08	07/08	11/08	07/09	11/08	01/09	07/09	01/09	07/09

In Table 1, significant results are evident:

- We can note once again a marked seasonal variation, with more important radon concentrations in colder periods (November/January) compared with the measured values in summer (in June/July). These results indicate that the stack effect is always present.
- Thanks to CNV, average radon concentrations are below 214 Bq/m^3 even in the cold period which is characterised by strong convective transfer of radon in the house, due to the stack effect.

- If CNV is combined with IMV, average radon concentrations never exceed 197 Bq/m^3. Thus these modifications have induced some really important radon concentrations decrease in both tested rooms.

C. Mechanical Ventilation with Double Flow (MVD)

To reduce the radon concentrations of a nursery school in the same village as the tested individual house, a Mechanical Ventilation with Double flow (MVD) was installed in this school. Then the radon concentrations were measured, between Friday 8th February (at 10:00 am) and Monday 18th 2013 (at 10:00 am) (Figure 2). However, as the MVD was considered by the school staff as too noisy, it was manually desactivated by this staff when the children (or staff) were present. Thus, on Monday, Tuesday, Wednesday, Thursday and Friday, the ventilation was cut between 8:00 am and 5:00 pm. So the MDV was on all nights and at week-end periods.

A very wide valuable range, going from 0 in 135 0 Bq/m3 was measured (figure 2). Although a priori the ventilation was on when children/staff were not present, we can note that the maximal radon concentration was measured on Wednesday (13/02/13) at midnight. A further investigation proved, that school staff omitted to restart the ventilation on Wednesday at 5:00. Then the ventilation was cut between Wednesday at 8:00 am until Thursday at 5:00 pm. The observed peaks were then in agreement with stopping and restarting the ventilation (increase of the contents observed after the morning stop and decrease after evening restart). The average concentration in radon on the 10 tested days was 198 Bq / m^3. However the concentration in radon calculated over the periods of presence of the children reached 398 Bq / m^3.

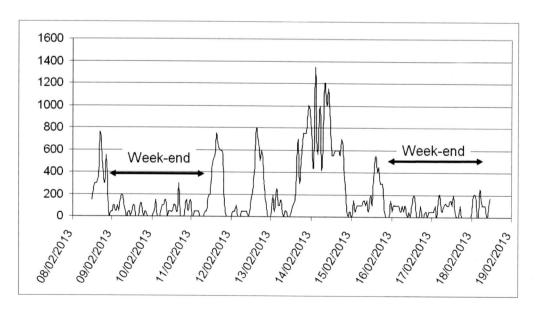

Figure 2. Concentrations in radon (Bq / m^3) in the nursery school from Friday (08/02/13 at 10:00 am) to Monday (18/02/13 at 10:00 am).

The efficiency of the system of MDV was highlighted when it is underway. However the increase of the concentration in radon when the ventilation was stopped, is problematic because it concerned the periods when the children were at school. Thus the utility of system is here doubtful.

CONCLUSION

The proportion of radon which escapes from the ground depends on geology, on the pedology of the ground and on weather conditions. In the tested houses (located in Calvados, France) radon concentrations could exceed 1000 Bq/m^3 while the external temperature is low. The PVN efficiency is evidenced by an important reduction in radon levels (closed to 0 Bq/m3), when the windows are opened. However, as soon as the windows are closed, at dusk, when the outside temperatures decrease, as a consequence of the stack effect, the convective transfer of radon, hence the radon concentration increases quickly. Thus this technique was not satisfactory.

Modifications (which consisted of CNV and IMV installations in the individual house and MVD installation in the nursery school) in houses aim:

- to increase aeration and so to dilute radon concentrations
- to decrease depression in houses in order to decrease the convective transfer of radon.

These modifications (CNV, IMV and MVD) were perfectly adapted and completely satisfactory. Indeed they allowed a very important decrease (from 80 to 95 %) of the radon average concentrations. Moreover, the radon concentrations peaks are consistently below 300 Bq/m^3, even during the cold periods which are characterised by strong convective transfer of radon in the buildings as a consequence of the stack effect.

The choice of the technique to be used must be reflected and correspond to a compromise between cost, efficiency, the ease of utilisation or the induced inconvenience.

ACKNOWLEDGMENTS

Thanks to Bernard Mazenc for his availability and the loan of the dosimeter allowing the measures and Yannick Reinert to have conceived and realized the works in the test house.

REFERENCES

Arrêté du 22 juillet 2004, relatif aux modalités de gestion du risque lié au radon dans les lieux ouverts au public, J.O. du 11 août 2004.

ASN (Autorité de sureté nucleaire) (2013) 'plan national d'action 2011-2015 pour la gestion du risque lié au radon'.

Baysson H., Tirmarche M. (2008) Risque de cancer du poumon après exposition au radon : état des connaissances épidémiologiques, *Archives des Maladies Professionnelles et de l'Environnement*, Vol. 69, Issue 1 58-66.

Baysson H., Tirmarche M., Tymen G., Gouva S., Caillaud D., Artus J.C., Vergnenegre A., Ducloy F. and Laurier D. (2005) Exposition domestique au radon et risque de cancer du poumon: Les résultats d'une étude épidémiologique menée en France, Indoor radon exposure and lung cancer risk. Results of an epidemiological study carried out in France, *Revue des Maladies Respiratoires*, Vol. 22, Issue 4 587-594.

Catelinois O., Rogel A., Laurier D., Billon S., Hemon D., Verger P., and Tirmarche M. (2006) Lung Cancer Attributable to Indoor Radon Exposure in France: Impact of the Risk Models and Uncertainty Analysis, *Environmental Health Perspectives*, vol. 114 1361-1366.

Collignan B., Sullerot B. (2008) Le radon dans les bâtiments, Guide pour la remédiation dans les constructions existantes et la prévention dans les constructions neuves, *CSTB*, 164 pages.

Darby S., Hill D., Auvinen A., Barros-Dios J.M., Baysson H., Bochicchio F. (2005) Radon in homes and risk of lung cancer: collaborative analysis of individual data from 13 European case-control studies. *BMJ* 330(7485):223; doi:10.1136/bmj.38308.477650.63.

Green, J.C.H. Miles, E.J. Bradley and D.M. Rees (2002) Radon atlas of England and Wales, NRPB Report NRPB-W26, National Radiological Protection Board, ISBN 0-85951-497-8.

Groves-Kirkby C.J, Denman A.R., Phillips P.S., Tornberg R., Woolridge A.C., Crockett R.G.M. (2008) Domestic radon remediation of U.K. dwellings by sub-slab depressurisation: Evidence for a baseline contribution from constructional materials, *Original Research Article, Environment International*, Vol. 34, 428-436.

Leleyter L., Baraud F., Riffault B. and Mazenc B. (2012) Ways to reduce radon concentration to improve house air quality, in *Handbook of Radon: Properties, Applications and Health*, Editors: Zachary Li and Christopher Feng, ISBN: 978-1-62100-177-5.

Leleyter L., Riffault B. and Mazenc B. (2010) *Concentration en radon dans une maison du Calvados Radon concentration in a house of Calvados*, C. R. Mecanique 338, pp 139–145.

Melloni B., Vergnenègre A., Lagrange P., Bonnaud F. (2000) Radon et exposition domestique, *Revue des Maladies Respiratoires*, N°17 1061-1071.

Rydock J. P., A. Næss-Rolstad, J. T. Brunsell, (2001) Diurnal variations in radon concentrations in a school and office: implications for determining radon exposure in day-use buildings, *Atmospheric Environment*, Vol.35, 2921-2926.

Singh M., K. Singh, S. Singh and Z. Papp, (2008) Variation of indoor radon progeny concentration and its role in dose assessment, *Journal of Environmental Radioactivity*, Vol. 99, 539-545.

Tirmarche M. (1998) Evaluation du risque de cancer lié à l'inhalation du radon, *Revue de l'ACOMEN*, vol4, 337-344.

Zmazek B., J. Vaupotič, (2007) Coping with radon problem in a private house, *Building and Environment*, Vol.42, 3685-3690.

In: Radon
Editor: Audrey M. Stacks

ISBN: 978-1-63463-742-8
© 2015 Nova Science Publishers, Inc.

Chapter 5

THE USE OF RADON AND THORON IN BALNEOTHERAPY

Fábio Tadeu Lazzerini and Daniel Marcos Bonotto[*]

Departamento de Petrologia e Metalogenia, Universidade
Estadual Paulista (UNESP), Rio Claro, São Paulo, Brasil

ABSTRACT

Radon (^{222}Rn, half-life 3.8 days) is a naturally occurring volatile noble gas formed from the normal radioactive decay series of ^{238}U, according to the following decay sequence: ^{238}U (4.49 Ga, α) $\rightarrow$ ^{234}Th (24.1 d, β^-) $\rightarrow$ ^{234}Pa (1.18 min, β^-) $\rightarrow$ ^{234}U (0.248 Ma, α) $\rightarrow$ ^{230}Th (75.2 ka, α) $\rightarrow$ ^{226}Ra (1622 a, α) $\rightarrow$ ^{222}Rn (3.83 d, α) $\rightarrow$ ... Thoron (^{220}Rn, half-life 56 seconds) is another naturally radioactive volatile noble gas formed in the ^{232}Th decay series according to the sequence: ^{232}Th (14.0 Ga, α) $\rightarrow$ ^{228}Ra (5.8 a, β^-) $\rightarrow$ ^{228}Ac (6.2 h, β^-) $\rightarrow$ ^{228}Th (1.9 y, α) $\rightarrow$ ^{224}Ra (3.7 d, α) $\rightarrow$ ^{220}Rn (55.6 s, α) $\rightarrow$ ... Radon and thoron are colorless, odorless, tasteless, chemically inert and radioactive gases produced continuously in rocks and soils through α-decay of ^{226}Ra and ^{224}Ra, respectively, with some atoms escaping to the surrounding fluid phase, such as groundwater and air. They are subjected to recoil at "birth", with the emanated fraction relatively to that produced in the solid phase being dependent on factors such as total surface area of solids and concentration/distribution of ^{238}U (^{226}Ra) in the minerals. ^{222}Rn decays to stable lead according to the sequence: ^{222}Rn (3.83 d, α) $\rightarrow$ ^{218}Po (3.05 min, α) $\rightarrow$ ^{214}Pb (26.8 min, β^-) $\rightarrow$ ^{214}Bi (19.7 min, β^-) $\rightarrow$ ^{214}Po (0.16 ms, α) $\rightarrow$ ^{210}Pb (22.3 a, β^-) $\rightarrow$ ^{210}Bi (5 d, β^-) $\rightarrow$ ^{210}Po (138.4 d, α) $\rightarrow$ ^{206}Pb. ^{220}Rn decays to stable lead according to the sequence: ^{220}Rn (55.6 s, α) $\rightarrow$ ^{216}Po (0.14 s, α) $\rightarrow$ ^{212}Pb (10.6 h, β^-) $\rightarrow$ ^{212}Bi (60.6 min, β^--64.1% or α-35.9%) $\rightarrow$ ^{212}Po (0.3 μs, α) or ^{208}Tl (3.0 min, β^-) $\rightarrow$ ^{208}Pb. High ^{222}Rn concentrations occur in groundwaters in many areas where wells are used for domestic water supply, inclusive in small rural water supplies. Some natural processes related to high concentration of radon in groundwater are: low transmissivity zones, uranium content of the source rock, severe chemical weathering, hydrothermal solution,

---

[*] Daniel Marcos Bonotto: Departamento de Petrologia e Metalogenia, Universidade Estadual Paulista (UNESP), Câmpus de Rio Claro, Av. 24-A No.1515, C.P. 178, CEP 13506-900, Rio Claro, São Paulo, Brasil. E-mail: danielbonotto@yahoo.com.br.

deposition, extensive fracturing and variations in stress in rocks associated with seismicity. Potential health hazards from radon in consuming water have been considered worldwide, especially when groundwaters are utilized for public water supplies, because ^{222}Rn concentrations in surface waters are often less than 3.7 Bq/L, while in groundwater the ^{222}Rn concentrations commonly are 10-100 times higher. Despite the concerns coupled to the health risks due to ingestion of dissolved radon and thoron in drinking water, these radioactive gases have been sometimes used in balneotherapy in view of attributed benefic physiological effects to human health. This chapter reports the results of investigations held elsewhere focusing some of these aspects.

INTRODUCTION

In many countries, spring waters have been extensively used for consumption purposes as an option to tap water, as many people believe they are healthy and/or can be utilized for health cures. Additionally, economic reasons have also favored their use as bottled waters so that the commercialization of mineral waters has widely increased. In Brazil, the production and commercialization of mineral waters is managed by the National Department of Mineral Production (DNPM). Traditionally, the mineral waters had been used or directly consumed in the springs, where touristic centers developed around. However, in the present days, the mineral water for consumption is distributed in vessels for ingestion distant from the springs, whereas hydrothermal spas for therapeutic baths and leisure exhibit infra-structure with hotels and many facilities for users.

Many spring waters in Brazil do not contain high concentrations of dissolved constituents, but the waters are considered mineral due to the radioactivity in them, chiefly due to the presence of dissolved radon (^{222}Rn) and thoron (^{220}Rn) [1]. ^{222}Rn (half-life 3.82 days) and ^{220}Rn (half-life 55.6 sec.) are noble gas isotopes belonging to the ^{238}U (4n+2) and ^{232}Th (4n) decay series, respectively. ^{222}Rn is produced by α-decaying ^{226}Ra and reaches the stable ^{206}Pb, whereas ^{220}Rn is generated by α-emitting ^{224}Ra, forming the stable ^{208}Pb [2, 3].

The ^{238}U radioactive decay series reaches ^{222}Rn according to the sequence [3]: ^{238}U (4.49 Ga, α) $\rightarrow$ ^{234}Th (24.1 d, β^-) $\rightarrow$ ^{234}Pa (1.18 min, β^-) $\rightarrow$ ^{234}U (0.248 Ma, α) $\rightarrow$ ^{230}Th (75.2 ka, α) $\rightarrow$ ^{226}Ra (1622 a, α) $\rightarrow$ ^{222}Rn (3.83 d, α) $\rightarrow$ ... Thoron (^{220}Rn, half-life 56 seconds) is another naturally radioactive volatile noble gas formed in The ^{232}Th radioactive decay series reaches ^{220}Rn according to the sequence [3]: ^{232}Th (14.0 Ga, α) $\rightarrow$ ^{228}Ra (5.8 a, β^-) $\rightarrow$ ^{228}Ac (6.2 h, β^-) $\rightarrow$ ^{228}Th (1.9 y, α) $\rightarrow$ ^{224}Ra (3.7 d, α) $\rightarrow$ ^{220}Rn (55.6 s, α) $\rightarrow$ ...

^{222}Rn decays to stable lead according to the sequence [3]: ^{222}Rn (3.83 d, α) $\rightarrow$ ^{218}Po (3.05 min, α) $\rightarrow$ ^{214}Pb (26.8 min, β^-) $\rightarrow$ ^{214}Bi (19.7 min, β^-) $\rightarrow$ ^{214}Po (0.16 ms, α) $\rightarrow$ ^{210}Pb (22.3 a, β^-) $\rightarrow$ ^{210}Bi (5 d, β^-) $\rightarrow$ ^{210}Po (138.4 d, α) $\rightarrow$ ^{206}Pb. ^{220}Rn decays to stable lead according to the sequence [3]: ^{220}Rn (55.6 s, α) $\rightarrow$ ^{216}Po (0.14 s, α) $\rightarrow$ ^{212}Pb (10.6 h, β^-) $\rightarrow$ ^{212}Bi (60.6 min, β^--64.1% or α-35.9%) $\rightarrow$ ^{212}Po (0.3 μs, α) or ^{208}Tl (3.0 min, β^-) $\rightarrow$ ^{208}Pb.

Some ^{222}Rn and ^{220}Rn fraction escapes from the rocks and minerals to the surrounding fluid phase, such as groundwater and air, whose proportion depends on factors such as total surface area of solids and the concentration (and distribution) of ^{238}U (^{226}Ra) and ^{232}Th (^{224}Ra) in them [4]. Potential health hazards from natural radionuclides in consuming water have been considered worldwide, with many countries adopting the guideline activity concentration for drinking water quality recommended by the World Health Organization [5].

In general, the recommendations apply to routine operational conditions of water-supply systems, however, special attention must be also given when groundwater is utilized for public water supplies, because ^{222}Rn concentrations in surface waters are often less than 3.7 Bq/L, while in groundwater the ^{222}Rn concentrations commonly are 10-100 times higher [5].

Despite the concerns coupled to the health risks due to ingestion of dissolved radon in drinking water, numerous papers are available on the determination of radon in mineral waters, some of them pointing to the use of treatments involving the intake of radon gas through inhalation or by transcutaneous resorption of radon dissolved in water in view of attributed benefic physiological effects to human health. For instance, natural and artificial grottoes carved in the rock bathing tubs found in some of the oldest spas in Central and Southern Europe have been considered by [6] who reported high ^{222}Rn levels of 2,000-4,000 Bq/L at the island of Ischia in the volcanic area near Naples, Merano and Lurisia spas in the alpine region (Italy). In Spain, radioactive waters (^{222}Rn activity concentration higher than 67.3 Bq/L) were recognized by [7] in 24 water sources that exhibited ^{222}Rn levels between 73 Bq/L (Caldas de Boí-Tartera, Lérida) and 1868 Bq/L (Arnedillo, La Rioja).

The medical spa Jáchymov, Czech Republic, is the oldest one in Bohemia and probably also in Europe. It utilizes the natural resources in a uranium mine in Jáchymov, which arose in 1518 and is still functional. There are springs of radon water occurring at a depth of 500 m under the ground, where, nowadays, four springs are used on the 12th level of the Svornost mine (Concord) (Table 1). A mix from all four springs is used for the baths that contains in the bath approximately 4.5-5.5 kBq/L [8].

The medical spa Jáchymov and other like the Radon Revital Bad, St. Blasien-Menzenschwand, Germany, points out that they cure by means of radon water that, unlike other resources, does not have chemical effects, but energetic ones [8]. The premise involved is that during a radon bath, the surface of the body is exposed to an energy shower of radon alpha particles. Then, the alpha particles trigger a chain of physiological reactions, which soothe pain, improve joint movement, blood circulation and healing, stimulate regeneration processes in the tissue of locomotive organs, harmonize vegetative functions, positively affect blood pressure, and they boost the immune system of the body [9]. It has been also claimed that the penetration of radon into blood circulation is negligible and its biological half-life (elimination of radon from the body) is 20 minutes (i.e., the duration of the radon bath) [8].

RADON (^{222}RN) EMANATED AT SOURCE

Radioactivity in thermal sources can occur as gaseous emissions originated mainly by radium and thorium, emanations dissolved in water or transferred from water to air, dissolved radioactive salts or colloids present in water and radioactive substances in country-rocks, soil, mining sites or caves.

Radon found in mean concentrations of 10 to 100 ppm is considered a trace gas on the crust and surface. However, it is the largest contributor to human exposure to radiation, amounting to 55% of total exposure. Probably due to its high solubility and gaseous form, it is typically the most common radionuclide and the one found in higher concentration in water, being considered of fundamental biochemical relevance. It has density 7.6 times greater than the air's one [10].

Table 1. Radon content in springs of the Svornost mine, Jáchymov, Czech Republic [8]

Spring name	Discharge (L/min)	Temperature (°C)	Radon content (kBq/L)
Curie	30	29	5
C1	30	29	11
Behounek	300	36	10
Agricola	10	29	20

Radon can be originated from natural sources, presenting a great spatial dispersion and occurring together with other natural radionuclides like ^{228}Ra, ^{226}Ra, ^{238}U, ^{230}Th, ^{210}Po, and ^{40}K. In most groundwaters, natural radioactivity is generally related to the presence of ^{40}K, whose abundance is only 0.012% of total potassium [11]. The main exceptions are aquifers near to uranium deposits and those related to granitic bodies, which are generally enriched in Th and U. Some minerals generated by their weathering are radiobarite, zirconite, monazite, columbite and fergusonite, among other [12].

Radon emanations sometimes are related to recent volcanic and seismic activity, oscillating even on hourly or daily scales [13]. Their ionizing potential has great influence on the physical and chemical composition of nearby aquifers, the behavior of atmospheric ions, as well as in total bioactivities [14]. Radon (^{222}Rn) and thoron (^{220}Rn) may occur in emanations of hydromineral sources, inducing radioactivity with ionizing power and potential physiological actions surveyed since a long time ago [15]. Although ^{220}Rn and ^{219}Rn [from ^{235}U (4n+3) decay series] have strong induced radiation, similar that of radium, their emanations are short-lived and present only while in the spring upwelling. Thus, ^{219}Rn was not here considered, while ^{220}Rn was but only as emanation of bioactivity in its place of occurrence. The policy concerning the classification of mineral waters in Brazil, with respect to temperature and radioactivity, obliges the use of the expression "at source", i.e., the waters will be classified as mineral thermal and/or radioactive if they are in their natural deposits (springs or wells). Although this chapter does not quantify the exact extent involved in this classification, we are inclined to follow the suggestion of the 40 meter radius surrounding the source. This is the area sampling parameter used for measuring bioactivity in visitors of the Sacred Springs on Bakreswar Temple, in India [16].

The minimum ^{222}Rn value of 52.5 Bq/L is indicated for potential bioactivity in therapeutic applications of radon emanated from hydromineral sources [17]. For therapeutic techniques through radon inhalation and nebulization, there is the recommendation of 67.1 Bq/L of ^{222}Rn [18]. Small levels up to 4.2 Bq/m^3 of ^{222}Rn have been reported in emanating environments of crenotherapy at Eger Thermal Baths (Turkey) [19].

Radon is quickly absorbed by inhalation, through mucous membranes, than by bathing, and whenever it is free in gaseous form it can contain concentrations two to twenty times higher than that of whenever it is dissolved in aqueous solution. In general, about 70% to 90% of the radon gas dissolved in water escapes into the air of environments of balneotherapy [20, 21]. The few samples compiled for the Brazilian database, where its presence is found, have indicated they are classified as potentially bioactive.

Much of the reviewed bibliography express emanated Radon in Bq/m^3. In waters with 85.9 Bq/L of dissolved ^{222}Rn it is not overlooked the influence of its concentration increase towards the ambient air [19].

However, there are some suggestions for estimating the contribution in the amount of radon found in the air, especially in indoor environments, due to their release out of the existing solutions. For each 1 Bq/L in the air, it is required an amount of water with 10,000 Bq/L of dissolved ^{222}Rn [22]. Or, also, an indoor environment with water containing 1,000 Bq/L of dissolved ^{222}Rn will have its amount in the air with a minimum of 100 Bq/m^3. The annual estimate effective dosage will be of 200 Bq/m^3 of ^{222}Rn [23].

As reference values of ^{222}Rn we can find that the estimated concentration in atmospheric air is 4 Bq/L [24], while its overall indoors measure is of 27.2 Bq/m^3. This value varies significantly from country to country: 75 Bq/m^3 in Italy, 67.1 Bq/m^3 in India, and 14.3 Bq/m^3 in Brazil. However, the maximum concentration permitted by the World Health Organization is of 100 Bq/m^3 [16, 20, 25].

The world average rate of exhalation of radon is 57.6 Bq/m^2/h and one of the largest concentrations is found in Bad Gestein Heilstollen Fountain (Germany), which uses 40,000 Bq/m^3 of emanated ^{222}Rn for intense therapeutic purposes. Given the linear non-threshold (LNT) hypothesis tested on positive biological effects to low radiation exposition to radon, indications of the minimum necessary therapeutic range between 148 and 500 Bq/m^3 of ^{222}Rn [26].

Recent experiments have been carried out in guinea pigs through inhalation for one day, with the following results obtained: improvement of hepatic antioxidants functions and inhibition of alcoholic toxicology in 4,000 Bq/m^3 of ^{222}Rn [27]; alleviation of symptoms of diabetics and enhancement of antioxidant enzymatic activity in 3,500 Bq/m^3 of ^{222}Rn [28] or in 18,000 Bq/m^3 of gaseous ^{222}Rn present at the Ikeda-Misasa Fountain environment (Japan) [29], as well as by aerosol produced with its water containing 13,000 Bq/m^3 of dissolved ^{222}Rn [30].

Similar researches that were made in humans, with lower concentrations, have demonstrated anti-inflammatory and edema's inhibition potential, by means of inhalations in environments with 2,000 Bq/m^3 of ^{222}Rn [28]. Longer term clinical trials (28 days) displayed anti-inflammatory and anti-oxidative properties in patients with bronchial asthma treated in daily inhalation sessions of 40 minutes in emanation rooms with 2,080 Bq/m^3 of ^{222}Rn [31].

Some physiological effects occur from radioactive emanations. For instance: the lipid level activity promotes the growth of healthy cells simultaneously to the inhibition of morbid cells; produces dieresis; stimulates digestive activity and alleviates constipation; increases excretion of uric acid; lowers blood pressure by dilating blood vessels and decreasing blood viscosity; increases sexual activity and ability to reproduce and modifies blood composition through reducing white blood cells and increasing red blood cells [15, 19, 32, 33].

The diseases which are indicated for emanation therapy are: ankylosing spondylitis; degenerative joint disease; spondylarthrosis; myofascial tissue syndrome; ovarian hypofunction; allergic bronchial asthma; chronic bronchitis; gout; chronic articular rheumatism; gonorrheal rheumatism; rheumatic arthritis; neuralgia; high blood pressure; premature aging and gynecology [8, 9, 18, 34].

Inhalation of its aerosols, vapors, natural dissolved or emanated gases have sedative, analgesic, decongestant, anti-inflammatory, desensitizing and neuro-vegetative restoring effects. Thus, they are also recommended in diseases of the respiratory system and its pathways, such as: *nasal* septum disorders; chronic rhinopharyngitis; rhinitis; adenoiditis; laryngitis; bronchitis; laryngeal nerve paralysis; bronchiectasis; emphysema; asthma; and other chronic lung diseases [35].

THORON (^{220}RN) EMANATED AT SOURCE

Thoron (^{220}Rn) is a gaseous radionuclide derived from the thorium (^{232}Th) decay series, whose half-life is short (55.6 seconds). The average Th concentration in soils is about 25 Bq/kg [36]. In general, its occurrence is associated with uranium and rare earths elements occurring in monazite, thorite, thorianite, uranothorite, zircon, spheno or allanite that are present mainly in granites, syenites, pegmatites; acidic intrusions and black shales [12]. Basalts, limestones and sandstones typically have low concentrations of this element [37]. A consultation at PubMed database for the keywords "mineral water" AND "^{220}Rn" or "thoron" exhibited only 1 result, indicating how scarce is the available information, thus justifying the importance of considering it in this review.

The levels of ^{220}Rn are determined by emanations from the soil, underlying ^{232}Th content, soil types, and atmospheric environmental conditions. Its average rate of exhalation is estimated at 3 Bq/m^2/s. Once it belongs to a radioactive series different than that of ^{222}Rn, it also has distinct environmental behavior, although with many similar chemical characteristics [10]. Its measurements have been used as tracers of various characteristics of natural environments, such as the levels of other radionuclides, proximity of prospective targets, typology and age of springs or mineral deposits [38].

The presence of thoron in soil is quite limited by the increase of the content of interstitial water and, unlike ^{222}Rn, its concentration in groundwater is little affected by physical agitation. The average thoron exhalation rate is estimated as 3 Bq/m^2/s. Its relative scarcity in aquifers is related to reduced interstitial spaces and slow water flow with respect to its short half-life. Because monazite is a typical thorium source, this mineral is much used under flowing water to produce thoron emanations, especially for hydrotherapy uses [39].

Thoron may be responsible for up to 10% of the total annual radiation dose received by the population of many countries. Some measurements of average thoron concentration present in the air of residential environments are (in Bq/m^3 of ^{220}Rn) [40]: 19 in Brazil; 40 in Korea; 53 in Ottawa (Canada); 98 in Hungary; 3,297 in Geiju Yunnan (China); 160 in Serbia; 37 in spas and 840 in caves of Slovenia.

In Japan, it is a common practice to attend to artificial spa baths whose emanations are produced by water flowing in radioactive tablets or minerals. In their surrounding environments, it can be observed levels exceeding 20,000 Bq/m^3 of ^{220}Rn and 700 Bq/m^3 of ^{222}Rn in the air [41].

Scarce data was found about the ^{220}Rn levels in hydromineral springs. The average is 0.1 to 0.2 Bq/L at the regional group of springs in Austria [42]. The ^{220}Rn activity concentration is 1.4 Bq/L in a hydromineral source in Switzerland [39]. Analyzes in 13 mineral waters bottled in France, Italy and Morocco exhibited levels between 0.91 and 3.4 Bq/L of ^{220}Rn and 4.2 to 8.6 Bq/L of ^{222}Rn. In five thermal resorts of Morocco, the average thoron concentration is circa 15 Bq/L [43]. The main physiological activities and therapeutic recommendations for thoron are similar to those of radon (^{222}Rn). Given the scarce bibliography with respect to this approach, we can highlight the clinical study of 3 weeks exposure to Japanese hot springs containing thoron, which demonstrated bioactivity on the reduction of lipid peroxides, assisting in treatment against hypertension and "mellitus" diabetes [44]. Another randomized analysis, taken with subjects exposed to one hour per day, during 2 weeks, in an emanatory environment of a Japanese thermal center that contains 4,900 Bq/L of ^{220}Rn dispersed within the air, has indicated antioxidant effects and potential therapeutic uses in the treatment of mellitus diabetes, rheumatoid arthritis and other age-related diseases [45].

So, even if this gas can emerge in dissolved form in water, its positive or negative biochemical activities are regarded as emanations contained in the air volume. Despite its brief existence, the majority of research on thoron focuses on the risks involved in indoor exposure to it. This has been quantified in Bq/m^3 of air and is originated in soil, building materials or pollution. There are also approaches focusing the risks resulting from radionuclides generated by its radioactive decay [40].

In this compilation of studies, the unique policy found with the classification of thoron-active waters is a Brazilian one, although thoron is also mentioned in the context of therapeutic techniques in onsen resorts at Japan [45]. The thermal spas were gradationally constructed in Brazil for therapeutic and leisure purposes, corresponding the period elapsed between the 1930s and 1950s to the most auspicious hydrothermal period in the country [18]. The Brazilian Code of Mineral Waters was established in this time, under French influence, by Register 7841 published on 8 August 1945 [1]. This rule is still in force without any actualization, focusing the mineral waters for spas and bottling uses, as well the potable waters for bottling [46]. Thus, a possible minimum value for potential biological activity of thoron taking into account the Brazilian legislation could be 26.92 Bq/L of ^{220}Rn [1].

It is noteworthy that few data on thoron levels were found in Brazilian hydromineral springs, most of them providing from old publications. Even the governmental laboratory responsible by hydrochemical analysis and official ratings for the sources exploitation has not performed such assessments. In a recent thoron survey in some spas in the Brazilian states of São Paulo and Minas Gerais, it has been observed values significantly lower than those obtained from previous estimates [47].

"HOURLY-RADIOACTIVITY"

A natural spring of discharging groundwater has its flow also accompanied by gases dissolved in solution and/or gases directly dispersed in the air. Among these gases, the radioactive ones (radon and thoron) have attracted the interest of researchers. This was due to the bioavailability increase in the vicinity of springs as a consequence of the ionizing properties that influence other elements and also due to the increase in the nearby air electrical conductivity [48].

Thus, the radioactive power at the hydromineral spring can be calculated by means of the concentration of radioactive gases dissolved in the volume of discharged water added to their flows emanated in the site. In the early 1900s, this property was named as "hourly-radioactivity" and it consisted as an attempt to explain the evident biological activities of waters containing low concentration of salts through natural exposure to radioactive emissions at renewing low doses and continuously through the body [49].

Discharges in spring waters and emanated gases can be leveraged in hydrotherapy in stored water (bathtub or pool) by means of increasing the influence of radioactivity, temperature and other properties, by absorbing larger amounts of alpha particles per exposure time. These are potential uses of "hourly-radioactivity" that can be expressed in $Bq/m^3/h$.

Initial studies showed that, in bathings utilizing radioactive waters, radiation emitted by alpha particles does not penetrate deeper than the outer layer of skin. Thus, beta and gamma emissions are responsible for the dose absorbed by the body in this type of exposure [32].

Currently, calculations for this type of exposure (mainly due to radon) seek to guarantee epidemiologic safety in ordinary home baths and the occupational limits in ventures that use radioactive hydrothermal sources (radon or radium hot springs), such as hotels, resorts, radioactive mineral water bottlers, etc.

Since such studies involve several variables, some estimates usually consider [50]:

- Radionuclides dissolved in bath waters that are also responsible for radon and thoron inhalation (emanations and decays);
- Radioactive gases moving in the air, in a healthy environment, with aerosols (favored in natural climatic conditions of evapotranspiration occurring at waterfalls and mists) and heat that influence differently the biochemical activities related to nature (ionic reactions); and
- Principle of spatial equilibrium, where the energy emitted per unit of water mass equals the energy absorbed per unit of body mass. In bathings of running water, the continuous natural flow allows for a significant increase in the renewable exposure area.

Though scarce, currently it is still possible to find works related to this property in "radon spas", for instance, in the Greek island of Lesvos [13]. This kind of estimate should be added to the dose calculations for occupational exposure, as it has been done in "radon hotels" in China [51].

The evaluation from radioactive gases emanating from hydromineral sources is complex and its analysis shows large variations. They can be considered potentially therapeutic according to the available quantity and to levels starting from 52.5 Bq/L of air [17].

Thus, the index of "hourly-radioactivity" emerged as an attempt to better estimate the radioactive gaseous phenomena in hydromineral sources and their effects. It considers the product of the levels of radiation by the flow rate of the gases during one hour.

Therefore, the radioactive power of a hydromineral spring consists of the "hourly-radioactive" sum of the gases that were spontaneously detached with those ones that were dissolved in running water.

Typically, the solubility coefficient of the gases dissolved in water, in ambient temperature, and in resort- therapeutic standard conditions is considered to be 0.25 [17].

Gioconda spring at Águas de São Pedro city, São Paulo State, Brazil, had detached gases (57.1 Bq/L). It was estimated a total "hourly-radioactivity" corresponding to 976,000 Bq/h, as shown in Table 2. The Cipó resort (in Bahia State, Brazil), whose flow rate is 32,220 L/h, exhibited a dissolved radon level of 187.9 Bq/L. Its gaseous flow is 2,215 L/h and its emanated radon level was 57 Bq/L . In a 20 minute bath, it was estimated that its hourly-radioactivity was corresponding to 1,331,208.8 Bq/h [52].

Table 2. "Hourly-radioactivity" of Águas de São Pedro (São Paulo State, Brazil)

"Spring"	Flow		Radon		"Hourly-radioactivity"		
	Liter/hour		Bq/L		Bq/h		
	Water	Gas	Water	Gas	Water	Gas	Total
Juventude	12,500	430	2.59	14.79	32,348.5	6,284.9	38,633.3
Gioconda	14,500	60	66.55	184.90	964,909.1	11,090.9	976,000
Almeida Sales	2,500	200	14.79	14.79	36,969.7	3,696.7	40,666.7

According to [53] Pupo (1940).

SOME STUDIES IN BALNEOTHERAPY

Several radioactive elements can be found in the water, such as uranium, radium and thorium. The average U concentration in basalts is 0.87 ppm, while its concentration in granites is 3 ppm; the Th concentration in basalts is 2.93 ppm, whereas its concentration in granites is 13 ppm. The waters are generally called "radioactive" due to the levels of dissolved radon (^{222}Rn) gas on it. In addition, in atmospheric waters, cosmic radiation is also an important source of radiation [36].

Radon is generated by the decay of radium (^{226}Ra), an alkaline earth metal which is present in minerals of easy lixiviation. The average radium content in Earth's crust is 40 Bq/kg, and under standard condition of saturated soil with porosity of 20%, this amount of radium tends to originate balanced concentrations of radon in groundwater (about 50 Bq/L). The radium's average concentration in river water is 10^{-7} mg/L, while for radon is 10^{-12} mg/L [23, 36].

Being an unstable and soluble gas, radon is the predominant radionuclide in almost all water types. Its concentration depends not only on the levels of its predecessor (^{226}Ra), but also on the emanation intensity of the substrate it is on [54]. Turbulence and physical impact facilitate the escape of this gas from the dissolved phase, and, therefore, it has not been recommended that its uptake occur by pumping or sudden breaks in its flow when radioactive springs are used for therapeutic purposes [18].

The following conversion factors are used to express the radioactivity caused by radon: 1 Bq/L = 0.075 Mache Unit = 0.027 nCi/L = 27.01 pCi/L. Some examples of radon levels in Brazilian water samples are [25]: 0.95 to 36 Bq/L for groundwater; 0.43 to 2.4 Bq/L for river water; 0.3 to 5.4 Bq/L for the sea water; and 0.39 to 0.47 Bq/L for public distributed water. The averaged sampling of public drinking water in the United States is between 37 to 7.4 Bq/L of ^{222}Rn [26]. In evaluating the hydric resources in Brazil, it is estimated an average level of 57.7 Bq/L of ^{222}Rn, and it can be noted that higher concentrations are to be found in the south and southeast regions, with approximate ^{222}Rn levels of 144.3 Bq/L [55].

High ^{222}Rn levels (300 to 3,200 Bq/L) occur in aquifers composed of granitic rocks, especially pegmatites. In sedimentary rocks its value reaches from 3 to 40 Bq/L, and can increase if in contact with peat and carbonates enriched in radioactive minerals [56]. The radon solubility increases with the increasing of pH and total dissolved solids (TDS), whereas the temperature (heating from 0°C to 75°C) can reduce up to 4.5 times its content [57].

Classifications for radioactive waters differ in each country. Some examples of minimum levels (in Bq/L of [222]Rn) are: Italy- 48; Cuba- 67.3; Japan- 110.7; Brazil- 134.2; Russia- 185; France- 370; Czech Republic- 1,192; and Germany- 6,885 [21, 58].

The maximum value for [222]Rn bioactivity through daily consumption of drinking water is regulated by the European Commission (EURATOM), which recommends the value of 1,000 Bq/L of [222]Rn [23]. This value must be observed when the effective dose absorbed by an adult is between 0.2 and 1.8 mSv/year, so that it does not exceed half of the maximum safe level. In respect to its minimum value, perhaps it could be recommended 32 Bq/L of [222]Rn that is appropriate for child's diet [59].

The radon crenotherapy by means of its ingestion has been recommended in cases of gastrointestinal diseases due to the positive effects it can offer: to sedate peristalsis and motility; to regulate neuroenteric plexus; to stimulate enzymatic activity and the digestive efficiency of gastric and pancreatic juices [60].

Some baths referred to "diuresis cure" have claimed improvements to kidney and urinary tract diseases (nephritis, nephrosis, calculi of lithiasis, cystitis, urethritis, prostatitis and other chronic diseases). This would happen due to the following actions [35]: hinder the oxalic precipitation; increase the dilution and alkalinization of urine; elimination of phosphoric acid; favor organic oxidation; and facilitate the drag of calculi by increasing the fluid flow.

The coefficient of physiological absorption of radiation is higher in low temperatures and in more dilute solutions, being able to thereby potentiate bioactivities of trace elements that are present. Thus, it is recommended the consumption of these waters as soon as possible, due to the short half-life of the bioactive gases and their decays originating lead, even if in minimum concentrations [32].

In several countries, it has been reported the use of radioactive hydric resources in spas, thermal fountains, resorts and sanatoriums. Such knowledge has been disseminated since ancient times, long before there was so much attention, as there is today, regarding the risks to its exposure. Potential beneficial physiological effects would occur primarily through immersion baths [22].

Almost all of the controlled clinical trials and medical assessments occur in local waters with a high content of dissolved radon (above 1,000 Bq/L). For instance, one study in the thermal waters of the Fountain Bad Gestein (Austria) (982 Bq/L) found that skin permeation in patients occurred after 10 daily balneotherapic sessions [61]. Another example is a clinical trial done on 60 patients with rheumatoid arthritis, by means of a rehabilitation program, in which a series of 15 baths in the waters of Bad Brambach Fountain (Germany) suggested the acquisition of long-term beneficial effects [62]. This fountain exhibited 1,300 Bq/L of [222]Rn and 1,600 mg/L of CO_2.

Controlled clinical studies, conducted with 42 people, through isothermal bathing in waters from Jáchymov spa (Czech Republic) containing 3,500 Bq/L of [222]Rn, showed that the treatment was effective in improving the condition of patients bearing rheumatoid arthritis [63]. The observation of 186 patients suffering from heart diseases indicated that balneotherapy has generated positive results in over 90% of the cases. This study used radioactive water from 2 Russian bathhouses containing 1,478 Bq/L and 4,436.4 Bq/L of [222]Rn [64]. Yet another example is a study involving 141 patients with ankylosing and negative-serum spondylitis, which demonstrated effective anti-inflammatory and analgesic activity, after they took immersion baths with water from 3 Russian fountains with different radon concentration: 1.500, 3.000 and 4.500 Bq/l of [222]Rn [65].

Regarding low levels of ^{222}Rn, no influence on urinary excretion was found when tests were carried out in baths with therapeutic waters (72.4 Bq/L of ^{222}Rn) [66]. A balneotherapic pilot study, carried within 15 days, using radioactive water in 27 patients with degenerative skeletal-muscular disorders showed no efficacy or activity on the endocrine system. The waters came from the Fountain of Eger (Turkey), and had 80 Bq/L of ^{222}Rn [67]. However, as an exception, higher antibacterial activity was found in radioactive medicinal waters of Poland, with 74 Bq/L of ^{222}Rn [68]. For this reason, the Polish legislation perhaps could be used as reference to minimum value of exposure to radon.

The lowest radon levels clinically evaluated as respiratory control and positive skin alterations were observed in patients with ankylosing spondylitis, after 3 weeks of submitting themselves to daily immersion baths with radioactive water with 415 Bq/L of ^{222}Rn [69]. This value perhaps could be considered the lower limit for potentially therapeutic bioactivity of radon dissolved in water. With a similar value of 500 Bq/L of ^{222}Rn, positive results were obtained in 40% of 148 patients diagnosed with cervical rheumatic disease, who received treatment with this radioactive water.

In the same study, when the patients received treatment in balneotherapy with radioactive waters containing 5,000 Bq/L of ^{222}Rn, the efficacy was slightly superior, reaching 55% of cases [34]. This amount is similar to that achieved by balneotherapic activity in spas in Italy [59]. Waters with up to 300 Bq/L of ^{222}Rn are considered healthy in Greece, due to the fact that total exposure dose of Greek spa´s workers are legally limited to 3 mSv/year [13].

Radioactive medicinal waters are widely used in traditional thermal centers of Poland, in order to obtain pain relief to joint diseases and spondilartrosis. There, one can highlight the following fountains: Heisig (800 Bq/L of ^{222}Rn), Skorepa (400 Bq/L of ^{222}Rn) and Swieradow (707 Bq/L of ^{222}Rn). The minimum reference value for ^{222}Rn bioactivity in baths perhaps could based on the quality criteria proposed by the European Association of Radon Spas (EURADON), that is 666 Bq/L of ^{222}Rn and has been supported by several researchers from the hydrotherapy industry.

Isothermal bathings (36°C to 38°C) of 10 to 20 minutes in radioactive mineral waters have been indicated in pathologies due to the increased diuresis, blood vessel dilation, as well as the decrease in blood pressure and oxygen or bradycardic consumption [35].

It should be noted that the reactions, accumulations and adjustments to radionuclides are quite different among living organisms [24].

The main biochemical characteristics of radon are derived from its alpha radiation, due to its ease of absorption through mucous membrane or skin, and by its good ionization or biochemical excitation potential.

However, its emitted energy of 5.49 MeV has penetrating capacity of only 41.1 μm in water, and 20 μm in human tissue. Its time of permanence in the human body is also short, being estimated that about 59% of the total absorbed is to be eliminated from 15 to 30 minutes after its ingestion. Its final decay is no longer detectable through analytical methods after three hours from the physiological absorption [6].

Despite the use of radontherapy being old and intense, its action principles are not completely known. The hormesis theory offers the best explanation relating ionizing radiation at low doses and occasional exposure [9, 33, 70]. The main therapeutic indications are [8]: osteo-articular and rheumatic diseases; sequelae of traumatisms; analgesic and antispasmodic action; gout; central nervous system; immune system; reproductive system; gynecological functions; dermatology and anti-aging.

CONCLUSION

The World Health Organization has pointed out the relevance of epidemiological studies which have clearly shown that long-term exposure to high radon concentrations in indoor air increases the risk of lung cancer. It has been also considered that radon ingested in drinking-water will give a radiation dose to the lining of the stomach. However, the majority of scientific studies have not shown a definitive link between consumption of drinking-water containing radon and an increased risk of stomach cancer. As the dose from radon present in drinking-water is normally received from inhalation rather than ingestion, it is more appropriate to measure the radon concentration in air than in drinking-water. Contrarily to these aspects, radon water has been sometimes used as a natural healing resource, where medical spas have pointed out the following indications of the radon therapy: improves the mobility of the joints, has an antiarthritic effect, improves the immunity of the organism, improves blood supply, operates on cellular chemical processes, stimulates repair mechanisms in the core of cells, reducing the need for analgesics or other medicines. In most cases the effect, even in the case of chronic illnesses, lasts longer than six months, being the treatment suitable for older patients. The radon therapy does not burden the organism in terms of thermal, circulatory or acid-base balance. It has been also used for post-operative and post-traumatic conditions which require intensive healing and regeneration of damaged tissues, such as total joint replacements, operations of the spine and areal scars resulting from burns. Diseases of peripheral vessels, such as vaso-neurosis, ischaemic disorders when suffering from atherosclerosis of limb arteries, conditions after vein inflammation or rheumatic vascullitis have responded well, too. The reported contraindications are for pregnant women, children and teenagers under 18, patients within one or two years after operation or another therapy of tumour disease if the oncologist does not permit radon therapy, and patients with any acute disease such as infection, unstable arterial hypertension, cardiac or respiratory complaints, unstable diabetes, untreated thyroid hyper-function, among other. Therefore, the noble gas radon has motivated intense debates in the scientific community as despite it has been widely recognized as a health threat other people have considered and used it as an important natural healing resource for the treatment of several human diseases.

ACKNOWLEDGMENTS

CNPq (National Council for Scientific and Technologic Development) in Brazil is thanked for the PhD scholarship to F.T.L.

REFERENCES

[1] DFPM (Division for Supporting the Mineral Production), *The mining code, the mineral waters code and how applying research in a mineral deposit*, 8th ed., DFPM, Rio de Janeiro (1966).
[2] C. G. Clayton, *Nuclear Geophysics*, Pergamon Press, Oxford (1983).

[3] S. Y. F. Chu, L. P. Ekström and R. B. Firestone, *The Lund/LBNL Nuclear Data Search*, http://nucleardata.nuclear.lu.se/nucleardata/toi/index.asp.

[4] S. Flügge and K. E. Zimens, *Zeitschrift fur Physikalische Chemie (Leipzig)* B42, 179 (1939).

[5] WHO (World Health Organization), *Guidelines for drinking water quality*, 4th ed., WHO Press, Geneva (2011).

[6] Z. Zdrojewicz and J. Strzelczyk, *Dose Response* 4, 106 (2006).

[7] F. M. Eyzaguirre, *Vademécum of Spanish mineral-medicinal waters*, Instituto de Salud Carlos III, Madrid (2003).

[8] Medical spa Jáchymov, *Czech medical spas*, Spa and wellness nature resorts, Jáchymov (2014).

[9] P. Deetjen, A. Falkenbach, D. Harder, H. Jöckel, A. Kaul, and H. von Philipsborn, *Radon als Heilmittel – Therapeutische Wirksamkeit, biologischer Wirkungsmechanismus und vergleichende Risikobewertung*, Verlag Dr. Kovac, Hamburg (2005).

[10] J. Vaupotic and N. Kávási, *Radiat. Prot. Dosim.* 141, 383 (2010).

[11] J. A. S. Adams and P. Gasparini, *Gamma ray spectrometry of rocks*, Elsevier, Amsterdam (1970).

[12] J. W. Gabelman, *Migration of uranium and thorium: exploration significance*, Studies in Geology No. 3, AAPG, Tulsa (1977).

[13] E. Vogiannis, M. Niaounakis and C. P. Halvadakis, *Environ. Int.* 30, 621 (2004).

[14] Sakoda, B. Hanamotoa, N. Harukic, T. Nagamatsua, and K. Yamaoka, *Appl. Radiat. Isotopes* 65, 50 (2007).

[15] Z. I. Kolar, *Czech. J. Phys.* 49, 43 (1999).

[16] H. Chaudhuri, N. K. Das, R. K. Bhandari, P. Sen, and B. Sinha, *Radiat. Meas.* 45, 143 (2010).

[17] J. F. Andrade Jr., *Boletim do Serviço Geológico e Mineralógico do Brasil*. Bol. 28, Ministério da Agricultura, Indústria e Comércio, Rio de Janeiro (1928).

[18] B. M. Mourão, *Medicina hidrológica – moderna terapêutica das águas minerais e estâncias de cura*. Secretaria Municipal de Educação, Poços de Caldas (1992).

[19] E. Deák and K. Nagy, *Proc. VII Hungarian Radon Forum and Radon in Environment Satellite Workshop* 1, 33 (2013).

[20] D. Desideri, M. R. Bruno and C. Roselli, *J. Radioanal. Nucl. Ch.* 261, 37 (2004).

[21] N. Voronov, *Environ. Geol.* 46, 630 (2004).

[22] M. N. Gómez and A. I. Martín-Megías, *Anales de Hidrologia Médica* 3, 109 (2010).

[23] EC (European Communities), *Official Journal of the European Communities* L330, 85 (2001).

[24] F. Besançon, *La Presse Thermale et Climatique* 127, 1 (1990).

[25] L. Marques, W. dos Santos, L. P. Geraldo, *Appl. Radiat. Isotopes* 60, 801 (2004).

[26] K. Becker, *Nonlinearity Biol. Toxicol. Med.* 1, 3 (2003).

[27] T. Toyota, T. Kataoka, Y. Nishiyama, T. Taguchi, and K. Yamaoka, *Mediators Inflammation* 382801, 1 (2012).

[28] T. Kataoka, J. Teraoka, A. Sakoda, Y. Nishiyama, K. Yamato, M. Monden, Y. Ishimori, T. Nomura, T. Taguchi, and K. Yamaoka, *Inflammation* 35, 713 (2012).

[29] T. Kataoka and K. Yamaoka, *ISRN Endocrinology* 292041, 1 (2012).

[30] K. Yamaoka, *J. Clin. Biochem. Nutr.* 39, 114 (2006).

[31] F. Mitsunobu, K. Yamaoka, K. Hanamoto, S. Kojima, Y. Hosaki, K. Ashida, K. Sugita, and Y. Tanizaki, *J. Radiat. Res.* 44, 95 (2003).

[32] D. Yu and J. K. Kim, *Chemosphere* 54, 639 (2004).

[33] M. Giacomino and D. Demichele, *Anales de Hidrologia Médica* 5, 147 (2012).

[34] K. Becker, *International Journal of Low Radiation* 1, 334 (2004).

[35] Frangipani, C. Ceriani, F. M. Flora, M. U. Filho, R. A. P. Simões, and T. C. Alvisi, *Termalismo no Brasil*. Sociedade Brasileira de Termalismo, Seção de Minas Gerais, Belo Horizonte (1995).

[36] J. Tölgyessy, *Chemistry and biology of water, air, and soil: environmental aspects*, Elsevier, Amsterdam (1993).

[37] T. V. Ramachandran, *Iran. J. Radiat. Res.* 8, 129 (2010).

[38] Y. Prasad, G. Prasad, V. M. Choubey, and R. C. Ramola, *Radiat. Meas.* 44, 122 (2009).

[39] S. Huxol, M. S. Brennwald, E. Hoehn, and R. Kipfer, *Chem. Geol.* 298, 116 (2012).

[40] J. McLaughlin, *Radiat. Prot. Dosim.* 141, 316 (2010).

[41] T. Ishikawa, M. Hosoda, A. Sorimachi, S. Tokonami, S. Katoh, and S. Ogashiwa, *J. Radioanal. Nucl. Ch.* 287, 709 (2011).

[42] S. Huxol, E. Höhn, H. Surbeck and R. Kipfer, *Geophysical Research Abstracts of the EGU General Assembly* 11, 2459 (2009).

[43] M. A. Misdaq, M. Ghilane, J. Ouguidi, and K. Outeqablit, *Radiat. Environ. Biophys.* 51, 375 (2012).

[44] T. Kataoka, Y. Aoyama, A. Sakoda, S. Nakagawa and K. Yamaoka, *Physiological Chemistry and Physics and Medical NMR* 38, 85 (2006).

[45] Y. Aoyama, T. Kataoka, S. Nakagawa, Y. Ishimori, F. Mitsunobu, and K. Yamaoka, *Iran. J. Radiat. Res.* 9, 221 (2012).

[46] S. H. Serra, *Águas minerais do Brasil*, Millenium Editora, Campinas (2009).

[47] D. M. Bonotto, *J. Environ. Radioactiv.* 132, 21 (2014).

[48] C. Cotar and V. Harley, *A treatise on the mineral waters of Vichy: for the use of practitioners*, H. K. Lewis Ed., London (1913).

[49] M. Piery and M. Milhaud, *Les eaux minerals radioactives*, G. Doin Ed., Paris (1924).

[50] D. S. Vinson, T. R. Campbell and A. Vengosh, *Appl. Geochem.* 23, 2676 (2008).

[51] G. Song, X. Wang, D. Chen, and Y. Chen, *J. Environ. Radioactiv.* 102, 400 (2011).

[52] J. F. Lobo, *Guia termal do Cipó*, Oficinas Gráficas da Imprensa Oficial da Bahia, Salvador (1961).

[53] J. A. Pupo, *Águas de São Pedro: suas indicações terapêuticas e seu plano de organização*, IPT-Instituto de Pesquisas Tecnológicas, São Paulo (1940).

[54] P. L. Brezonik and W. A. Arnold, *Water chemistry: an introduction to the chemistry of natural and engineered aquatic systems*, Oxford University Press, Oxford (2011).

[55] J. M. Godoy and M. L. Godoy, *J. Environ. Radioactiv.* 85, 71 (2006).

[56] D. Banks, B. Frengstad, A. K. Midtgard, J. R. Krog, and T. Strand, *Sci. Total Environ.* 222, 71 (1998).

[57] J. Soto, P. L. Fernández, L. S. Quindós, and J. Gómez-Arozamena, *Sci. Total Environ.* 162, 187 (1995).

[58] J. R. Fagundo, A. Cima and P. González, *Revision bibliografica sobre classificación de las aguas minerales y mineromedicinales*, Centro Nacional de Termalismo "Victor Santamarina", Havana (2001).

[59] C. Nuccetelli, F. Bochicchio and G. Ruocco, *Natural radioactivity in mineral and spa waters: the current regulatory approach in Italy*, Istituto Superiore di Sanità, Roma (2002).

[60] M. C. Albertini, M. Dacha, L. Teodori, and M. E. Conti, *Int. J. Environmental Health* 1, 153 (2007).

[61] H. Tempfer, W. Hofmann, A. Schober, H. Lettner, and A. L. Dinu, *Radiat. Environ. Biophys.* 49, 249 (2010).

[62] Franke, L. Reiner, H. G. Pratzel, T. Franke, and K. L. Resch, *Rheumatology* 39, 894 (2000).

[63] F. Zolzer, Z. Hon, Z. F. Skalicka, R. Havrankova, L. Navratil, J. Rosina, and J. Skopek, *Int. Arch. Occup. Environ. Health* 12, 1 (2012).

[64] S. V. Klemenkov, O. B. Davydova, I. F. Levitskiĭ, O. G. Atrashkevich, I. V. Kubushko, and V. A. Makarenko, *Vopr Kurortol Fizioter Lech Fiz Kult.* 6, 6 (1999).

[65] V. V. Barnatskiĭ, V. D. Grigoreva and E. N. Kaliushina, *Vopr Kurortol Fizioter Lech Fiz Kult.* 4, 13 (2005).

[66] N. Kávási, T. Kovács, J. Somlai, V. Jobbágy, K. Nagy, E. Deák, I. Berhés, T. Bender, T. Ishikawa, and S. Tokonami, *Radiat. Prot. Dosim.* 146, 27 (2011).

[67] K. Nagy, I. Berhés, T. Kovács, N. Kávási, J. Somlai, and T. Bender, *Radiat. Environ. Biophys.* 48, 311 (2009).

[68] C. Serrano, M. Romero, L. Alou, D. Sevillano, I. Corvillo, F. Armijo, and F. Maraver, *Journal of Water and Health* 10, 400 (2012).

[69] Falkenbach, J. Kovács, A. Franke, K. Jörgens, and K. Ammer, *Reumatol. Int.* 25, 205 (2005).

[70] H.-Y. Thong and H. I. Maibach, *Dose Response* 6, 1 (2008).

In: Radon
Editor: Audrey M. Stacks

ISBN: 978-1-63463-742-8
© 2015 Nova Science Publishers, Inc.

Chapter 6

THE POSSIBLE APPLICATIONS OF RADON INHALATION TREATMENT AS ANTIOXIDANT THERAPY FOR HEPATOPATHY

Takahiro Kataoka[1], Akihiro Sakoda[2], Reo Etani[1], Yuu Ishimori[2], Fumihiro Mitsunobu[3] and Kiyonori Yamaoka[1, *]

[1]Graduate School of Health Sciences, Okayama University, Shikata-cho 2-chome, Kita-ku, Okayama-shi, Okayama, Japan
[2]Ningyo-toge Environmental Engineering Center, Japan Atomic Energy Agency, Kagamino-cho, Tomata-gun, Okayam, Japan
[3]Misasa Medical Center, Okayama University Hospital, Yamada, Misasa-cho, Tohaku-gun, Totori, Japan

ABSTRACT

The possibility of antioxidant therapy has been reported for several diseases such as ischemic stroke. The therapy could also be applied for diseases caused by reactive oxygen species (ROS). It has been reported that ROS or free radicals may cause various types of hepatopathy, including alcoholic liver disease. Low dose (0.5 Gy) X- or γ-irradiation activates the antioxidative functions of the mouse liver and inhibits ROS- or free radical-induced hepatopathy. Radon therapy is performed mainly for pain-related diseases in Japan and Europe. Several clinical studies have been reported, but the possible mechanisms of the beneficial effects remain unknown. Recently, we have reported that the possible mechanism of radon therapy is the activation of antioxidative functions following radon inhalation. For example, radon inhalation inhibits and alleviates chronic constriction injury induced pain or inflammatory pain in mice due to the activation of antioxidative functions. In addition, although hepatopathy is not the main indication for radon therapy, our recent studies suggested that radon inhalation inhibits hepatopathy caused by ROS or free radicals. In this chapter, based on

---

* Corresponding author: Graduate School of Health Sciences, Okayama University, 5-1 Shikata-cho 2-chome, Kita-ku, Okayama-shi, Okayama 700-8558, Japan, Phone/Fax: +81-86-235-6852, E-mail: yamaoka@md.okayama-u.ac.jp.

experiments with mice, we reviewed the possible applications of radon inhalation as an antioxidant therapy for hepatopathy from the viewpoints of recent antioxidant therapy, hepatopathy induced by ROS or free radicals, and the beneficial effects of radon inhalation for hepatopathy.

1. INTRODUCTION

We previously reviewed the activation of bio-defense systems by exposure to radon and possible applications of lifestyle diseases in a chapter of a book published by NOVA publishers [1]. In the chapter, we indicated that radon inhalation activates antioxidative functions in the brain, lung, liver, and kidney of mice and inhibits alcohol [2] or carbon tetrachloride (CCl_4) [3] induced hepatopathy in mice. Radon inhalation also has possibilities of the prevention of hypertension, diabetes and pain in rabbits due to the increase in the level of α-atrial natriuretic polypeptide (ANP), which decreases blood pressure by relaxation of the vascular smooth muscle, the level of insulin which promotes glycogen synthesis, and the level of β-endorphin and M-enkephalin with morphine-like analgesic actions [4]. In addition, radon therapy is performed for pain [5] or respiratory- related diseases [6] in the Misasa Medical Center Okayama University Hospital, Japan.

We recently reported several studies on the activation of antioxidative functions in the liver of mice and the inhibition of oxidative stress in the liver [2,3,7-9]. Although hepatopathy is not the main indication for radon therapy, our results demonstrated that radon inhalation clearly inhibited oxidative damage to the liver. In this chapter, we reviewed the possible applications of radon inhalation treatment as an antioxidant therapy for hepatopathy from the viewpoints of recent antioxidant therapy, hepatopathy induced by reactive oxygen species (ROS) or free radicals, and the beneficial effects of radon inhalation for hepatopathy.

2. CLINICAL STUDIES OF RADON THERAPY

2.1. Treatment for Pain Related Diseases

Radon therapies using spas, caves, or galleries are performed for various diseases in Japan, [5,6], Russia, and the USA, as well as several countries in Europe [10,11]. These therapies are performed in combination with ordinary treatment. Specifically, it appears to be effective against pain related diseases. For example, radon therapy for patients with osteoarthritis [5] is performed at Misasa Medical Center, Okayama University Hospital, Japan. Participants do not bathe in, but only stay in a bathroom with a concentration of 2080 Bq/m^3, about 100-fold higher than the background level. The room temperature is 42°C and at high humidity (90%). Every 2 days, nasal inhalation of radon is performed for 40 minutes per day, a total of 9 to 12 times over 3 or 4 weeks. This treatment enhanced the antioxidative and immune functions, which can contribute to the prevention of osteoarthritis related to peroxidation reactions and immune depression. This study also suggested that the changes in vasoactive and pain-associated substances following radon therapy play a role in alleviating pain.

Although radon therapy is widely performed in Russia, as well as central Europe, papers are written in various languages, such as Russian, German, or Dutch. Falkenbach chose five randomized controlled clinical trials with a total of 338 patients suffering from rheumatic diseases and examined whether radon therapy was associated with a significant alleviation of pain compared to control interventions [10]. The trials included radon bath treatment (three trials; radon concentration in the baths, 0.8–3 kBq/L) or radon speleotherapy (two trials; radon concentration in the air, 37–160 kBq/m^3) with control intervention in degenerative spinal disease (two trials), rheumatoid arthritis (one trial), and ankylosing spondylitis (two trials). Although the authors mentioned the limitations of this study, which may arise from the fact that pain caused by various rheumatic diseases was pooled for meta-analysis, the results suggested that radon therapy significantly reduced pain at 3 or 6 months after treatment.

Although radon therapy is also performed in Montana [12], USA, it is slightly different from the therapy in Japan and Europe. In Japan and Europe, medical doctors prescribe radon therapy, while in Montana, it is a self-treatment option in which people simply enter the mine and spend one or more hours inhaling radon. They sit on chairs and padded benches and they may read, sleep or sit and talk with one another or play cards or board games. Approximately 500 people visit the mine every year and many of these repeatedly visit for years. They are likely to spend anywhere from 30 to 60 hours over ten days. The mean radon concentration in the mine is approximately 55000 Bq/m^3. To evaluate the radon effects in Montana, the finger, shoulder or neck flexibility of 21 participants, suffering from osteoarthritis, rheumatoid arthritis or both, were measured before, during and after a course of radon therapy [12]. The results indicated overall patterns of improvement in joint flexibility.

2.2. Treatment for Respiratory Related Diseases

Radon therapy also has curative effects on respiratory diseases such as bronchial asthma [6]. Radon therapy significantly increased the forced expiratory volume in one second (%FEV1), and activated antioxidant enzymes such as catalase and superoxide dismutase (SOD), while radon therapy significantly decreased the lipid peroxide level. ROS induced by inflammatory cells is thought to be involved in the pathophysiology of asthma. These findings suggest that radon and thermal therapy improved the pulmonary function of asthmatics by increasing the reduced activities of antioxidant functions.

2.3. Comparison between Radon Effects and Thermal Effects

As described above, radon therapy activates antioxidative functions. It has been also reported that SOD increases by in vivo hyperthermia (39^{0}C) [13]. To clarify which treatments (radon or hyperthermia) contribute to activating antoxidative functions, we compared the differences between radioactive effects and thermal effects under sauna or hot spring conditions with a similar chemical component [14]. In this study, the subjects were divided into 3 groups: control, radon, and thermal groups. Subjects in the radon group stayed in a hot bathroom with a high concentration of radon at the Misasa Medical Center, Okayama University Hospital (temperature: 36°C, radon concentration: 2080 Bq/m^3). Subjects in the thermal group (temperature: 48°C, radon concentration: 54 Bq/m^3) and the control group (temperature: 36 °C, radon concentration: 54 Bq/m^3) visited a local sauna. The thermal group, as well as the radon group, showed increased activities of SOD and catalase and inhibited lipid peroxidation and total cholesterol. This study also examined the changes in several parameters such as the concanavalin A (ConA)-induced mitogen response, the percentage of CD4-positive cells, which is a marker of helper T cells, the percentage of CD8-positive cells, which is a common marker of killer T cells and suppressor T cells, the atrial natriuretic polypeptide, which relaxes the vascular smooth muscle, the β-endorphin with morphine-like analgesic actions, and adrenocorticotropic hormone (ACTH), insulin, and glucose-6-phosphate dehydrogenase. The results were on the whole better in the radon group than in the thermal group. These findings indicate that radon inhalation is more effective than thermal treatment against some diseases.

3. OXIDATIVE DAMAGE TO THE LIVER AND ANTIOXIDANTS

3.1. Sources of ROS and Endogenous Antioxidants

The production of ROS is from not only exogenous sources but also endogenous sources. The former includes irradiation, food, drugs, pollutants, and toxins. The latter includes production from neutrophils, mitochondria, xanthin oxidase, and the ischemic process [15]. Exposure to radiation is one of the major exogenous sources of the production of ROS, such as hydroxyl radicals. Although it is thought that these radicals are produced from the ionization of intracellular water, radiation also induces mitochondrial ROS production [16]. These findings may indicate that radiation is from not only exogenous sources but also endogenous sources for activating endogenous ROS production.

We previously wrote about endogenous antioxidants [1]. SOD is an antioxidtive enzymes that changes superoxide anion into hydrogen peroxide (H$_2$O$_2$). Catalase and glutathione peroxidase (GPx) are also antioxidative enzymes, and these enzymes detoxify H$_2$O$_2$ into H$_2$O. Reduced glutathione (GSH) directly reacts with H$_2$O$_2$ and hydroxyl radicals by the catalyzing role of GPx. Oxidized glutathione (GSSG) is produced by the oxidation of GSH. However, glutathione reductase (GR), which is also an enzyme, catalyzes the regeneration of GSH from GSSG. Thus, GR and GPx are both enzymes in the glutathione-regenerating pathway. Therefore, the increase in only an antioxidative enzyme is insufficienct to protect against oxidative damage.

3.2 Hepatopathy Induced by ROS

It is widely accepted that the overproduction of ROS causes the development of liver diseases such as alcoholic liver diseases [17, 18] and non-alcoholic steatohepatitis (NASH) [19]. Numerous studies have found that ROS production following alcohol administration contributes to the development of liver damage. Specifically, alcohol administration has been found to cause the production of superoxide anion, hydroxyl radicals, and hydrogen peroxide and to increase lipid peroxide in the liver [20]. Several studies have also suggested that oxidative stress plays an important role in the pathogenesis of NASH. Specifically, it has been reported that superoxide anions play a pivotal role in the development of the disease [18]. These findings indicate that inhibition of ROS production or ROS detoxification prevents the development of liver diseases.

3.3. Protection of the liver from Oxidative Damage

Numerous studies have been reported that antioxidants protect the liver from ROS or free radicals. For example, quercetin, a flavonoid antioxidant, prevents against ethanol-induced oxidative stress in mouse liver [21]. Pretreatment with quercetin increased antioxidative related substances such as SOD, catalase, GPx, GR, and GSH in comparison to the ethanol group. In this study, the main role of quercetin appears not to be to act as a radical scavenger, but to activate endogenous antioxidants. Quercetin also decreases liver damage in mice with NASH. Quercetin treatment showed a significant inhibition of the increase in hepatic damage enzymes, lipoperoxidation, and deoxyribonucleic acid (DNA) damage [22]. Another report indicated that ascorbic acid protects from hepatotoxicity and oxidative stress caused by CCl_4 in the liver of rats [23]. This study also indicated that ascorbic acid treatment activates antioxidative functions in the liver and protects the liver against CCl_4 -induced oxidative stress. In addition, the combination of antioxidants is more effective against oxidative stress. The combination of melatonin, which is a direct free radical scavenger, and vitamin C dramatically enhanced the protective effects against iron-induced lipid peroxidation [24]. The mechanism proposed was that that vitamin C may recycle melatonin. As a result, synergistic antioxidative actions were observed.

4. ANTIOXIDANT THERAPY

4.1. Clinical Use of Antioxidants

Numerous clinical or animal studies have reported that antioxidants are useful against ROS-related diseases. Antioxidants are used for the purpose of the providing protection against ROS-induced damage. Firuzi reviewed the current status of antioxidant therapy including antioxidants and their clinical use, such as edaravone, idebenone, N-aceylcystein, α-lipoic acid, micronized purified flavonoids fraction, 0-β-hydroxyethyl-rutosides, silibinin, baicalein and catechines [25]. For example, edaravone is one of antioxidants that protects against ischemic stroke. ROS play an important role in the development of ischemia-reperfusion injury in stroke patients, so antioxidants are likely to provide beneficial effects. Several studies demonstrated that edaravone administration improved the functional outcome of patients [26] and reduced brain edema [27]. In addition, the adverse effect of edaravone has been observed at only 0.1% [28]. However, functional improvement was observed only in patients with mild cardioembolic stroke and not in moderate to severe stroke [29]. On the whole, antioxidant therapy seems to be ineffective when disease is already well-established. However, earlier antioxidant treatment appears to be more effective for preventing the development of diseases. Another possibility pointed out is that the combination of different antioxidants might be more effective than one single antioxidant [25].

4.2 Antioxidant Therapy for the Liver

Ashwani reviewed antioxidants as therapeutic agents for liver disease [30]. The review summarizes the results of clinical trial use of antioxidant therapy for chronic hepatitis C, alcoholic liver disease and NASH. Several clinical studies on chronic hepatitis C virus infection showed that antioxidants therapy improved hepatic function parameters, such as the alanine aminotransferase (ALT) level [31-33], whereas it did not show any improvement in others [34-36]. This seems to indicate that treating with a high dose antioxidant or combination of antioxidants for long periods provides clinical effects. A study also showed the improvement of ALT level in suffering from NASH [37], but antioxidant therapy for alcoholic liver diseases did not improve the ALT level [38-40]. Although animal studies and diseases mechanisms clearly showed that ROS or free radicals contribute to the development of liver diseases, further studies are still needed to establish antioxidant therapy for liver diseases.

5. ACTIVATION OF ANTIOXIDATIVE FUNCTIONS BY LOW-DOSE X- OR γ -IRRADIATION OR RADON INHALATION

5.1. X- or γ-Irradiation

Low-dose X- or γ-irradiation induces various effects, especially the activation of antioxidative functions in several organs of mice, including the spleen [41], liver [42], and brain [43]. In the case of low-dose X- or γ-irradiation, the optimum dose for activation of SOD activity appears to be 0.5 Gy [44]. Moreover, the peak in SOD activity occurred after around 4 hours of exposure to X-irradiation and kept a slightly high activity rate for several weeks [44]. Low-dose irradiation increases not only SOD activity but also the GSH level due to glutathione synthesis [41-43].

Kelch-like ECH-associated protein 1 (Keap1) – nuclear factor erythroid 2-related factor 2 (Nrf2) system regulates antioxidant response element (ARE) - mediated the expression of genes encoding antioxidants [45]. This Keap1-Nrf2 system acts as a sensor to ROS for protecting from oxidative stress. To investigate the mechanism of activation of antioxidative functions, some studies have been conducted. Low-dose γ-irradiation induced a translocation of Nrf2 from cytoplasm to the nucleus in mouse macrophage RAW264.7 cells [46]. These findings indicate that production of ROS by low-dose irradiation induced activation of antioxidative functions.

5.2. Radon Inhalation

Radon inhalation at a concentration of 400 or 4000 Bq/m^3 also increases antioxidative enzymes such as SOD and catalase in the brain, lung, liver, and kidney of mice [47]. To clarify the radon concentration dependency and inhalation time dependency of activation of antioxidative functions in the plasma, liver, pancreas, heart, thymus, kidney, brain, small intestine, lung, and stomach of mice, the mice inhaled radon at concentrations of 250, 500,

1000, 2000, or 4000 Bq/m^3 for 0.5, 1, 2, 4, or 8 days. Our results suggested that continuous exposure to radon increases SOD activity in most organs, except for the stomach, within 2 days, but SOD activity transiently decreased after the peak of the activation. These findings indicates that radon inhalation protects mice from oxidative damage in most organs [48].

6. BENEFICIAL EFFECTS OF LOW-DOSE X- OR γ- IRRADIATION OR RADON INHALATION FOR HEPATOPATHY

6.1. Inhibition of Low-Dose Irradiation on Ferric Nitrilotriacetate (Fe^{3+}-NTA) Induced Hepatopathy

We previously reported activation of the biodefense system by low-dose irradiation or radon inhalation inhibits diabetes and hepatopathy in mice [49]. Specifically, the report suggested that low-dose irradiation including radon inhalation activates antioxidative functions in several organs of mice or rats. Therefore, it is highly possible that the main indications of low-dose irradiation therapy are ROS or free radicals related diseases. For example, Fe^{3+}-NTA administration induced acute hepatopathy due to production of free radicals. Iron level in serum significantly increased in sham-irradiated rats as well as prior 0.5 Gy irradiated rats, and no significant differences in iron level were observed between sham irradiation and 0.5 Gy irradiation, suggesting that prior 0.5 Gy irradiation does not inhibit free radical production. However, only irradiated rats significantly inhibited the hepatopathy. These results indicate that low-dose irradiation does not inhibit the production of free radicals by Fe^{3+}-NTA administration, but dismutates free radicals. In fact, prior 0.5 Gy irradiation significantly inhibits the production of the thiobarbituric acid reacting substances (TBARSs), which show the level of oxidant injury, in the liver [50]. In the same manner, post 0.5 Gy irradiation inhibited Fe^{3+}-NTA induced hepatopathy [51].

6.2. Inhibition of Irradiation on CCl$_4$ Induced Hepatopathy

We also examined the inhibitory effects of 0.5 Gy irradiation on hepatopathy induced by another free radical inducer, CCl$_4$ [52]. Trichloromethyl radical or trichoromethyl peroxy radical contribute to the development of hepatopathy [53, 54]. However, it is likely that activation of antioxidative functions by low-dose irradiation reduces oxidative damage induced by CCl$_4$ because trichloromethyl radical or trichoromethyl peroxy radical are detoxified. In this study, the irradiation of γ-rays was initiated 24 hours after the injection of CCl$_4$. Therefore, the results of this study show the curative effects of low-dose irradiation.

The results showed that hepatopathy improved 3 days after irradiation, and that the activities of GR, GPx, and the total glutathione content (t-GSH) of the irradiation group elevated after irradiation.

6.3. Inhibition of Prior or Post X-Irradiation on CCl$_4$ Induced Hepatopathy in Acatalasemic Mice

Acatalasemic mice had lower catalase activity, one of the antioxidative enzymes, than normal mice [55]. Therefore, acatalasemic mice are likely to be sensitive to oxidative stress. Our results also showed that catalase activity in the liver of acatalasemic mice was a half lower than that of normal mice. However, GPx activity in the liver of acatalasemic mice was higher than that of normal mice and the lipid peroxide level, which shows the oxidative damage level, in the liver of acatalasemic mice was at the same level as that of normal mice, suggesting that acatalasemic mice are not damaged by ROS under no treatment condition. CCl$_4$ administration induced hepatopathy in normal and acatalasemic mice. On the other hand, low-dose (0.5 Gy) irradiation significantly increased SOD, catalase, and t-GSH in the liver of normal mice, and catalase of acatalasemic mice, suggesting that low-dose irradiation activates antioxidative functions in the liver. In mice injected with CCl$_4$, the activities of glutamic oxaloacetic transaminase (GOT) and glutamic pyruvic transaminase (GPT), which show hepatic functions, in the serum of normal and acatalasemic mice were significantly higher than that of irradiated mice, suggesting that pre-treatment with low-dose irradiation inhibits CCl$_4$ induced hepatopathy in normal and aatalasemic mice due to activation of the antioxidative functions [56]. In addition, no differences in the inhibitory effects were observed between normal and acatalasemic mice. These findings suggested that that the effects of CCl$_4$ are properly neutralized by high GPx activity and low-dose irradiation in the acatalasemic mouse liver.

Effects of post low-dose irradiation were also examined [57]. Post treatment with low-dose irradiation accelerated the rate of recovery from CCl$_4$ induced hepatopathy in normal and acatalasemic mice probably due to activation of the antioxidative functions in the liver. The recovery of the acatalasemic mice from CCl$_4$ induced hepatopathy was lower than that of the normal mice. This is probably because catalase plays an important role in the recovery from CCl$_4$ induced hepatopathy.

6.4. Inhibition of Radon Inhalation on CCl$_4$ Induced Hepatopathy

We have reported several studies of the effects of CCl$_4$ induced hepatopathy in mice to clarify the mechanism of radon effects including preventive or curative effect and the antioxidative ability against CCl$_4$ induced liver damage. We attempted two kinds of inhalation conditions; high radon concentration (18,000 Bq/m^3) - short inhalation time (6 hours) [3] and low radon concentration (2,000 Bq/m^3) - long inhalation time (24 hours) [9].

Radon inhalation at a concentration of 18000 Bq/m^3 for 6 hours increases antioxidative substances such as t-GSH, GPx, and GR and decreases the lipid peroxide level in the liver of mice, and inhibits CCl$_4$-induced hepatopathy [3]. These findings indicate that radon inhalation has similar antioxidative effects to low-dose X- or γ- irradiation.

Furthermore, we compared the inhibitory effects of prior or posterior radon inhalation on CCl$_4$-induced oxidative damage in the liver of mice [8]. Mice inhaled radon at a concentration of 18000 Bq/m^3 for 6 hours before or after CCl$_4$ administration. Although CCl$_4$ administration significantly increased the lipid peroxide level in the liver, the level was significantly lower in the prior or posterior radon inhaled mice than in the CCl_4 administered mice. These findings suggest that not only 0.5 Gy irradiation but also radon inhalation has preventive and curative effects on acute hepatopathy induced by free radicals. In addition, to assess the antioxidant activity following radon inhalation, we compared the antioxidant activities of antioxidant vitamins, such as ascorbic acid (vitamin C) or α-tocopherol (vitamin E), and radon from the results of the inhibition of CCl_4-induced hepatopathy based on hepatic function-associated parameters, oxidative damage-associated parameters and histological changes [9]. Results showed that radon inhalation has an anti-oxidative effect against CCl_4-induced hepatopathy that is comparable to treatment with ascorbic acid at a dose of 500 mg/kg weight or α-tocopherol treatment at a dose of 300 mg/kg weight. Radon therapy is performed for various diseases at Misasa Medical Center, Okayama University Hospital. Although hepatopathy is not the main indication for radon therapy, these reports demonstrated that radon inhalation clearly inhibited oxidative damage to the liver. Furthermore antioxidative activity following radon inhalation is effective as can be seen from the comparison between antioxidant vitamins and radon.

6.5. Inhibition of Radon Inhalation on Alcohol Induced Acute Hepatopathy

We have also reported that radon inhalation inhibits hepatopathy induced by alcohol administration in mice [7]. As described above, production of ROS contributes to the development of alcohol-induced hepatopathy [17, 18]. Alcohol administration also leads to GSH depletion in the liver [58, 59], due to the direct conjugation of GSH with acetaldehyde and reactive intermediates of alcohol oxidation. For example, radon inhalation at a concentration of 4000 Bq/m^3 for 24 hours inhibits not only acute alcohol-induced oxidative damage, but also hepatopathy and fatty liver in mice. The activation of antioxidative functions in the liver of mice following radon inhalation may detoxify ROS induced by alcohol metabolism.

7. OTHER ANIMAL STUDIES OF RADON EFFECTS

7.1. Radon Inhalation Inhibits Oxidative Damages of Several Organs in Mice

We previously reported that radon inhalation has antioxidative and anti-inflammatory effects and pain relief in mouse. Specifically, these effects are induced by activation of antioxidative functions in mouse organs following radon inhalation. For example, the increase in antioxidative enzymes and substances following radon inhalation protects the mouse kidney from CCl_4 induced oxidative damage [3, 9]. Considering the renal function results, the antioxidative effects against CCl_4 were equivalent to the administration of $300 - 500$ mg/kg body weight of α-tocopherol [9]. These results indicate that radon inhalation is effective against CCl_4 induced damages for not only the liver but also the kidney.

Another study shows the protective effects for streptozotocin (STZ) induced type-1 diabetes in mice [60]. STZ selectively destroys pancreatic β cells, which produce insulin, through the generation of ROS and alkylation of DNA [61, 62]. Since the production of ROS in the pancreas of human is one of the causes of the development of type-1 diabetes, STZ is widely used in studies of experimental type-1 diabetes. As described above, the increasing of antioxidative functions in the pancreas by radon inhalation probably inhibits STZ induced type-1 diabetes. Results showed that radon inhalation inhibits STZ induced type-1 diabetes in mice. The possible mechanisms of the inhibitory effects indicated that activation of the antioxidative functions by radon inhalation suppressed β cell apoptosis and improved insulin secretion.

Radon inhalation also inhibits transient global cerebral ischemic injury in gerbils [63]. Production of ROS following cerebral ischemia–reperfusion is an important underlying cause of neuronal injury [64]. Specifically, ROS production following global cerebral ischemia has been linked to delayed neuronal death in affected neurons of the CA1 region of the hippocampus [65, 66]. Results showed that transient global cerebral ischemia induced neuronal damage in hippocampal CA1. However, radon inhalation inhibited the neuronal damage. SOD activity in the radon-treated gerbil brain was more significantly activated than that in sham-operated gerbils. These findings indicated that radon inhalation dismutates ROS due to the activation of antioxidative functions by radon inhalation. Thus, it is likely that radon inhalation would be useful against oxidative damage for not only the liver, but also the kidneys, pancreas, and brain.

7.2 Radon Inhalation Has Anti-Inflammatory Effects in Mice

Radon inhalation also has anti-inflammatory effects in mice. We previously reported that radon inhalation inhibits carrageenan-induced inflammatory paw edema in mice [67]. Although carrageenan administration to the paw significantly increased serum tumor necrosis factor-alpha (TNF-α) and nitric oxide (NO), which are mediators in the inflammatory response, and leukocyte infiltration in the paw. However, radon inhalation inhibits these inflammatory responses.

Since Cuzzocrea suggests that these inflammatory responses are mediated by ROS, antioxidative functions have an important role in the inhibition of carrageenan-induced inflammatory paw edema. [68] The SOD and catalase activities in the paws of radon inhaled mice were significantly higher than those of the carrageenan administered mice. These findings indicated that radon inhalation activated antioxidative functions and inhibited carrageenan-induced inflammatory paw edema.

Radon inhalation also suppresses dextran sulfate sodium (DSS)-induced colitis in mice [69]. In DSS treated mice, the disease activity index (DAI) score, which includes diarrheal stool score and bloody stool score and provides an indicator of the severity of the colitis, was significantly elevated.

However, the DAI score of radon inhaled and DSS treated mice were significantly lower than that of DSS treated mice, suggesting radon inhalation DSS-induced colitis. Since this model also attributes the development of colitis to the production of ROS, activation of antioxidative functions in colon by radon inhalation has an important role in the inhibition of DSS-induced colitis. SOD activity and t-GSH content of radon inhaled and DSS treated mice was significantly higher than those of DSS treated mice.

Thus, it is likely that radon inhalation has anti-inflammatory effects due to detoxification of ROS by activation of antioxidative functions.

7.3. Radon Inhalation Provides Pain Relief in Mice

Although we demonstrated that radon inhalation has anti-inflammatory effects, we did not evaluate whether radon inhalation has antinociceptive effects. Therefore, we examined whether radon inhalation has antinociceptive effects against formalin-induced transient inflammatory pain in mice [70]. The formalin-induced inflammatory pain model is appropriate to examine whether radon inhalation inhibits inflammatory pain, because mice show two kinds of specific pain-related behaviors such as a licking response after formalin administration to the paw. The first pain-related behavior is a direct stimulation of formalin to sensory nerve ending indicating acute pain. The second behavior is an inflammatory response indicating persistent pain [71, 72]. Results showed that radon inhalation inhibits only the second phase, suggesting that radon inhalation had anti-inflammatory and antinociceptive effects.

We also examined whether radon inhalation produces a remission of chronic constriction injury (CCI)-induced neuropathic pain [73]. To examine the preventive and curative effects, mice inhaled radon before or after CCI surgery. Neuropathic pain was evaluated by a von Frey test [74]. Although pain-like behavior following CCI surgery was significantly inhibited by pre-treatment or post treatment of radon, pretreatment with radon had longer-lasting effects on neuropathic pain than post-treatment. This is probably because the SOD activity of pre-treated mice was greater than that of post-treated mice.

Thus, it is likely that radon inhalation prevents and cures pain in mice.

FUTURE PROSPECTS

As describe above, radon therapy is clearly effective against pain-related diseases. Although hepatopathy is not the main indication for radon therapy, our recent studies suggested that radon inhalation inhibits hepatopathy induced by several kinds of ROS or free radicals. On the other hand, a combination of different antioxidants can be more effective than one single antioxidant [25]. Since radon inhalation increases several kinds of antioxidative substances such as SOD, catalase, t-GSH, GPx and GR, it is possible that radon therapy is useful as antioxidant therapy. In addition, a number of clinical studies indicate that long-lasting pain relief and a reduction in the use of the analgesic dose [75]. Therefore, the combination of normal care and radon therapy may enhance therapeutic efficacy of liver disease treatments. To our knowledge, no clinical studies have reported any curative effects on liver diseases. Further clinical studies of radon therapy are needed.

REFERENCES

[1]   Kataoka T., Yamaoka K. Activation of bio-defense system by radon inhalation and its applicable possibility for the treatment of lifestyle diseases, Radon: Properties, Applications and Health Hazards. Nova Science Publishers, Inc. 335-356, 2012.

[2]   Kataoka T., Sakoda A., Yoshimoto M., Toyota T., Yamamoto Y., Ishimori Y., Hanamoto K., Kawabe A., Mitsunobu F., Yamaoka K. A comparative study on the

effect of continuous radon inhalation on several-time acute alcohol-induced oxidative damage to liver and brain in mouse. *Radiati. Safety Manag.* 10:1-7, 2012.

[3] Kataoka T., Nishiyama Y., Toyota T., Yoshimoto M., Sakoda A., Ishimori Y., Aoyama Y., Taguchi T., Yamaoka K. Radon inhalation protects mice from carbon-tetrachloride-induced hepatic and renal damage. *Inflammation* 34:559-567, 2011.

[4] Yamaoka K., Komoto Y. Experimental study of alleviation of hypertension, diabetes and pain by radon inhalation. *Physiol. Chem. Phys. Med. NMR*, 28:1-5, 19996.

[5] Yamaoka K., Mitsunobu F., Hanamoto K., Mori S., Tanizaki Y., Sugita K. Study on biologic effects of radon and thermal therapy on osteoarthritis. *J. Pain* 5:20-25, 2004.

[6] Mitsunobu F., Yamaoka K., Hanamoto K., Kojima S., Hosaki Y., Ashida K., Sugita K., Tanizaki Y. Elevation of antioxidant enzymes in the clinical effects of radon and thermal therapy for bronchial asthma. *J. Radiat. Res.* 44:95-99, 2003.

[7] Toyota T., Kataoka T., Nishiyama Y., Taguchi T., Yamaoka K. Inhibitory effects of pretreatment with radon on acute alcohol-induced hepatopathy in mice. *Mediators Inflamm.* 2012(Article ID 382801):1-10, 2012.

[8] Nishiyama Y., Kataoka T., Teraoka J., Sakoda A., Ishimori Y., Yamaoka K. Inhibitory effects of pre and post radon inhalation on carbon tetrachloride-induced oxidative damage in mouse organs. *Radioisotopes* 61:231-241, 2012.

[9] Kataoka T., Nishiyama Y., Yamato K., Teraoka J., Morii Y., Sakoda A., Ishimori Y., Taguchi T., Yamaoka K. Comparative study on the inhibitory effects of antioxidant vitamins and radon on carbon tetrachloride-induced hepatopathy. *J. Radiat. Res.* 53:830-839, 2012.

[10] Falkenbach A., Kovacs J., Franke A., Jörgens K., Ammer K. Radon therapy for the treatment of rheumatic diseases—review and meta-analysis of controlled clinical trials. *Rheumatol. Int.* 25:205-210, 2005.

[11] Van Tubergen A., Boonen A., Landewé R., Rutten-Van Mölken M., Van Der Heijde D., Hidding A., Van Der Linden S. Combined spa-exercise therapy is effective in ankylosing spondylitis patients: a randomised controlled *trial. Arthritis Rheum.* 45:430-438, 2001.

[12] Erickson B. E. Range of motion assessment of elderly arthritis sufferers at Montana (USA) Radon Health Mines. *Int. J. Low Radiation* 3:325-336, 2006.

[13] Hass M. A., Massaro D. Regulation of the synthesis of superoxide dismutase in rat lungs during oxidant and hyperthermic stresses. *J. C. B.* 26:776-781, 1998.

[14] Yamaoka K., Mitsunobu F., Hanamoto K., Shibuya K., Mori S., Tanizaki Y., Sugita K. Biochemical Comparison between radon effects and thermal effects on humans in radon hot spring therapy. *J. Radiat. Res.* 45:83-88, 2004.

[15] Kohen R., Nyska A. Oxidation of biological systems: oxidative stress phenomena, antioxidants, redox reactions, and methods for their quantification. *Toxicol. Pathol.* 30:620-650, 2002.

[16] Yamamori T., Yasui H., Yamazumi M., Wada Y., Nakamura Y., Nakamura H., Inanami O. Ionizing radiation induces mitochondrial reactive oxygen species production accompanied by upregulation of mitochondrial electron transport chain function and mitochondrial content under control of the cell cycle checkpoint. *Free Radic. Biol. Med.* 53:260-270, 2012.

[17] Defeng W., Arthur I. Cederbaum. Oxidative stress and alcoholic liver disease. *Semin. Liver Dis.* 29:141-154, 2009.

[18] Wu D., Cederbaum A.I. Oxidative stress and alcoholic liver disease. *Seminars in liver disease* 29:141-154, 2009.

[19] Laurent A., Nicco C., Tran Van Nhieu J., Borderie D., Chéreau C., Conti F., Jaffray P., Soubrane O., Calmus Y., Weill B., Batteux F. Pivotal role of superoxide anion and beneficial effect of antioxidant molecules in murine steatohepatitis. *Hepatology.* 39:1277-1285, 2004.

[20] Dey A., Cederbaum A. I., Alcohol and Oxidative Liver Injury. *Hepatology* 43:S63-74, 2006.

[21] Molina M.F., Sanchez-Reus I., Iglesias I., Benedi J. Quercetin, a flavonoid antioxidant, prevents and protects against ethanol-induced oxidative stress in mouse liver. *Biol. Pharm. Bull.* 26:1398-1402, 2003.

[22] Marcolin E., Forgiarini L.F., Rodrigues G., Tieppo J., Borghetti G. S., Bassani V. L., Picada J. N., Marroni N. P. Quercetin decreases liver damage in mice with non-alcoholic steatohepatitis. *Basic Clin. Pharmacol. Toxicol.* 112:385-391, 2013.

[23] Ozturk I. C., Ozturk F., Gul M., Ates B., Cetin A. Protective effects of ascorbic acid on hepatotoxicity and oxidative stress caused by carbon tetrachloride in the liver of Wistar rats. *Cell Biochem. Funct.* 27:309-315, 2009.

[24] Gitto E., Tan D. X., Reiter R. J., Karbownik M., Manchester L. C., Cuzzocrea S., Fulia F., Barberi I. Individual and synergistic antioxidative actions of melatonin: studies with vitamin E, vitamin C, glutathione and desferrioxamine (desferoxamine) in rat liver homogenates. *J. Pharm. Pharmacol.* 53:1393-1401, 2001.

[25] Firuzi O., Miri R., Tavakkoli M., Saso L. Antioxidant therapy: current status and future prospects. *Curr. Med. Chem.* 18:3871-3888, 2011.

[26] Edaravone-Acute-Infarction-Study-Group. Effect of a novel free radical scavenger, edaravone (MCI-186), on acute brain infarction. Randomized, placebo-controlled, double-blind study at multicenters. *Cerebrovasc. Dis.* 15:222-229, 2003.

[27] Suda S., Igarashi H., Arai Y., Andou J., Chishiki T., Katayama Y. Effect of edaravone, a free radical scavenger, on ischemic cerebral edema assessed by magnetic resonance imaging. *Neurol. Med. Chir.* 47:197-202, 2007.

[28] Watanabe, T., Tahara M., Todo S. The novel antioxidant edaravone: frombench to bedside. *Cardiovasc. Ther.* 26:101-114, 2008.

[29] Inatomi Y., Takita T., Yonehara T., Fujioka S., Hashimoto Y., Hirano T., Uchino M. Efficacy of edaravone in cardioembolic stroke. *Intern. Med.* 45:253-257, 2006.

[30] Singal A. K., Jampana S. C., Weinman S. A. Antioxidants as therapeutic agents for liver disease. *Liver International.* 31:1432–1448, 2011.

[31] Matsuoka S., Matsumura H., Nakamura H., Oshiro S., Arakawa Y., Hayashi J., Sekine N., Nirei K., Yamagami H., Ogawa M., Nakajima N., Amaki S., Tanaka N., Moriyama M. Zinc supplementation improves the outcome of chronic hepatitis C and liver cirrhosis. *J. Clin. Biochem. Nutr.* 45:292–303, 2009.

[32] von Herbay A., Stahl W., Niederau C., Sies H. Vitamin E improves the aminotransferase status of patients suffering from viral hepatitis C: a randomized, double-blind, placebo-controlled study. *Free Radic. Res.* 27:599–605, 1997.

[33] Gabbay E., Zigmond E., Pappo O., Hemed N., Rowe M., Zabrecky G., Cohen R., Ilan Y. Antioxidant therapy for chronic hepatitis C after failure of interferon: results of phase II randomized, double-blind placebo controlled clinical trial. *World J. Gastroenterol.* 13:5317–5323, 2007.

[34] Ideo G., Bellobuono A., Tempini S., Mondazzi L., Airoldi A., Benetti G., Bissoli F., Cestari C., Colombo E., Del Poggio P., Fracassetti O., Lazzaroni S., Marelli A., Paris B., Prada A., Rainer E., Roffi L. Antioxidant drugs combined with alpha-interferon in chronic hepatitis C not responsive to alpha-interferon alone: a randomized, multicentre study. *Eur. J. Gastroenterol. Hepatol.* 11:1203-1207, 1999.

[35] Gordon A., Hobbs D. A., Bowden D. S., Bailey M. J., Mitchell J., Francis A. J., Roberts S. K. Effects of Silybum marianum on serum hepatitis C virus RNA, alanine aminotransferase levels and well-being in patients with chronic hepatitis C. *J. Gastroenterol. Hepatol.* 21:275-280, 2006.

[36] Hawke R. L., Schrieber S. J., Soule T. A., Wen Z., Smith P. C., Reddy K. R., Wahed A. S., Belle S. H., Afdhal N. H., Navarro V. J., Berman J., Liu Q. Y., Doo E., Fried M. W., SyNCH Trial Group. Silymarin ascending multiple oral dosing phase I study in noncirrhotic patients with chronic hepatitis C. *J. Clin. Pharmacol.* 50:434–449, 2010.

[37] Pamuk G. E., Sonsuz A. N-acetylcysteine in the treatment of non-alcoholic steatohepatitis. *J. Gastroenterol. Hepatol.* 18:1220–1221, 2003.

[38] Lieber C. S., Weiss D. G., Groszmann R., Paronetto F., Schenker S. II. Veterans Affairs Cooperative Study of polyenylphosphatidylcholine in alcoholic liver disease. Alcohol Clin. Exp. Res. 27:1765–1772, 2003.

[39] Ferenci P., Dragosics B., Dittrich H., Frank H., Benda L., Lochs H., Meryn S., Base W., Schneider B. Randomized controlled trial of silymarin treatment in patients with cirrhosis of the liver. *J. Hepatol.* 9:105-113, 1989.

[40] Mezey E., Potter J. J., Rennie-Tankersley L., Caballeria J., Pares A. A randomized placebo controlled trial of vitamin E for alcoholic hepatitis. *J. Hepatol.* 40:40–46, 2004.

[41] Yamaoka K., Kojima S., Takahashi M., Nomura T., Iriyama K. Change of glutathione peroxidase synthesis along with that of superoxide dismutase synthesis in mice spleens after low-dose X-ray irradiation. *Biochim. Biophys. Acta* 1381:265-270, 1998.

[42] Kojima S, Matsuki O, Nomura T., Kubodera A., Honda Y., Honda S., Tanooka H., Wakasugi H., Yamaoka K. Induction of mRNAs for glutathione synthesis-related proteins in mouse liver by low doses of γ-rays. *Biochim. Biophys. Acta* 1381:312-318, 1998.

[43] Kojima S., Matsuki O., Nomura T., Shimura N., Kubodera A., Yamaoka K., Tanooka H., Wakasugi H., Honda Y., Honda S., Sasaki T. Localization of glutathione and induction of glutathione synthesis-related proteins in mouse brain by low doses of γ-rays. *Brain Res.* 808:262-269, 1998.

[44] Yamaoka K., Edamatsu R., Mori A. Increased SOD activities and decreased lipid peroxide levels induced by low dose X irradiation in rat organs. *Free Radic. Biol. Med.* 11:299-306, 1991.

[45] Cho H. Y., Kleeberger S. R. Nrf2 protects against airway disorders. *Toxicol. Appl. Pharmacol.* 244:43-56, 2010.

[46] Tsukimoto M., Tamaishi N., Homma T., Kojima S. Low-dose gamma-ray irradiation induces translocation of Nrf2 into nuclear in mouse macrophage RAW264.7 cells. *J. Radiat. Res.* 51:349-353, 2010.

[47] Nakagawa S., Kataoka T., Sakoda A, Ishimori Y., Hanamoto K., Yamaoka K. Basic study on activation of antioxidation function in some organs of mice by radon inhalation using new radon exposure device. *Radioisotopes* 57:241-251, 2008.

[48] Kataoka, T., Sakoda A., Ishimori Y., Toyota T., Nishiyama Y., Tanaka H., Mitsunobu F., Yamaoka K. Study of the response of superoxide dismutase in mouse organs to radon using a new large-scale facility for exposing small animals to radon. *J. Radiat. Res.* 52:775-781, 2011.

[49] Kataoka T., Yamaoka K. Review: Activation of bio-defense system by low-dose irradiation or radon inhalation and its applicable possibility for treatment of diabetes and hepatopathy. *ISRN Endocrinol.* 2012(Article ID 292041):1-11, 2012.

[50] Yamaoka K., Nomura T., Iriyama K., Kojima S. Inhibitory effects of prior low dose X-ray irradiation on Fe^{3+}-NTA-induced hepatopathy in rats. *Physiol. Chem. Phys. Med. NMR* 30:15–23, 1998.

[51] Yamaoka K., Kojima S., Nomura T. Inhibitory effects of post low dose γ-ray irradiation on ferric- nitrilotriacetate induced mice liver damage. *Free Radic. Res.*32:213–221, 2000.

[52] Nomura T., Yamaoka K. Low-dose γ-ray irradiation reduces oxidative damage induced by CCl_4 in mouse liver. *Free Radic. Biol. Med.* 27:1324–1333, 1999.

[53] Recknagel, R. O., Ghoshal, A. K. Lipoperoxidation as a vector in carbon tetrachloride hepatotoxicity. *Lab. Invest.* 15:132-148, 1966.

[54] Durk, H., Frank H. Carbon tetrachloride metabolism in vivo and exhalation of volatile alkanes: dependence upon oxygen parital pressure. *Toxicol.* 30:249-257, 1984.

[55] Wang D. H., Masuoka N., Kira S. Animal model for oxidative stress research—catalase mutant mice. *Environ. Health. Prev. Med.* 8:37-40, 2003.

[56] Yamaoka K., Kataoka T., Nomura T., Taguchi T., Wang D. H., Mori S., Hanamoto K., Kira S. Inhibitory effects of prior low-dose X-ray irradiation on carbon tetrachloride-induced hepatopathy in acatalasemic mice. *J. Radiat. Res.* 45:89-95, 2004.

[57] Kataoka T., Nomura T., Wang D. H., Taguchi T., Yamaoka K. Effects of post low-dose X-ray irradiation on carbon tetrachloride-induced acatalasemic mice liver damage. *Physiol. Chem. Phys. Med. NMR* 37:109-126, 2005.

[58] Vogt B. L., Richie J. P. Glutathione depletion and recovery after acute ethanol administration in the aging mouse. *Biochem. Pharmacol.* 73:1613–1621, 2007.

[59] Strubelt O., Younes M., Pentz R. Enhancement by glutathione depletion of ethanol-induced acute.

[60] Nishiyama Y., Kataoka T., Teraoka J., Sakoda A., Tanaka H., Ishimori Y., Taguchi T., Mitsunobu F., Yamaoka K. Suppression of streptozotocin-induced type-1 diabetes in mice by radon inhalation. *Physiol. Res.* 62:57-66, 2013.

[61] Lenzen S. The mechanisms of alloxan- and streptozotocin-induced diabetes. *Diabetologia* 51:216-226, 2008.

[62] Szkudelski T. The mechanisms of alloxan and streptozotocin action in B cells of rat pancreas. *Physiol. Res.* 50:537-546, 2001.

[63] Kataoka T., Etani R., Takata Y., Nishiyama Y., Kawabe A., Kumashiro M., Taguchi T., Yamaoka K. Radon inhalation protects against transient global cerebral ischemic injury in gerbils. *Inflammation* 37:1675-1682, 2014.

[64] Chan P.H., Kawase M., Murakami K., Chen S.F., Li Y., Calagui B., Reola L., Carlson E., Epstein C. J. Overexpression of SOD1 in transgenic rats protects vulnerable neurons against ischemic damage after global cerebral ischemia and reperfusion. *J. Neurosci.* 18:8292–8299, 1998.

[65] Wang Q., Sun A.Y., Simonyi A., Jensen M. D., Shelat P. B., Rottinghaus G. E., MacDonald R.S., Miller D. K., Lubahn D.E., Weisman G.A., Sun G.Y. Neuroprotective mechanisms of curcumin against cerebral ischemia-induced neuronal apoptosis and behavioral deficits. *J. Neurosci. Res.* 82:138–148, 2005.

[66] Urabe T., Yamasaki Y., Hattori N., Yoshikawa M., Uchida K., Mizuno Y. Accumulation of 4-hydroxynonenal-modified proteins in hippocampal CA1 pyramidal neurons precedes delayed neuronal damage in the gerbil brain. *Neurosci.* 100:241–250, 2000.

[67] Kataoka T., Teraoka J., Sakoda A., Nishiyama Y., Yamato K., Monden M., Ishimori Y., Nomura, T. Taguchi T., Yamaoka K. Protective effects of radon inhalation on carrageenan-induced inflammatory paw edema in mice. *Inflammation* 35:713-722, 2012.

[68] Cuzzocrea S., Zingarelli B., Hake P., Salzman A. L., Szabó C. Antiinflammatory effects of mercaptoethylguanidine, A combined inhibitor of nitric oxide synthase and peroxynitrite scavenger, in carrageenan-induced models of inflammation. *Free Radic. Biol. Med.* 24:450–459, 1998.

[69] Nishiyama Y., Kataoka T., Yamato K., Taguchi T., Yamaoka K. Suppression of dextran sulfate sodium-induced colitis in mice by radon inhalation. *Mediat. Inflamm.* 2012:(Article ID 23961)1-11, 2012.

[70] Yamato K., Kataoka T., Nishiyama Y., Taguchi T., Yamaoka K. Antinociceptive Effects of Radon Inhalation on Formalin-induced Inflammatory Pain in Mice. *Inflammation* 36:355-363, 2013.

[71] Dubuisson D., Dennis S. G. The formalin test: A quantitative study of the analgesic effects of morphine, meperidine, and brain stem stimulation in rats and cats. *Pain* 4:161–174, 1977.

[72] Viggiano E., Monda M., Viggiano A., Viggiano A., Aurilio C., De Luca B. Persistent facial pain increases superoxide anion production in the spinal trigeminal nucleus. *Mol. Cell. Biochem.* 339:149-154, 2010.

[73] Yamato K., Kataoka, T., Nishiyama Y., Taguchi T., Yamaoka K. Preventive and curative effects of radon inhalation on chronic constriction injury-induced neuropathic pain in mice. *Eur. J. Pain* 17:480-492, 2013.

[74] Funakubo M., Sato J., Honda T., Mizumura K. The inner ear is involved in the aggravation of nociceptive behavior induced by lowering barometric pressure of nerve injured rats. *Eur. J. Pain* 14:32–39, 2010.

[75] Franke A., Reiner L., Resch, K .L. Long-term benefit of radon spa therapy in the rehabilitation of rheumatoid arthritis: a randomised, double-blinded trial. *Rheumatol. Int.* 27:703-713, 2007.

In: Radon
Editor: Audrey M. Stacks

ISBN: 978-1-63463-742-8
© 2015 Nova Science Publishers, Inc.

Chapter 7

RADON IN TAP WATER IN THE TERRITORY OF TBILISI CITY

Nodar Kekelidze[1,2,3], Teimuraz Jakhutashvili[1], Eremia Tulashvili[1], Manana Chkhaidze[1], Zaur Berishvili[1], Lela Mtsariashvili[1] and Irina Ambokadze[1]*

[1]Iv. Javakhishvili Tbilisi State University, Tbilisi, Georgia
[2]F. Tavadze Institute of Metallurgy and Materials Science, Tbilisi, Georgia
[3]Georgian Technical University, Tbilisi, Georgia

ABSTRACT

Content of radioactive gas radon – Rn-222 in tap water of municipal water-supply system in various territorial sites of Tbilisi city – capital of Georgia has been investigated. Within the framework of study there were analyzed the water resources used for supply of urban population by drinking water. It is shown, that now water supply is made from 11 sources of natural water which can be divided on two essentially various groups - sources in which underground waters (basically, from artesian wells) are used, and sources in which surface waters (river and from water reservoirs) are used. Water samples were selected in the residential buildings located in the main territorial sites of the city – in total 52 territorial entities have been allocated. Researches were carried out in the period January-December, 2013. Total amount of control points has made 118 points. Samples in nearby settlements (10 control points) were selected for comparison. In many control points sampling and the control of radon content was carried out monthly. Modern radon detector RAD7 was used for determination of radon content. It was established that radon content in water considerably changes depending on sampling time (that connects with possible changes of specific conditions of water transport to the consumer – distance from intermediate storage reservoirs, duration of stay in water mains, etc.) as well as on location of control point (that connects with primary prevalence in certain territories of water transport from surface sources – in this case activity of

---

* Corresponding author's email: n.kekelidze@tsu.ge.

samples corresponded to group with very low radon content (<0.3 Bq/L) and low (0.3 - 1.0 Bq/L), or from underground sources – in this case activity of samples corresponded to group with typical radon content (1.0 - 3.0 Bq/L) and above typical (3.0 - 10.0 Bq/L)). Based on the received data (more than 700 results) there was issued radon map of tap water in the city territory. Comparison with literary data has been carried out, in particular it is noticed, that the received values do not exceed recommended reference levels and are not dangerous for the population.

Keywords: Radon, activity concentration, drinking (tap) water, Tbilisi

INTRODUCTION

Radon is one of three main gases which are a part of so-called "terrestrial breath" during which argon, helium and radon constantly escape from the Earth interior into atmosphere. Only radon is radioactive among these all gases [1].

Radon – inert gas without color and odor, 7.5 times heavier than air, it is well dissolves in water. Its half-time is rather small (3.8 days), and it constantly escapes in soil layers (and then into the atmosphere). Its short-lived decay products - Po-218, Pb-214, Bi-214, Po-214 and rather "long-lived" Pb-210 (half-time of 22.3 year) mainly contribute to natural radio-activity of various environmental structures, in particular, natural (including drinking) water.

Radon enters into water from surrounding soil, and also from granites, basalts, sands which adjoin with auriferous layers. Radon concentration in usually used water is small, but water from some deep wells and artesian wells can contain a lot of radon - from 100 pCi/L up to 1000000 pCi/L (3.7 - 37000 Bq/L) [2].

Dissolved in water radon operates doubly. On the one hand, together with water radon enters into digestive system, and on the other hand, people inhale radon allocated from water during its utilization.

Radon inhalation can cause lung cancer. Radon decays into radioactive elements which can subside in lungs during inhalation. Because they decay later too, these particles radiate energy impulses that can damage lung tissue and increase probability of development of lung cancer during all life. Drinking water containing radon also represents risk of development of cancer lumps of internal organs, first of all stomach cancer [3].

Thus, studying of radon content in various environmental structures (in the first place in drinking water and so-called "indoor" air in living accommodation) and determining the effect of various external factors on them is of great importance for health protection of public.

Researches of radon content level in drinking water are carried out within many decades. The beginning of systematic studying of radon radiation should be considered 70th years of 20 centuries when in the territory of Helsinki wells with very high radon concentration in water have been found out. Further, in view of growth of attention to the radon problem, especially radon dissolved in water, similar researches are carried out in many countries.

Results of numerous researches show, that radon content in water in the various countries changes in the big ranges. For example, radon concentration in tap water of Kulachi city, Pakistan, changed in the interval 0.333-0.903 Bq/L with average value 0.602 Bq/L [4]. Average value of radon concentration in tap water of Mashhad city, Iran, makes up 11.44

Bq/L [5], and size of changing of radon content in tap water of the capital of Iran – Teheran is 2.70 – 6.0 Bq/L with average value 3.70 Bq/L [6]. Interval of radon content in tap water of Lahore city, Pakistan, is 2.0 – 7.9 Bq/L with average value 4.5 Bq/L [7].

In the work [8] there were studied samples of tap water in various settlements of Bulgaria, and it was shown that radon concentration in them varied from 17±0.40 Bq/L up to 185.5±10.4 Bq/L. In the 1980s, a number of national studies of radon in public water supplies in the United States were carried out; for example, beginning in November 1980, Environmental Protection Agency of the USA (US EPA) systematically sampled the 48 contiguous states, focusing on water supplies that served more than 1,000 people, and as a result it was shown that radon concentration ranged from 0 to over 500 Bq/L, the average was 12.6 Bq/L/L [9].

In the present work it was studied radon content in drinking tap water (selected from municipal water-supply system) in various territorial sites of Tbilisi city – capital of Georgia.

TASKING

Modern water-supply system of Tbilisi has begun to be created actually in 1933. Main water resources of drinking water are located in 20 - 40 km to the north from the city (in so-called Kartli artesian basin), and are presented by 11 sources which can be divided into 2 groups:

- surface sources (Jinvali reservoir, Bodorna buffer pool, Bulachauri water pipe, Saguramo water pipe, Tbilisi reservoir, Samgori cleaning construction, Grmagele cleaning construction);
- underground sources (Choport-Misaktsieli water pipe, Natakhtari water pipe, New Natakhtari water pipe, Mukhrani artesian water pipe).

The whole of city territory is divided into 4 specific zones which water supply is provided, accordingly, by four zone watercourses. Zone watercourses, basically, are located parallel to another, on the left and right coast of the river Mtkvari (running practically in the city center). Water from zone watercourses arrives in zone reservoirs, pump stations and further in city water highways, whence arrives directly to the consumer. All highways represent uniform water-supply system that allows to supply in case of need water to the consumer from different zone systems.

In the present work there was a task in view of research of radon content in tap water on various territorial sites for the purpose of an establishment of its quantitative values, and also dependence on a territorial location of control points.

RESEARCH OBJECTS

Research objects were the samples of drinking tap water selected in various control points practically in the whole territory of the city. Based on the available information materials, 52 territorial entities (including 28 – on the right-bank part and 24 – on the left-bank part of the city) have been allocated in city. They have been identified on the basis of informal "historical" territorial sites (are connected in many cases with local settlements earlier existing in these territories which further at the expansion of city territory became its part). Names of these territories (see column T, Table 1) are widely used by the population for the identification of location of those or other objects (places of residing, location of public and office buildings, etc.).

Table 1. Distribution of territories (T) of Tbilisi (C - city; R - right bank, L - left bank) and nearby settlements (## 14 - 18) by the groups of activity (GA) of radon in tap water

#	C	T	GA			
			I	II	III	IV
1	Tbilisi, R	Patara Digomi (PD)			▓	▓
2	"-"	Digomi housing unit (DHU)	▓	▓		
3	"-"	Agrarian University (AU), Didi Digomi (DD), Village Digomi (VD), Vashlijvari (Vj), Gotua (G), Vedzisi (Vz), Nutsubidze's Microdistricts (NM), Vaja-Pshavela's Quarters (VPQ), Kazbegi (Kz), Saburtalo str. (Sb), Bakhtrioni (B), Kostava (K), Bagebi (Bg), Vake (Vk), Vera (Vr), Mtatsminda (Mt), Sololaki (S), Ortachala (Ort)			▓	
4	"-"	Kvemo Ponichala (KP)		▓		
5	"-"	Zemo Ponichala (ZP), Village Ponichala (VP)	▓			
6	"-"	Tskhneti (Tsk), Kojori (Kj), Tabakhmela (Tb), Tsavkisi (Ts), Okrokana (Ok)			▓	
7	Tbilisi, L	Zahesi (Z)				▓
8	"-"	Avchala (Avch), Gldani (Gl)			▓	
9	"-"	Mukhiani (M), Temka (T), Sanzona (Sz), Nadzaladevi (Nd)		▓		
10	"-"	Zemo Chugureti (ZCh), Svanetisubani (Sv), Elia (E), Zemo Avlabari (ZA), Isani (I), Vazisubani (Vz), Varketili (Vrk), Dampalo (Dm), Orkhevi (Ork)	▓			
11	"-"	Didube (D), Kvemo Chugureti (KCh)		▓		
12	"-"	Avlabari (Avl), Navtlugi (Nv), Samgori (Sm), Alekseevka (A), Settlement Airport (SA)	▓			
13	"-"	Lilo (L)		▓		
14	Rustavi, Gardabani, Mtskheta	Rustavi, Village Mtisdziri, Mtskheta				▓
15	"-"	Mtskheta			▓	
16	"-"	Nataxtari, Tserovani				▓
17	Dusheti	Tsitelsopeli				▓
18	"-"	Bulachauri		▓		

The total amount of control points has made 118 points (in a number of control points water was selected monthly).

For comparison water samples was selected also in some nearby cities and settlements (in total 7) in which water selection was carried out in 10 control points.

METHODOLOGY

Sampling was carried out in special glass containers in capacity 250 mL. Containers were filled with water up to the top and densely closed by a cover. Then the selected water samples were transported to the laboratory for the analysis.

Electronic radon detector RAD7 was used for determination of radon content in water. In the device RAD7 it is used the method of detecting of alpha particles (created during decay process of radon and its products) based on the use of semi-conductor solid state detector. Protocol of measurement Wat-250 was used. Software Capture was used for processing of the received results.

Measurement error of the device does not exceed ±5 %. Dynamic range is 0.004 – 400 Bq/L. Detection limit of radon activity in water is estimated on the level of 0.03 - 0.04 Bq/L. For an estimation of real radiation background special repeated measurements of distilled water samples were carried out. The results received on two various devices have shown background activity in the range of 0.03 - 0.22 Bq/L (with arithmetic mean of 0.09 Bq/L).

For the investigated points taking into account reference level of 11 Bq/L recommended by US EPA [10] and the received results 5 groups of radon activity level in water have been established, in particular:

- 1st group (I) – control points in which value of radon concentration is very low - did not exceed 0.3 Bq/L (i.e., close to the background);
- 2nd group (II) – control points in which it is possible to consider that value of radon concentration is low - in the range of 0.3 - 1.0 Bq/L;
- 3rd group (III) – control points in which value of radon concentration can be designated conditionally as typical - in the range of 1.0 - 3.0 Bq/L;
- 4th group (IV) – control points in which it is possible to consider that value of radon concentration is above typical - in the range of 3.0 - 10.0 Bq/L;
- 5th group (V) – control points in which it is possible to consider that value of radon concentration is high - in the range of 10 - 30 Bq/L.

For the characteristic of time history (stability) of radon activity in control points depending on the period (month) of measurements the value of standard relative deviation of the received results was used which was calculated for specific control point by the received results in the course of year. For values of relative standard deviation no more than 50 % corresponding average value conditionally was accepted as stable (constant) value of activity concentration (and, accordingly, group of activity) for the given control point. If value of relative standard deviation exceeded this size, value of activity concentration (and, accordingly, group of activity) for the given control point was considered as the unstable.

RESULTS

For the whole period of observations it has been received more than 700 values of radon concentration in the water samples selected in various control points in the territories of Tbilisi and some settlements in its geographical area.

In the generalized view distribution of territories of Tbilisi (the order of location of territories, basically, was defined by their "geographical vicinity") and nearby settlements by the groups of radon activity in drinking water for the period January - December, 2013 is given in Table 1.

Radon map of tap water in the territory of Tbilisi is given in Figure 1 (the shaded areas correspond to populated territories).

In Table 2 there are given average values of radon concentration and distribution of control points by groups of activity in Tbilisi and nearby settlements for the period January - December, 2013.

Figure 2 shows distribution of quantity (N) of territories of Tbilisi by the intervals of radon activity (R, Bq/L).

In Table 3 there are given monthly average, minimal and maximal values of radon activity, their generalized values, and also relative standard deviation (RSD) in various control points in territorial entities of Tbilisi and nearby settlements (in the range $\leq 0.3 \div 1.0$ Bq/L with primary water supply from surface (*srf*) sources and $1.0 \div 10.0$ Bq/L with primary water supply from underground sources (*und*)).

Note. Given distribution of territories by the types of water supply was carried out based on the analysis of radon activity values in drinking water on corresponding territorial entities (see section Analysis).

Figure 3 shows monthly dependence of averaged activity of radon in the territories of Tbilisi with primary water supply from surface (*srf*) and underground (*und*) sources for the period January - December, 2013.

Apparently from the received results, it is possible to note following features;

- average activity of radon in water in the territories of Tbilisi changes in sufficiently wide range – from background values and close to them (0.03 - 0.3 Bq/L) up to 2.2 Bq/L and more (maximal value of 6.54 Bq/L, territory Zahesi);
- in nearby settlements the size of changing of average radon-in-water activity is a little another – from 0.33 Bq/L (in settlement Bulachauri) up to 8.93 Bq/L and more (maximal value of 13.1 Bq/L in Rustavi);
- the whole territory of Tbilisi sufficiently conditionally can be divided into 2 big groups - territories, where radon activity is low (0.3 - 1.0 Bq/L) or very low (<0.3 Bq/L) with average value of 0.45 Bq/L (minimal value of 0.03 Bq/L, maximal value of 3.2 Bq/L), and territories, where radon activity is typical (1.0 - 3.0 Bq/L) or above typical (3.0 - 10.0 Bq/L) with average value of 1.82 Bq/L (minimal value of 0.1 Bq/L, maximal value of 6.54 Bq/L); it is necessary to notice, that the greatest relative density (47.5 %) is made by control points with typical values of activity; thus in nearby settlements the condition also differs a little - the greatest relative density (70 %) is made by points with activity above the typical;
- in seasonal dependence it is possible to note some tendency to decrease of activity concentration in summer months;
- stability of activity in various control points also varies in sufficiently wide limits, thus its average value (from 40% to 56%) in the first group is a little bit more in comparison with the second group (from 10% to 57%); it is possible to note that

value of stability in individual control points depending on the period changed from 16 up to 150 % in the first group and from 7 to 87% in the second group.

Table 2. Average values (A_{av}) of radon concentration in tap water, distribution of quantity of control points (N_p) by the group of activity (*GA*), and their ratios ($R_N = N_p/N_t$ %; N_t – total quantity of control points) in the territory (*T*) of Tbilisi (*Tb*) and nearby settlements (*Tb-A*)

#	T	GA	A_{av}, Bq/L	N_p	R_N, %
1	Tb	I	0.1	37	31.4
2	"-"	II	0.6	22	18.6
3	"-"	III	1.8	56	47.5
4	"-"	IV	3.9	3	2.5
5	Tb-A	II	0.3	1	10.0
6	"-"	III	2.9	2	20.0
7	"-"	IV	4.2	7	70.0

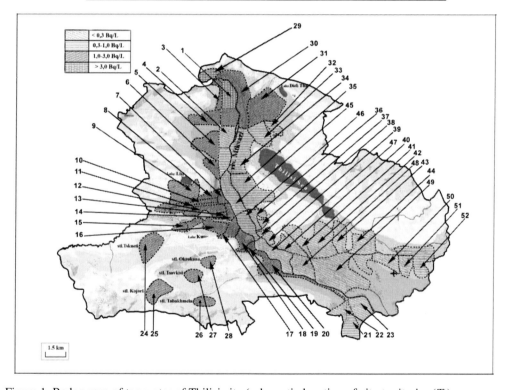

Figure 1. Radon map of tap water of Tbilisi city (schematic location of city territories (*T*)).

#	T	#	T	#	T	#	T	#	T	#	T	#	T	#	T	#	T
1	PD	7	G	13	B	19	S	25	Kj	31	Gl	37	Sv	43	Dm	49	Sm
2	DHU	8	Vz	14	K	20	Ort	26	Tb	32	M	38	E	44	Ork	50	A
3	AU	9	NM	15	Bg	21	KP	27	Ts	33	T	39	ZA	45	D	51	SA
4	DD	10	VPQ	16	Vk	22	ZP	28	Ok	34	Sz	40	I	46	KCh	52	L
5	VD	11	Kz	17	Vr	23	VP	29	Z	35	Nd	41	Vz	47	Avl		
6	Vj	12	Sb	18	Mt	24	Tsk	30	Avch	36	ZCh	42	Vrk	48	Nv		

Table 3. Monthly average activity of radon (A) in tap water, their averaged (A_{av}), minimal (A_{mn}) and maximal (A_{mx}) values, relative standard deviation (*RSD*), and also their generalized values (Prm - *aver, min, max*) in the territories of Tbilisi (*Tb*) and nearby settlements (*Tb-A*) with primary water supply from surface (*srf*) and underground (*und*) sources (S)

#	C	S	AR Bq/L	Prm	A, Bq/L												A_{av} Bq/L	$A_{mn,}$ Bq/L	$A_{mx,}$ Bq/L	RSD, %
					Jan.	Feb.	Mar.	Apr.	May.	Jun.	Jul.	Aug.	Sept.	Oct.	Nov.	Dec.				
1	*Tb*	*srf*	≤0.3 ÷ 1	aver	0.44	0.62	0.48	0.50	0.57	0.49	0.16	0.19	0.77	0.54	0.30	0.30	0.45	0.16	0.77	40
2				min	0.03	0.06	0.06	0.03	0.03	0.04	0.03	0.03	0.04	0.1	0.04	0.08	0.05	0.03	0.12	56
3				max	1.26	2.36	1.63	3.20	2.20	2.00	0.84	0.81	2.69	1.87	1.20	1.57	1.80	0.81	3.20	41
4		*und*	1 ÷ 10	aver	1.95	2.06	1.76	2.20	2.01	1.70	1.65	1.62	1.72	1.64	1.80	1.73	1.82	1.62	2.20	10
5				min	0.54	0.75	0.28	0.10	0.37	0.20	0.39	0.56	0.21	0.26	0.11	0.36	0.34	0.10	0.75	57
6				max	4.45	3.86	4.63	6.13	6.54	6.09	5.17	3.33	4.34	4.74	4.92	5.77	5.00	3.33	6.54	20
7	*Tb-A*	*srf*	≤0.3 ÷ 1	aver				0.62	0.46	0.05	0.31	0.18	0.31	0.21	0.47	0.35	0.33	0.05	0.62	53
8		*und*	1 ÷ 10	aver	6.01	4.77	4.86	2.03	4.05	4.71	5.52	4.32	5.74	3.59	8.93	6.22	5.06	2.03	8.93	33
9				min	4.44	1.36	3.37	0.67	0.91	3.05	3.06	3.47	4.49	2.88	6.14	4.75	3.22	0.67	6.14	51
10				max	9.14	13.1	7.06	3.47	7.82	7.25	8.62	5.28	8.13	4.51	11.7	8.98	7.92	3.47	13.1	35

Note. AR – activity range.

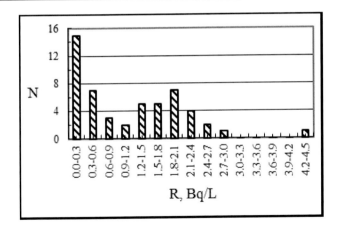

Figure 2. Distribution of quantity (N) of territorial entities of Tbilisi by the intervals of radon activity (R, Bq/L).

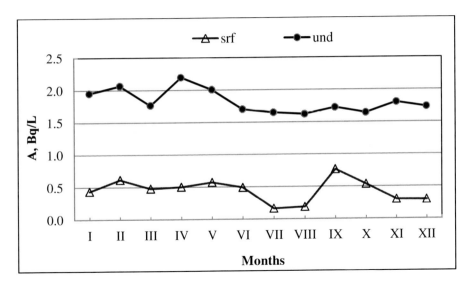

Figure 3. Monthly dependence of averaged activity of radon (A, Bq/L) in tap water in various control points in the territories of Tbilisi with primary water supply from surface (*srf*) and underground (*und*) sources.

ANALYSIS

Apparently from the received data, in Tbilisi the big size of changing of radon activity in drinking (tap) water is observed – the basic data array is within the limits from 0.1 and less up to 6.5 Bq/L. Size of changing in nearby settlements is a little another - from 0.33 up to 13.1 Bq/L.

It is necessary to consider that observable big size of changing of radon activity in drinking water in the territory of Tbilisi is connected, first of all, with a considerable quantity and a variety of characteristics of natural sources of water resources. As it was marked above, water supply of Tbilisi is provided with natural sources of two types - surface (basically, river water) and underground (basically, artesian waters). It is possible to consider, that the first type of waters differs by the small values of radon activity, and the second - higher. It proves to be true the results received for settlements, being in immediate proximity from target systems of maintenance of quality of drinking water. So, for example, in settlement Bulachauri (being in immediate proximity from a surface source of drinking water Bulachauri water pipe) value of activity in water system laid in the range from 0.05 up to 0.6 Bq/L (with average value of 0.33 Bq/L), and in settlement Natakhtari (being in immediate proximity from an artesian source of drinking water Natakhtari water pipe) value of activity in water system laid in the range from 1.1 up to 6.3 Bq/L (with average value 4.0 Bq/L). It is possible to consider these values, in certain degree, as "reference" at the estimation of type of natural source providing water supply of corresponding territory (or control point). For example, in the territory Zahesi (# 29, Figure 1) which among the all city territories is most close located to Natakhtari water pipe (and approximately on the same distance, as well as settlement Natakhtari), values of activity in water system also were the highest in comparison with other territories - from 1.6 up to 6.54 Bq/L (with average value 4.7 Bq/L, group IV), thus activity was sufficiently stable (RSD within 30-39 %). It gives the reason to consider, that this territory (and also some other territories in which there are observed typical and above typical values of radon activity in drinking water - more than 1.0 Bq/L) is provided by water from underground sources (for example, Natakhtari water pipe). Such feature of water-supply system the city explains presence of two maxima on the histograms of distribution of territories depending on the radon activity in drinking water (Figure 2).

In is necessary to note that values of radon activity in tap water in the territories which are sufficiently far from sources (for example, ## 36 - 52, Figure 1), correspond to low or very low groups (i.e., less than 1.0 Bq/L). These territories are located on the left bank of the river Mtkvari and, basically, are areas of new buildings. For their supply with drinking water resources of artesian sources were not adequate (which have been mastered in an initial stage of creation of modern water-supply system of Tbilisi and, basically, provided historically "older" territories located on the right bank) and additional water resources on the basis of surface sources began to be created.

It is interesting to notice, that territory of Digomi housing unit (# 2) is located on the right bank, but also characterized by low values of activity. It, apparently, is connected with that, in this territory one of the first areas of new buildings is located, and it is possible, use of surface sources has begun with it for supply with drinking water. Thus in the northern part of the territory #1 (directly adjoining to the territory # 29) values of radon activity in water also correspond to the group IV, and in its southern part (which directly adjoining to the territory # 2) basically water with smaller values of activity (group III) prevails. In the territory # 2 values of activity continue to decrease (group II, and further group I). On the other hand nearby (also on the right bank) there are located territories (# 3, 4 etc.) which also were areas of new buildings, however radon activity in these territories is considerably more (more than 1.0 Bq/L). Apparently, these areas (and also northern part of the territory # 1 and territory # 29) laid close to the first zone watercourse by which water run from Natakhtari water pipe, and for their supply water was taken from this highway.

It is necessary to note that on the left bank likewise there are located historically "old" territories (for example, # 47, Avlabari), but they are characterized by low values of radon activity. Apparently, they have been translated in due course by the supply from surface sources of water. Considering, that now the water supply system is united in a uniform network of water highways, in separate local sources can remain (probably, periodically) water delivery from underground sources. So for example, in one of control points in this territory average value of activity has made 1.7 Bq/L that is connected with such case. Such cases take place also in other territories (on the left bank), also being "old" areas, in particular Didube, Kvemo Chugureti (## 45, 46). The raised quantity of "unstable" points (i.e., points in which value of activity varies in sufficiently big range during period of year) are observed in these districts. Apparently, in these cases depending on the period (month) of observation prevails one type of water, or another.

It is necessary to notice, that, of course, fluctuations of radon activity can take place also because of various features of control points of sampling, for example, distances from the basic water lines, selection time, etc.

The observable tendency to decrease of radon activity in summer months, basically, apparently, is connected with increase in temperature of water in accumulator reservoirs therefore speed of decontamination of radon from water can increase.

Proceeding from the received results by the values of radon concentration in drinking water, the majority of the investigated control points (territories) in Tbilisi was in the groups with low (0.3 - 1.0 Bq/L) or typical (1.0 - 3.0 Bq/L) values which are noticeably below recommended reference levels. This circumstance, in certain degree, allows considering the general condition with radon activity not so dangerous. It is necessary to notice nevertheless, that insignificant quantity of control points (up to 4%) is in the group with rather high values of activity (3.0 - 10.0 Bq/L). It is possible also to notice, that the results received for nearby settlements are considerably above, than within Tbilisi. The majority of the investigated control points was in the groups with typical (1.0 - 3.0 Bq/L) or above typical values of activity (3.0 - 10.0 Bq/L), and in one case (Rustavi) activity of 13.1 Bq/L corresponding to the group with high values (10 - 30 Bq/L) has been registered. It is possible to consider, that this circumstance is connected by that water supply in these settlements (except settlement Bulachauri) is provided by natural sources of the second type. Thus it is necessary to notice, that in Rustavi water supply is provided by own sufficiently close located underground source, that, apparently, causes higher values of radon activity in tap water [11].

Comparison of the received results with some literary data is carried out in Table 4.

Table 4. Average, minimal and maximal values of radon concentration (A) in tap water in various regions of the world

#	Region, country	A, Bq/L			Ref.
		average	minimal	maximal	
1	Kulachi, Pakistan		0.602	1.218	[4]
2	Mashhad, Iran	16.238			[5]
3	Tehran, Iran	3.7			[6]
4	Lahore, Pakistan		2.0	7.9	[7]
5	Various regions of Bulgaria (Haskovo, Smolian, Kustendil, Blagoevgrad, Sofia)		1.17±0.40	185.5±10.4	[8]
6	Various states of USA	12.6	0	>500	[9]
7	Tbilisi, Georgia				Present work
	srf	0.45	0.03	3.2	
	und	1.82	0.1	6.54	

Apparently from the data, the received results by the average values of radon activity are close to the values received in works [4, 6].

Received results testify about topicality of researches of radiation background of natural waters (drinking, surface, etc.) and necessity of their further studying.

ACKNOWLEDGMENTS

This publication was supported by the project # 5644 "Radon background in drinking and surface waters in areal of capital of Georgia – Tbilisi" of Science and Technology Center in Ukraine and Shota Rustaveli National Science Foundation.

CONCLUSION

1. It was carried out analysis of water-supply system of Tbilisi and it was noted its some features; there were discussed methodological approaches on objects of research and processing of results.
2. Within January-December, 2013 there were carried out more than 700 measurements of radon activity in tap water samples selected in 118 control points (52 territories) in Tbilisi and 10 control points in nearby settlements.
3. It was established, that radon activity in water in the territories of Tbilisi changes in sufficiently wide range – from background values and close to them (0.03 - 0.3 Bq/L) up to 2.2 Bq/L and more (maximal value of 6.54 Bq/L); in nearby settlements size of changing of radon activity in water is a little another - from 0.33 Bq/L up to 8.93 Bq/L (maximal value of 13.1 Bq/L).
4. It was shown, that the whole territory of Tbilisi sufficiently conditionally can be divided into 2 big groups - territories, where radon activity is low (0.3 - 1.0 Bq/L) or very low (<0.3 Bq/L) with average value of 0.45 Bq/L and territories, where radon activity is typical (1.0 - 3.0 Bq/L) or above typical (3.0 - 10.0 Bq/L) with average value of 1.82 Bq/L; the greatest relative density (47.5 %) is made by control points with typical values of activity; thus in nearby settlements the condition also differs a little - the greatest relative density (70 %) is made by points with activity above the typical.
5. It was carried out analysis in which it was shown, that the received results, basically, are connected with features of water resources of drinking water, in particular with existence of qualitatively differing sources – surface and underground; comparison with literary data has been carried out.
6. Based on the received results radon map of the territory of Tbilisi has been created.

REFERENCES

[1] Utkin V.I. Radon problem in ecology, *Soros Educational Journal*, vol. 6, issue 3, p. 73, 2000.

[2] USGS. Radon-222 in the Ground Water of Chester County, Pennsylvania. By Lisa A. Senior. *Water-Resources Investigations* Report 98-4169. Lemoyne, Pennsylvania, 1998.

[3] EPA. Regulatory Information. Basic Information about Radon in Drinking Water: http://water.epa.gov/lawsregs/rulesregs/sdwa/radon/basicinformation.cfm.

[4] Tabassum N., Mujtaba Sh. Measurement of Annual Effective Doses of Radon from Drinking Water and Dwellings by CR-39 Track Detectors in Kulachi City of Pakistan. *Journal of Basic & Applied Sciences*, vol. 8, pp. 528-536, 2012.

[5] Binesh A., Mowlavi A.A., Mohammadi S. Estimation of the effective dose from radon ingestion and inhalation in drinking water sources of Mashhad, Iran. *Iran. J. Radiat. Res.*, vol. 10, issue 1, pp. 37-41, 2012.

[6] Alirezazadeh N. Radon concentrations in public water supplies in Tehran and evaluation of radiation dose. *Iran. J. Radiat. Res.*, vol. 3, issue 2, pp. 79-83, 2005.

[7] Manzoor F., Alaamer A.S., Tahir S.N.A. Exposures to 222Rn from consumption of underground municipal water supplies in Pakistan. *Radiat. Prot. Dosim.*, vol. 130, issue 3, pp. 392-396, 2008.

[8] Totzeva R., Kotova R. Radon and Radium-226 Content in Some Bulgarian Drinking Waters. Balwois-2010, Ohrid, Republic of Macedonia, 25-29 May, 2010.

[9] National Academy of Sciences. Committee on the Risk Assessment of Exposures to Radon in Drinking Water, Board of Radiation Effects Research, Commission on Life Sciences, National Research Council. *Risk Assessment of Radon in Drinking Water.* National Academy Press, Washington, DC, 1999.

[10] Federal Register. July 18, 1991. Part II. Environmental Protection Agency. 40 CFR Parts 141 and 142. National Primary Drinking Water Regulations; Radionuclides; *Proposed Rule.* (56 FR 33050).

[11] Kekelidze N.P., Kajaia G., Jakhutashvili T.V., Tulashvili E.V., Mtsariashvili L.A., Berishvili Z.V. Comparative Measurements of Radon Content in Tap Water in Tbilisi and Rustavi Cities. F.F. Quercia and D. Vidojevic (eds.), *Clean Soil and Safe Water*, NATO Science for Peace and Security Series C: Environmental Security, DOI 10.1007/978-94-007-2240-8_6. © Springer Science+Business Media B.V., pp. 65-76, 2012.

In: Radon
Editor: Audrey M. Stacks

ISBN: 978-1-63463-742-8
© 2015 Nova Science Publishers, Inc.

Chapter 8

RADON IN WATER — HYDROGEOLOGY AND HEALTH IMPLICATION

*N. Todorović[*1], J. Nikolov[1], T. Petrović Pantić[2], J. Kovačević[2], I. Stojković[3] and M. Krmar[1]*

[1]University of Novi Sad, Faculty of Sciences,
Department of Physics, Novi Sad, Serbia
[2]Geological Survey of Serbia, Belgrade, Serbia
[3]University of Novi Sad,
Faculty of Technical Sciences, Novi Sad, Serbia

ABSTRACT

Radon presence in the environment is associated mainly with trace amounts of uranium and radium in rocks and soil. Underground rock containing natural uranium continuously releases radon into water in contact with it (groundwater). When groundwaters reach the surface, in spas, wells or springs, the radon concentrations decrease sharply with the water movement and with purification treatment. But if the water is consumed directly from the point of emergence, as is habitual in rural sites, the time is often not long enough to prevent the health risks associated with its short-lived daughters. Hence, there is a need to determine the radon activity concentrations in groundwaters used directly (or indirectly through irrigation) and to estimate the doses received by the public consuming these waters. The risk due to radon in drinking-water derived from groundwater is typically low compared with that due to total inhaled radon but is distinct, as exposure occurs through both consumption of dissolved gas and inhalation of released radon and its daughter radionuclides. Moreover, the use of radon-containing groundwater supplies not treated for radon removal (usually by aeration) for general domestic purposes will increase the levels of radon in the indoor air, thus increasing the dose from indoor inhalation.

Radon analyses of groundwater samples can – beside the health implications – supply useful information for hydrogeological and hydrological purposes, groundwater

---

[*] Corresponding author's email: natasa.todorovic@df.uns.ac.rs.

quality (missing or existence of a protective soil cover), infiltration and exfiltration of groundwater, and age determinations of groundwater after seepage.

INTRODUCTION

Radon is a naturally occurring volatile gas formed from the alpha radioactive decay of radium. It is colorless, odorless, tasteless, chemically inert, and radioactive. Of all the radioisotopes that contribute to natural background radiation, radon presents the largest risk to human health. Most ^{222}Rn enters homes via migration of soil gas. Radon has also been identified as a public health concern when present in drinking water. Surface water contains very small amounts of dissolved radon. Typically, concentrations in surface waters are less than 4000 Bq/m^3. Water from wells can have high radon concentrations. Because radon is relatively insoluble in water, water use releases radon into the indoor air and contributes to the total indoor-airborne radon concentration. Ingestion of radon in water may also pose a direct health risk through irradiation of sensitive cells in the gastrointestinal tract and other organs once it is absorbed into the bloodstream. Thus, radon in drinking water could potentially produce adverse health effects in addition to lung cancer [1].

Radon levels in groundwater vary significantly over smaller distance scales. Local differences in geology tend to greatly influence the patterns of radon levels observed at specific locations. Over small distances, there is often no consistent relationship between measured radon levels in groundwater and radium levels in the groundwater or in the parent bedrock [2]. Radon volatilizes rapidly from surface water, and measured radon levels in surface water supplies are generally insignificant compared to those found in groundwater.

Waters originating from granite formations have been identified as having elevated concentrations of radon and other radionuclides in the uranium and thorium series [3].

Measurements of ^{222}Rn in groundwater have been performed in connection with geological, hydrogeological and hydrological surveys and health hazard studies. The widespread occurrence of radon and its relatively high specific activity in natural water, as well as simple instrumentation for its measurement, have resulted in a number of applications to hydrogeological and engineering problems, frequently in combination with analysis of other nuclides [4]. The content of radon in water samples must therefore be determined by reliable methods, and since radon is a very mobile gas and can easily escape from water during sampling and transportation, careful sample preparation is necessary.

This study was conducted in order to discuss the following subjects: significance of radon determination in waters for geology and hydrogeology considerations; radon occurrence in thermal springs and spas; as well as health impact and annual effective dose assessment from ingestion and inhalation depending on measured radon levels in drinking waters.

RADON, GEOLOGY AND HYDROGEOLOGY

Total amount of radon present in the earth crust can be determined using the following parameters: average concentration of radon in the earth crust is 1,5 ppm and uranium/radon ratio is 1: 2.15· 10^{-12}. As the mass of the earth crust (with 40 km average depth) is 5.7 ·10 ^{19}

tons, this gives mass of total uranium of $8.5 \cdot 10^{13}$ tons, and respectively the mass of total radon present in the earth crust of only 175 tons. This relatively small amount can have a large influence on the earth surface.

Geological settings are the most important factor having influence on the source and distribution of radon in soil, rocks and waters. Radionuclides can have different distribution in different geologic units because of variability in source areas and depositional processes in sedimentary rocks or different forming process of igneous and metamorphic rocks [5]. The release of radon from rocks and soils is controlled largely by the types of mineral in which uranium and radium occur [6].

Radon emanation coefficient is the fraction of the total amount of radon produced by radium decay that escapes from the soil particles and gets into the pores of the medium. The United Nations Scientific Committee on the Effect of Atomic Radiation UNSCEAR (2000) [7] reported that the emanation coefficients of rock and soil typically range from 5% to 70% with a representative value of 20%. The average value of emanation coefficient for soil is about 40% and for rocks is about 5% [8], for granit 6-33%, granodiorit 17-40%, gneiss 1-14% [9]. Emanation coefficients of radon for some minerals are: 0,29-4,17 % for monazite; 0,46-1,04 for zircon; 0,3-0,76 for uraninite; 1,34-22,5 for thorite; 16,8-22,9 for cerite [10].

Radon easily emanates from the rocks so high values of radon can be found in groundwater. It is usually present in groundwater that is in contact with granite rocks, gneiss, schist, shales, but is also common to find high concentrations of radon in groundwater that is in contact with sandstone, limestone or marls.

Radon dissolves relatively well in water, and due to its short lifetime ($T_{1/2}$ =3.82 days) it can migrate with groundwater over a long distances along faults and fractures depending on the velocity of fluid flow [6]. The migration of Rn in groundwater to the surface is controlled by:

- presence of ^{226}Ra and ^{228}U;
- lithology of the area;
- presence of faults, fractures and the circulation paths of the groundwater;
- hydrogeological properties of rocks (porosity, permeability, yield);
- presence of CO_2 or some other dissolved gas;
- temperature and pressure.

Radon is gaseous product of uranium decay series, respectively, radioactive decay of ^{226}Ra which are naturally present in rock and soil. The highest ^{226}Ra concentrations are observed in shale, volcanic and phosphate rocks followed by granites, clay rocks, sandstone then limestone and other carbonate rocks. The high ^{226}Ra levels in shale are likely due to associations of clay rich material of organic origin, while phosphate rocks of sedimentary origin are well known as minerals rich in uranium [11]. Uranium is mainly found in igneous rocks, mostly in intrusive rocks, then in volcanic rocks, black shales, crystalline schist, Permian sediments (red sandstone), phosphatic rocks and sometimes in carbonate rocks. The highest contents of Rn are mostly associated with granite and shale which have the highest content of uranium. Concentration of uranium in shale is 3,7 ppm, in granite 3 ppm, in limestone 2,2 ppm, in basalt 1 ppm, in sandstone 0,45 ppm, while in ultramafic rocks is 0,003 [9]. In Europe, uranium deposits of the fissure-filling type are mostly in Permo-Triassic rocks, which form a large part of the Carpatho-Balkan metallogenic province extending from central Europe, across eastern Balkans to the Black Sea in Turkey [12]. In sandstone, uranium is absorbed into Fe-O of the rock matrix or its weathering products. Clay, siltstone and sandstone with more organic material in matrix, have more radioactive elements than marls and limestone. The difference in radioactivity of sedimentary rocks is in close relation with their genetic properties. Emission of radon from sandstones is restricted by the relatively low specific surface area of the uranium minerals and appears to be more dependent on rock fractures. In rocks that have a higher surface area in contact with groundwater, release of radon is higher [6].

Radioactivity of igneous rock increases with the percentage of SiO_2 in the rock, respectively, from lowest radioactivity present in ultrabasic (peridotite) and basic rocks (gabro), over the intermediate rocks (diorite) to acid igneous rocks (granite) with a highest radioactivity.

For concentration of radioactive elements in rocks, significant are the hydrothermal, pneumatolytic and other post-igneous processes that are the reason for secondary alteration of rocks. Products of these processes are generally found nearby the tectonic zones, joints and faults, where the radioactive elements are concentrated in the geochemical barriers [13]. High radon concentration is noticed in waters nearby the tectonic zones, where the fracture and faults are the most common path for the Rn migration to the surface.

As the Rn is inert gas, the presence of CO_2 in the groundwater would tend to induce migration of radon [6], with a consequence of having water discharged on surface with high CO_2 and Rn concentrations.

The radon concentration in the water increases with the higher rate of spring discharge. A larger volume of water can dissolve radon from a larger volume of rocks in the outflow zone and simultaneously carry the gas faster to the discharge area [14]. Rn is also transported faster through more porous and permeable rocks.

Although ^{222}Rn is a gaseous progeny of ^{226}Ra, high ^{222}Rn content in water does not always lead to presence of ^{226}Ra [15] and ^{238}U, which is proved on analyzed content of ^{226}Ra, ^{222}Rn and ^{238}U in thermal and mineral groundwater samples of Serbia (Figure 1 and Figure 2).

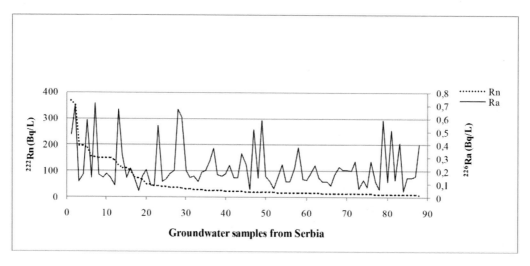

Figure 1. Relationship of ^{222}Rn and ^{226}Ra content in thermal and mineral waters of Serbia, number of samples n=88 (according to data from [25]).

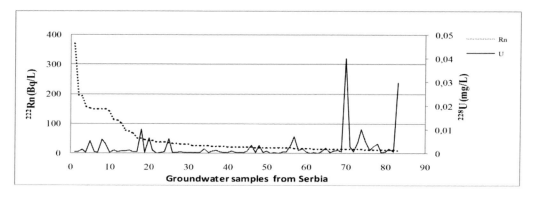

Figure 2. Relationship of ^{222}Rn and ^{228}U content in thermal and mineral waters of Serbia, number of samples n=83 (according to data from [25]).

Figure 1 and Figure 2 show that concentration of Rn in groundwater is not in correlation with presence of Ra and U. Radon is commonly present in groundwater in the concentrations that are greater than any of its parents (Ra, U), because solubility limits of these radionuclides are greater than those for radon in most groundwaters [5]. Low to moderate radium concentrations will lead to high radon concentrations if radium is adsorbed on the surface of the mineral grains [16].

Waters in Serbia with increased concentration of Rn are by ionic composition predominantly HCO$_3$, while according to cation composition they vary from Na to Na-Ca to Ca-Na or Ca-Mg. No correlation can be made between Rn and majority of ions found in water, as this is generally the most present type of water in Serbia. Also, according to study of Rn in groundwater from Chester County, Pennsylvania [5], no correlation was found between Rn and majority of dissolved ions in groundwater, like correlation with pH and other chemical constituents.

The average concentration of ^{222}Rn is about 20 Bq/L found in groundwater sources [17]. Content of Rn in groundwater is in range from 3 to nearly 80000 Bq/L [6]. Content of ^{226}Ra in thermal groundwater is in range from 0,74 to 5550 mBq/L [18]. According to U.S. Environmental Protection Agency regulation [2], the upper limit for ^{222}Rn in drinking water is 11,1 Bq/L, while according to World Health Organization, recommended as safe limit for drinking purpose is 100 Bq/L of radon in water, content of ^{226}Ra up to the 3,7 Bq/L and for uranium 15 μg/L is maximum acceptable contaminant level [17]. Water with content of ^{222}Rn more than 100 Bq/L is classified as radon water, while water with ^{226}Ra more than 3,7 Bq/L is classified as radium water [19]. Although waters containing radon are not recommended for drinking, these waters can be used in medical purposes and they have the important role in balneology (with Rn content more than 74 Bq/L).

Highest Rn concentrations are usually found in thermal waters. With the increase of groundwater temperature solubility of radon increases rapidly. The consequence of the radioactive U, Th and K decay (which are mostly concentrated in the acid igneous rocks) is the temperature emission and heating of groundwater, with Rn emission as the consequence of U decay as well.

Globally, ^{222}Rn contents up to 1868 Bq/L are measured in thermal spring in Spain [20], in Lądek Zdrój, Poland, in range from 122 to 1284 Bq/L [21] and in thermal water from Bad Gastein, Austria, with radon content of 749 Bq/L [22]. Several springs on mt. Kitka, south of

Skopje (FYROM), have content of Rn from 3000-15000 Bq/L. The area is characterized by horst-anticline in Precambrian made by schist with intruded granitic rocks [23].

In Serbia, the highest content of ^{222}Rn is measured in spring Cerska Slatina. Content of ^{222}Rn was measured in range from 350 to 3600 Bq/L, while ^{226}Ra content is constant, about 0,7 Bq/L, and ^{228}U content is low [24]. This spring is located on south slopes of mt. Cer granitoid, in contact with nearby Tertiary clastites (sandstones and conglomerates) of the hill Iverak. Atmospheric waters are flowing through highly permeable limestone formation underplaying the Tertiary sediments, located on higher altitude than Cerska Slatina spring. Groundwater circulates up along the contact of granites and overlying clastites (Figure 3), and it is enriched with radioactive elements from the uranium mineralization located nearby [24].

At the spring, value of ^{222}Rn was measured in range of 402 to 691 Bq/L. On the south slope of Iverak hill Banja Badanja spa is located, having spring water with Rn concentration of 188,7 Bq/L [25].

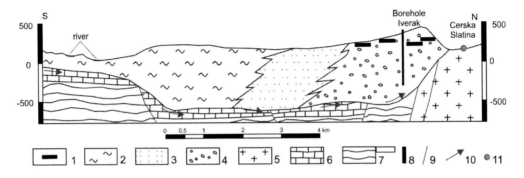

Figure 3. Geological cross-section line of Iverak area (according to data [24]). Legend: 1. uranium mineralizations; 2. clay; 3. sandstone; 4. conglomerate; 5. granitoid; 6. limestone; 7. metasandstone and schist; 8. borehole; 9. fault; 10. direction of groundwater flow; 11. spring with high concentrations of ^{222}Rn.

Figure 4. Conceptual model for high radon production in groundwater (according to [16]).

Another significant spring in Serbia is Hajdučica, located in the area of mt. Bukulja (granite), and having Rn concentration from 1000 to 1445 Bq/L [26]. In the area of mt. Bukulja number of uranium mineralization zones are registered, and two uranium ore bodies.

The high content of ^{222}Rn is also measured in water at Niška Banja spa, especially at the spring "Školska česma" where content of ^{222}Rn is 648 Bq/L [27]. This area is characterized by complex geology and tectonics. Travertine is formed around thermal water by outgassing of CO_2 from Ca-bicarbonate solution. Its thickness reaches 25 m, and content of ^{226}Ra is 470±40 Bq/kg [27]. Upwelling anoxic thermal water mixed with cold, oxygen rich water leads to iron hydroxide precipitation scavenging of ^{226}Ra from the thermal water and thus forming an efficient radon source (Figure 4 [16]). The good example of that process is Ribarska Banja spa (Serbia). Contents of ^{222}Rn in water from deep boreholes (depth 852 m and 1543 m) are 42-54 Bq/L, while in water extracted from fault zone in crystalline schist (depth 163 m), where the thermal water is mixed with shallow cold water, is 104 Bq/L [28,29]. Rn is monitored and evaluated during the prospection of mineral ore bodies in mining exploration. Increase of Rn in water was also shown to be a precursor of earthquake [30]. As a consequence of pressure and stress increase, larger fractures are opened in the rocks, and this leads to enrichment of groundwater with Rn. From hydrogeological aspect, presence of Rn in water is very significant as a water origin indicator. Rn enriched waters usually have high mineralization or temperature or increased concentration of CO_2 compared to the groundwater of the surrounding area [31]. The analysis of radon concentrations provides useful information on the groundwater's dynamic characteristics, the effects of pumping on an aquifer, the phases of the hydrological cycle [32], and the present joints, fracturing and type of porosity in the rocks.

RADON IN THERMAL SPRINGS AND SPAS

In Serbia, there are numerous thermal springs and air spas: Ribarska banja spa, Sijarinska banja spa, Vranjska banja spa and Bujanovačka banja spa, that contribute to the general health of people. The south-eastern part of Serbia represents an area with a lot of well-known thermal water sources [29].

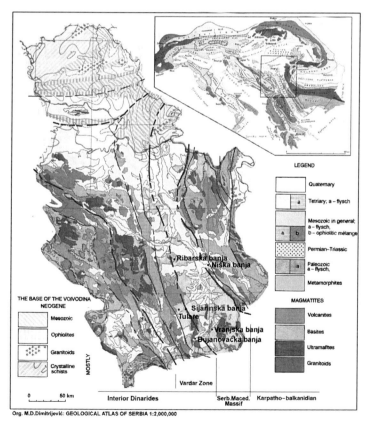

Figure 5. A geological map of Serbia (according to [29]).

The analyzed waters from Ribarska banja spa, Sijarinska banja spa, Vranjska banja spa and Bujanovačka banja spa belong to geotectonic unit Serbo-Macedonian massif. Niška Banja spa is located between two geotectonic units: Serbo-Macedonian massif and Carpatho-Balkanides, which results in its complex geology and tectonics [29]. Figure 5 represents a geological map of Serbia, in which the sampling locations are marked. As it can be seen, in the southern part of Serbia, the Serbo-Macedonian massif consists of crystalline schists that are intruded by granitoids and andezite.

According to their chemical composition, all analyzed waters in this geotectonic unit are mineralized, HCO_3–Na, and in waters from Ribarska banja spa and Vranjska banja spa there is also a higher content of SO_4.

RIBARSKA BANJA SPA

Waters from Ribarska banja are taken from three boreholes. At borehole CRB-1, the temperature of the water is 388 °C from fault zone from 53 to 145 m, and the other two deep boreholes are RB-4, with temperature 41.6 °C at depth 852 m, and RB-5, with temperature 548 °C at depth 1543 m. The geology of this area is very complex. The terrain is dominated by green schists divided into three groups. The majority of thermal waters belong to the group of poorly metamorphosed and non-metamorphosed rocks of clastic character, presented by phyllite, metamorphosed sandstone, meta-alevrolite, conglomerates and sandstone [29].

SIJARINSKA BANJA SPA

Sijarinska banja has ~18 springs and boreholes with thermal waters, with a temperature from 188 °C to 788 °C (borehole B-4). The deepest borehole is B-4 with a depth of 1232 m, which goes through crystalline schists (gneiss and amfibolite schists). Fifty meters away from B-4 is the borehole Aragon, bored through andesite with interlayers of marble onix, with a depth of 40 m. The water from this borehole has a temperature of 668 °C. The main touristic attraction in this spa is the borehole 'Gejzer' (geyser). From this borehole thermal water appears with high pressure from a depth of 8.5 m; the height of the water fountain had even been 8 m, but after a while it decreased to a stable height of 4 m above the ground. The temperature of the water from 'Gejzer' is 718 °C. Thermal waters from Sijarinska banja are related to a big dislocation, Tupalska, which represents the western border of the Serbo-Macedonian massif [29].

On the western side, 15 km away from Sijarinska banja spa is the small village of Tulare. In this village, there is a borehole with the depth of 300 m with thermal water at a temperature of 268 °C. This borehole is bored through andesite and gneiss.

VRANJSKA BANJA SPA

Vranjska banja spa is the spa with the warmest water in Serbia. The temperature of the water in some springs in this spa is around 958 °C, and the water from one borehole (1603 m deep) in this spa is at a temperature of 1118 °C. In this spa, there are also two more deep boreholes that were not available for sampling.

The sampling locations were 'Stara kaptaža' (with a temperature from 708 °C to 908 °C) and borehole B-1 (with a temperature of 838 °C) from which the water is conducted to the fountain in the spa park and is used for drinking purposes. The borehole B-1 was drilled through pyritizated biotite schists and in some places graniodiorite. The geological structure of the wider area of this spa is dominated by gneiss, crystalline schists and metamorphosed basic volcanics, and younger granitoide rocks are embedded through the whole observed complex. In the wider area, there are bursts of dacite, andesite and quartz latite [29].

BUJANOVAČKA BANJA SPA

Twenty kilometres south from Vranjska banja there is the Bujanovačka banja spa. The oldest rocks on this terrain are crystalline schists that are represented by gneiss, amphibolites and quartzite, in which Bujanovac granitoid is embedded. Cold tap water is flowing from granite through Parizanska česma, and from hydrothermal alerted granite there is a borehole A-3 with water at a temperature of 208 °C. In the spa area through granite and schists the granite gruss are formed, which is presented by sand and gravel with the presence of pieces and blocks of granite and schists. From this horizon, the sample A-2 of the warmest water in this spa (temperature of 468 °C) was taken; this water is used for balneotherapy. In the central part of the Bujanovačka valley, Miocene sandy shales and marls have been deposited and from this area the sample (YU-1) is taken (temperature of 308 °C); this water is used as bottled water for drinking purposes [29].

GEOLOGY AND HYDROGEOLOGY OF NIŠKA BANJA SPA

In Serbia there is only one radon spa called Niška Banja, a well-known health resort in south-east of Serbia. The Niška banja spa is located between two geotectonic units: Serbo-Macedonian massif and Carpatho-Balkanides, which results in its complex geology and tectonics.

Lithologically, the spa is located within the Neogene basin whose thickness is ~500 m. Mesozoic limestones, dolomites, dolomite limestones and clastic-carbonate sediments are found beneath these sediments. Permian and Carboniferous sandstones, conglomerates, alevrolite and shale are widely present over the larger area of the terrain [33]. On the terrain there are argillaceous schists, conglomerate, alevrolite, sandstone and limestone rocks, which date from the Devonian. Thermal springs in this spa have low-mineralised HCO_3-Ca water.

In radon spas, ^{222}Rn, considered as a hazardous carcinogen in the atmosphere of mines, mills and dwellings, is used at elevated levels for the deliberate exposure of children and adults for medical reasons. This paradoxical situation is largely caused by the unresolved issue of the shape of the dose-effect relationship at low level doses. Radon and its decay product nuclei which are present in the indoor environment of spa facilities have been identified as an agent of additional radiation burden both for bathers and working personnel [33].

There are a lot of papers dealing with this region of Serbia. According to them, Niška Banja was identified as a high natural radiation area [34]. A preliminary (screening) survey was conducted for indoor radon/thoron, the soil gas concentration ranges were from 63.7 to 1300 kBq/m^3 for radon [35]. Another detailed survey was conducted for radium in soil, radon in soil gas, and the gamma dose rate [36]. It was concluded that the estimated annual effective dose may exceed 50 mSv with a regional average of about 30 mSv for inhabitants in small Niška Banja town [37].

Geological map of the wider area of the Niška Banja spa is given in Figure 6.

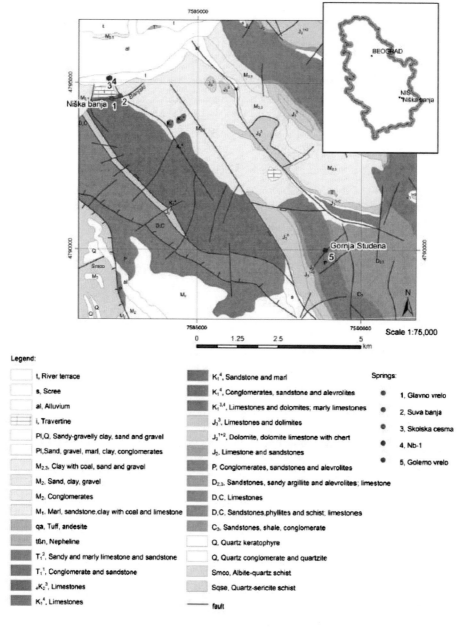

Figure 6. Geological map of the wider area of Niška Banja spa (according to [27]).

In the exploration area there are numerous joints and faults, spreading NW-SE and NNW-SSE. These structures are the most important for groundwater circulation.

Thermal water of the spring Glavno vrelo appears along the fault Banjski rased (Figure 7), on smashed fault zone between K^4_1 limestone and Miocene sediments. Travertine is formed around Glavno vrelo by outgassing of CO_2 from Ca-bicarbonate solution. Its thickness reaches 25 m. By exploring ^{226}Ra content in travertine in wider area of Niška Banja spa it has been concluded that travertine from Niška Banja has high content of ^{222}Rn, while other travertine from wider area of the spa have low values [33].

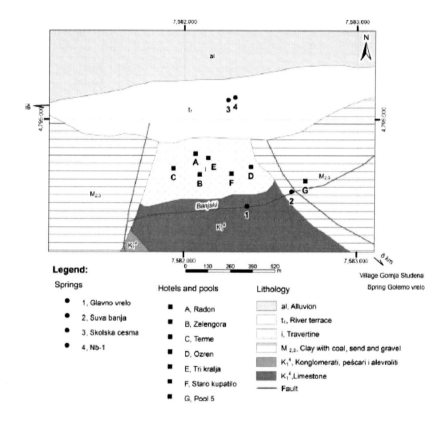

Figure 7. Geological scheme of Niška Banja spa with position of springs, hotels and pool (according to [27]).

Suva banja springs from breccias fault zone of K^4_1 limestone. The original leakage of spring Suva banja was within the cave opening in breccias zone. The spring used to dry out from time to time so in the period 1931-1956 it was lowered by 16.5 m which resulted in its constant outflow [38]. Today the spring S. banja is located about 300m north-east from the spring Glavno vrelo. Its water is used for the supply of the open summer pool 5 and the remaining water goes through pipelines to the spring capturing of Glavno vrelo, where the two waters mix. From there, through pipelines, the water goes to the Institute "Niška Banja" which has three stationeries ("Radon", "Zelengora" and "Terme"), "Staro kupatilo" with two pools, and it also supplies the hotel "Ozren" (Figure 7).

In the village Gornja Studena there is a karst spring Golemo vrelo used for water supply of Niška Banja and the city Niš. The tap "Tri kralja" located in the park uses this water, too.

"Školska česma" springs on river terrace near the school. Water flows on six pipes and is used for drinking. The borehole NB-1 510 m deep was made 50 m from the "Školska česma". Thermal water with the yield of 100 L/s and temperature 36.9 °C is reached on 475 m depth [33].

The borehole is closed and according to the data of the earlier explorations ^{222}Rn content in the water is 3.7 Bq/L [31].

Limestone is not considered to be a radon source [33]. Aquifer recharges by infiltration of rainfalls and river flows on area with karstified Jurrasic and Cretaceous rocks [33]. Water sinks to the contact between Permian sandstone and Paleozoic schist on depth of 1500 m [33] and under hydrostatic pressure springs along fault zone in Niška Banja spa. Due to deep circulation the heating of water is enabled, while on its way to the surface is cooled due to the influx of colder water from karst. Subthermal or thermal springs may get their radon from radium adsorbed on shallow iron hydroxide deposits that have been formed by the interaction of deep anoxic waters with young karst water rich in oxygen [16]. In Niška Banja spa, Permian sandstone contains ferric oxides which have given them red color, like sandstone of Low Jurrasic. According to their chemical composition waters are low mineral, with electroconductivity ranging from 540 to 571 mS/cm, and predominantly HCO_3 and Ca (Figure 8), with the exception of the borehole Nb-1, where higher content of potassium and magnesium has been recorded. Occasionally there is higher water turbidity and even though the water content stays stable, its radioactivity becomes higher. According to [31] this is the result of red Permian sandstone erosion. S. banja is characterized by temperature oscillations from 12.5 to 37.7 °C, which indicates that there is mixing of cold and thermal waters. The highest temperature recorded in Niška Banja area is measured in the Glavno vrelo (39.6 °C), whiles the lowest one is in the "Školska česma" (21°C).

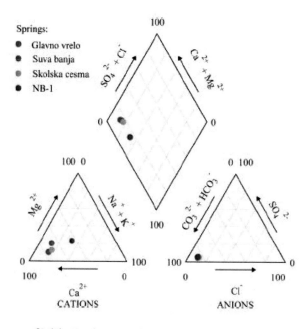

Figure 8. Piper diagram of Niška Banja spa springs (according to [27]).

HYDROGEOLOGY OF VOJVODINA

The Vojvodina region in the Pannonian Plain, on the site of the ancient Pannonian Sea, is a region with possibility of elevated radon levels in water because of numerous underground hot springs and sources of natural gas, as well as some crude oil reservoirs [39]. Hydrogeological map of Vojvodina province is presented on Figure 9.

The Vojvodina Province is characterized by four groundwater systems. The upper one (maximum thickness: about 2000 m) extends from the ground surface down to the top of the Lower Pontian; it includes Quaternary, Paludinian and Upper Pontian sediments [40]. The second groundwater system (maximum thickness: about 1500 m) comprises Lower Pontian and Pannonian sandstones of lower porosity than those of the overlying first system. The third one (maximum thickness: about 1500 m) consists of Miocene, Palaeogene, Jurassic and Cretaceous rocks (i.e., limestones, sandstones, conglomerates, and breccias). The fourth one is the lowest groundwater system (less than 500 m thick), which comprises Triassic and Paleozoic igneous, metamorphic and sedimentary rocks [40].

Most of the thermal waters in Vojvodina originate from the first, second and third groundwater systems hosted in Quaternary and Neogene formations, with a very small amount coming from the older geologic units.

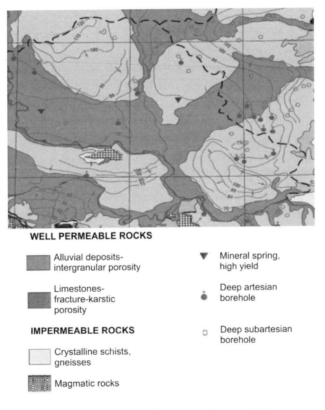

Figure 9. Hydrogeological map of Vojvodina province (according to [40]).

These groundwater systems occur at varying depths throughout the study area. For example, some waters of the first system are found at 2000 m depth, and some aquifers of northeastern Banat are hosted in Upper Miocene sandstones, at depths ranging between 200 and 1700 m, while some of the hot water aquifers in the northern and central parts of Banat and Bačka are in Upper Miocene and Pliocene sandstones (at 750–1700 m depth).

In southern Srem, eastern Srem and western Bačka, some hot aquifers are hosted in fractured and karstic carbonates and dolomites of mostly Triassic age at depths ranging from 500 to 1300 m [40].

Since hot water aquifers are widespread in Central Banat and Bačka, these areas are of great importance for water supply and for balneological and energy applications. The water-bearing formations in the 11 Central Bačka boreholes consist mainly of gravel and sands of high primary (intergranular) porosity (40–43%), having clay layers as aquitards.

The water samples from Vojvodina region were collected from public drinking fountains in Fruška Gora orthodox monasteries, where the faithful come believing that this water heals and they use it for drinking and face-washing.

HEALTH IMPACT

Biologic Basis of Risk Estimation. The biologic effects of radon exposure under the low exposure conditions found in domestic environments are postulated to be initiated by the passage of single alpha particles. The alpha particle tracks produce multiple sites of DNA damage that result in deletions and rearrangements of chromosomal regions and lead to the genetic instabilities implicated in tumor progression.

Because low exposure conditions involve cells exposed to single tracks, variations in exposure translate into variations in the number of exposed cells rather than in the amount of damage per cell. This mechanistic interpretation is consistent with a linear, no-threshold relationship between high-linear energy transfer (LET) radiation exposure and cancer risk. However, quantitative estimation of cancer risk requires assumptions about the probability of an exposed cell becoming transformed and the latent period before malignant transformations are complete. When these values are known for singly hit cells, the results might lead to reconsideration of the linear no-threshold assumption used at present [1].

Ingestion Risk. The cancer risk arising from ingestion of radon dissolved in water has to be derived from the calculated dose absorbed by the tissues at risk because no direct studies have quantified the risk. Studies of the behavior of radon and other inert gases have established that they are absorbed from the gastrointestinal tract and readily eliminated from the body through the lungs. The stomach is of particular concern. The range of an alpha-particle emitted when radon decays is such that alpha-particles emitted within the stomach are unable to reach the cells at risk in the stomach wall. Thus, the dose to the wall depends heavily on the extent to which radon diffuses to and into the wall. Once radon has entered the blood, through either the stomach or the small intestine, it is distributed among the organs according to the blood flow to them and the relative solubility of radon in the organs and in blood. Radon dissolved in blood that enters the lung will equilibrate with air in gas-exchange region and be removed from the body [1].

Inhalation Risk. Lung cancer arising from exposure to radon and its decay products is bronchogenic. The alpha-particle dose delivered to the target cells in the bronchial epithelium is necessarily modeled on the basis of physical and biologic factors. The dose depends particularly on the diameter of the inhaled ambient aerosol particles to which most of the decay products attach. These particles deposit on the airway surfaces and deliver the pertinent dose, and the dose can vary because of changes in particle size for a factor of ~2 in normal home conditions. The dose from ^{222}Rn gas itself is smaller than the dose from decay products on the airways, mainly because of the location of the gas in the airway relative to the target cells. The dose from ^{222}Rn gas that is soluble in body tissues is also smaller than the decay-product dose. Two of the underground-miner studies showed no statistically significant risk of cancer in organs other than the lung due to inhaled radon and radon decay products. The dosimetry supports that observation, although there is a need to continue the miner observations [1].

NATURE OF HEALTH IMPACTS

Exposure to radon and its progeny is believed to be associated with increased risks of several kinds of cancer [2]. When radon or its progeny are inhaled, lung cancer accounts for most of the total incremental cancer risk. Ingestion of radon in water is suspected of being associated with increased risk of tumors of several internal organs, primarily the stomach. Environmental Protection Agency (EPA) subsequently calculated the unit risk of inhalation of radon gas to 0.06 percent of the total risk from radon in drinking water, using radiation dosimetry data and risk coefficients [2].

Risk estimates indicate that inhalation of radon progeny accounts for most (approximately 89 percent) of the individual risk associated with domestic water use, with almost all of the remainder (11 percent) resulting from ingestion of radon gas. Inhalation of radon progeny is associated primarily with increased risk of lung cancer, while ingestion exposure is associated primarily with elevated risk of stomach cancer. Ingestion of radon also results in slightly increased risk cancer of the colon, liver and other tissues [2].

HEALTH IMPLICATIONS

Radon is an unstable radionuclide that disintegrates through short lived decay products before eventually reaching the end product of stable lead. The short lived decay products of radon are responsible for most of the hazard by inhalation.

Radon progeny rather than ^{222}Rn gas deliver the actual radiation dose to lung tissues. The solid airborne ^{222}Rn progeny, particularly ^{218}Po, ^{214}Pb, and ^{214}Bi, are of health importance because they can be inspired and retained in the lung. The radiation released during the subsequent decay of the alpha-emitting decay products ^{218}Po and ^{214}Po delivers a radiologically significant dose to the respiratory epithelium. The ratio of progeny to ^{222}Rn gas ranges from 0.2-0.8 with a typical value of 0.4. The ratio between progeny and ^{222}Rn gas is called the equilibrium factor [41].

After decay of the ^{222}Rn gas, a high percentage of the decay products attaches to ambient aerosols. A small percentage of the decay products remain unattached; others increase their diameter through chemical and physical processes. The percent of the attached products depends on numerous factors, including the size and concentration of the airborne particles.

The size and density of a particle determine its behavior in the respiratory tract. The unattached particle fraction with a 1nm diameter is generally removed in the nose and mouth during breathing and has limited penetration of the bronchi. Maximal deposition occurs as the particles with diameters ranging from 3-10 nm increase their rate of penetration through the mouth and nose, ultimately depositing in the bronchial region. The deposition rate decreases for particles as their diameter increases toward 100 nm and larger because the particles are less able to diffuse to the airway surface. However, particle deposition into the respiratory tract through impaction starts to increase again for particles above 500 nm. Larger particles with a diameter exceeding 3.5 μm deposit predominantly in the nose and mouth during inhalation and do not reach the sensitive respiratory epithelium [2].

Ninety-five percents (95%) of exposure to radon is from indoor air; about one percent (1%) is from drinking water sources. Most of this 1% drinking water exposure is from inhalation of radon gas released from running water activities, such as bathing, showering, and cleaning. Only 0.1% of exposure is from swallowing drinking water contaminated with radon gas. Although the effects of ingested radon are not fully understood, calculations suggest that the great majority of the radiation dose from such exposure is in the stomach [2].

In many countries, some homes obtain drinking water from groundwater sources (springs, wells, adits, and boreholes). Because of its volatility, radon contamination occurs primarily in ground, not surface, water sources of drinking water. Underground, this water often moves through rock containing natural uranium that releases radon to the water. That is why, water from wells normally has much higher concentrations of radon than surface water such as rivers, lakes, and streams [2].

ACTION LEVEL OF RADON CONCENTRATION IN THE WATER

The United States set a Maximum Contaminant Level for radon in drinking water of 150 Bq/L for radon concentration in drinking water from private water supplies, The European Union Commission recommended Action Level of 1000 Bq/L. This is set so that the risk to a typical person drinking such water is similar to the risk from breathing air which contains radon at the Action Level of 200 Bq/m^3 [2]. In Table 1 domestic radon concentrations and Acton Level in different countries are presented.

Table 1. Domestic radon concentrations and Acton Level in different countries (according to data from [2])

Country	Average radon concentration in homes (Bq/m^3)	Action Level (Bq/m^3)
Czech Republic	140	200
Finland	123	400
Germany	50	250
Ireland	60	200
Israel	/	200
Lithuania	37	100
Luxembourg	/	250
Norway	51-60	200

Country	Average radon concentration in homes (Bq/m^3)	Action Level (Bq/m^3)
Poland	/	400
Russia	19-250	/
Sweden	108	400
Switzerland	70	1000
United Kingdom	20	200
European Community	/	400
USA	46	150
Canada	/	800

ANNUAL EFFECTIVE DOSE ASSESSMENTS

Radon has been recognized as a radiation hazard causing excess incidence of lung cancer among underground miners. In Europe, the annual effective dose from all sources of radiation in the environment is estimated at 3.3 mSv with indoor doses from radon, thoron and their short-lived progeny accounting for 1.6 mSv of this value. ^{226}Ra and its decay product ^{222}Rn in water contribute to human exposure by ingestion (drinking water) and inhalation (using thermal waters in health purposes) [29].

The radon concentration of drinking water is an important issue from the dosimetry aspect, because more attention is paid to the control of public natural radiation exposure. The radiation dose due to radon can be divided into two parts, the dose from ingestion and the dose from inhalation [42]. On some occasions, water is consumed immediately after leaving the faucet before its radon is released into the air. This water goes directly to the stomach. Before the ingested water leaves the stomach, some of the dissolved radon can diffuse into and through the stomach wall. During that process, the radon passes next to stem or progenitor cells that are radiosensitive. These cells can receive a radiation dose from alpha particles emitted by radon and radon decay products that are created in the stomach wall. After passing through the wall, radon and decay products are absorbed in blood and transported throughout the body, where they can deliver a dose to other organs [43].

For the ingestion part, radon and its daughters in drinking water impact a radiation dose to the stomach. The committed annual effective dose contribution of citizens, taking ^{222}Rn concentrations in account, was calculated according to the [44], using the formula:

$$E = K \cdot C \cdot KM \cdot t \tag{1}$$

where E is the committed effective dose from ingestion (Sv), K is the ingesting dose conversion factor of ^{222}Rn (10^{-8} Sv/Bq for adults, and 2×10^{-8} Sv/Bq for children [35]), C is the concentration of ^{222}Rn (Bq/L), KM is the water consumption (2 L/day), t is the duration of consumption (365 days) [35]. For the dose calculations per year, a conservative consumption of 2 L/day for "standard adult" drinking the same water and directly from the source point was assumed [45, 46]. The Rn concentration of drinking waters decreases during storage, processing, etc., so by the evaluation of dose, the consumption of interest is that of water taken directly from the tap [44].

Exposures to radon come mainly from the inhalation of the decay products of radon, which deposit inhomogeneously within the human respiratory tract and irradiate the bronchial epithelium [7]. According to the UNSCEAR report [7], 1 Bq/m^3 of radon in air, with an equilibrium factor of 0.4 and occupation factor of 0.8, gives an effective dose to the lung of 25 μSv/year. Assuming the ratio between the radon concentrations in air released from water to that in water to be 10^{-4}, the conversion factor from unit concentration of radon at equilibrium is 2.8 μSv m^3/Bq [34].

The contribution of drinking-water to total exposure is typically very small and is due largely to naturally occurring radionuclides in the uranium and thorium decay series [42]. For drinking water radioactivity total indicative dose TID (effective dose from radionuclides in drinking water except ^{3}H, ^{40}K, radon and radon progenies) is 0.1 mSv/year [42].

The guidance levels for radionuclides in drinking-water are presented in [43] for radionuclides originating from natural sources or discharged into the environment as the result of current or past activities. The guidance levels for radionuclides in drinking water were calculated by the equation:

$$GL = \frac{IDC}{h_{ing} \cdot q}$$

(2)

where GL is guidance level of radionuclide in drinking water, IDC is individual dose criterion; equal to 0.1 mSv/year, h_{ing} is dose coefficient for ingestion by adults (mSv/Bq) and q is annual ingested volume of drinking water, assumed to be 730 L/year [43]. The guidance level for ^{238}U, ^{226}Ra, ^{232}Th, and ^{137}Cs is presented in Table 2.

Table 2. Guidance levels GL for radionuclides in drinking water (according to data from [43])

Radionuclide	GL (Bq/L)
^{238}U	10
^{226}Ra	1
^{232}Th	1
^{137}Cs	10

An estimate of committed effective dose for each radionuclide should be made and the sum of these doses determined. If the following additive formula is satisfied, no further action is required:

$$\sum_i \frac{C_i}{GL_i} \leq 1$$

(3)

where C_i is the measured activity concentration of radionuclide i, GL_i is the guidance level value (Table 2).

Radon in Water

The total dose is the sum of the dose contributions of the single radionuclides (GD_i), which are calculated from the activity concentrations (C_i) with the legal valid dose conversion factors ($h(g_i)$) for adults, respectively, and an annual consumption (KM) of 730 L/year [47]:

$$GD = \sum_i GD_i = \sum_i h(g_i) \cdot C_i \cdot KM$$

(4)

Sampling technique is generally the major source of error in measuring the radon content in water [39]. The water sample must be representative of the water being tested and such that it has never been in contact with air. According to Environmental Protection Agency (EPA) [48], samples were collected from a non-aerated faucet of spigot to minimize desorption of radon: a 1L glass beaker was filled with water from the source, allowing to gently overflow, and the sample container was opened and submerged inverted into a beaker. Then, the sample container was rotated so that it was filled with water. The Teflon-lined cap was applied with the sample container still under water to eliminate headspace. Analysis would begin within 3 d after receipt of sample in the laboratory. The measurement results of radon ^{222}Rn and radium ^{226}Ra in thermal water samples are presented in Table 3.

Table 3. Activity concentration of ^{222}Rn and ^{226}Ra in thermal waters (according to data from [29])

Location	Activity concentration of ^{222}Rn (Bq/L)	Activity concentration of ^{226}Ra (Bq/L)
Bujanovacka banja		
A-2	30±11	0.47±0.20
A-3	52±13	0.32±0.17
Partizanska česma	46±9	0.21±0.09
YU-1	10.4±0.9	0.48±0.09
Sijarinska banja		
B-4	32±7	0.37±0.08
Aragon	52±9	0.41±0.18
Gejzir	48±6	0.45±0.09
Tulare	52±10	0.35±0.18
Ribarska banja		
RB-4	42±7	0.32±0.19
RB-5	54±8	0.48±0.18
CRB-1	104±15	0.26±0.08
Vranjska banja		
B-1	48±5	0.27±0.07
Stara kaptaža	26±4	0.35±0.16
Niška banja		
Glavno vrelo	61±5	0.49±0.18
Tri kralja	69±10	0.54±0.09
Školska česma	648±38	1.13±0.17

According to US Environmental Protection Agency regulation the upper limit for ^{222}Rn in drinking water is 11.1 Bq/L [2]. This means that almost all the water sources that are analyzed should not be used for drinking purposes, except YU-1, which is used as bottled water. from the obtained results [29], [34], it can be concluded that just one water sample from public fountain 'Školska česma' in radon spa Niška banja has a really high activity concentration of ^{222}Rn, more than six times higher than EU recommendations, in which a radon level, 100 Bq/L is acceptable [49] and there is no need for any interventions for the activity concentrations that are below this limit. Glavno vrelo and Školska česma are natural water springs in Niška Banja spa; they are settled on open air space and open for public all day. Water from this public fountain inhabitants use for everyday drinking. For ^{226}Ra, the EU recommendation limit is 0.5 Bq/L [50]. In the sample from the same source, the result of ^{226}Ra activity concentration measured here is 1.13(0.17) Bq/L, which exceeds the EU limit mentioned above. Moreover, the same conclusion that the presence of radon is not correlated with that of radium in the sample was derived. Other obtained activity concentrations of ^{226}Ra are on the recommended EU limit or below [35].

Radon is soluble in water and its route of exposure is important if high concentrations are found in drinking water [51]. Citizens of the Niška Banja use water from public fountains (Školska česma and Tri kralja) and also from natural springs (Glavno vrelo) for drinking and they often consume water immediately after leaving the faucet before its radon is released into the air. By drinking the water with high radon concentration, radon and its daughters impact a radiation dose to stomach. For ingestion, the committed annual effective dose contribution to citizens, taking ^{222}Rn concentration in account, was calculated according to Eq(1).

Table 4. Annual effective dose for citizens in Niška banja spa from public drinking fountains (according to data from [27])

Location	Annual effective dose – ingestion (mSv/year)
Glavno vrelo	0.45±0.04
Tri kralja	0.50±0.07
Školska česma	4.7±0.3

Table 5. Mapping of ^{222}Rn activities (with 2σ counting uncertainty, 95% CL) and annual effective doses in drinking waters in Fruška Gora mountain (according to data from [39])

Sample location	Activity concentration of ^{222}Rn (Bq/L)	Annual effective dose – ingestion (mSv/year)	Annual effective dose – inhalation (μSv/year)
Jazak	10.3±0.7	0.075±0.005	25.6±1.7
Ravanica	5.6±0.5	0.041±0.004	13.9±1.2
Hopovo	4.1±0.5	0.030±0.003	10.1±1.2
Kuveždin	22.2±1.1	0.162±0.008	55.4±2.7
Petkovica	17.2±1.0	0.125±0.007	43.0±2.5
Šišatovac	6.7±0.6	0.049±0.004	16.9±1.5
S. Karlovci	6.8±0.5	0.049±0.004	16.9±1.2
V. Remeta	1.32±0.24	0.010±0.010	3.3±0.6
Krušedol	23.9±0.9	0.172±0.006	59.0±2.2
Rakovac	0.85±0.20	0.0060±0.0014	2.1±0.5

The spring Školska česma could be translated as School's fault, and as its name said it is a spring near to the school in Niška Banja. The water from this spring is often used for drinking by adults but also by children. The calculated annual effective dose from ingestion for the water from this spring is very high, which was expected considering the measured activity concentrations of ^{222}Rn in water and in the air near to this spring, Table 4.

Calculated activities, as well as an assessment of effective doses received by population from drinking waters from fountain waters in Fruška Gora monasteries, are displayed in Table 5. Comparing the measured ^{222}Rn concentrations with the reference level of 100 Bq/L [40], it can be concluded that the drinking waters from Fruška Gora mountain do not contain elevated concentrations of radon.

CONCLUSION

Of all the radioisotopes that contribute to natural background radiation, radon presents the largest risk to human health. Although radon presence in the environment comes from uranium decay series, i.e., radioactive decay of ^{226}Ra, naturally present in rocks and soil, high ^{222}Rn content in groundwaters is not always consistently correlated to the increased levels of ^{226}Ra and ^{238}U. Geological settings are the most important factor having influence on the source and distribution of radon in soil, rocks and waters. Radionuclides can have different distribution in different geologic units because of variability in source areas and depositional processes in sedimentary rocks or different forming process of igneous and metamorphic rocks. The release of radon from rocks and soils is controlled largely by the types of mineral in which uranium and radium occur. It is confirmed that local differences in geology cause significant variations in radon levels in groundwater over smaller distance scales.

From hydrogeological aspect, level of Rn presence in groundwater is very significant as its origin indicator, its dynamic characteristics, the effects of pumping on an aquifer, the phases of the hydrological cycle, and the present joints, fracturing and type of porosity in the rocks. Content of ^{222}Rn in groundwater sources is in range from 3 to nearly 80000 Bq/L, with the average concentration of 20 Bq/L.

Highest Rn concentrations are usually found in thermal waters, where radon solubility increases with the groundwater temperature. Although waters containing radon are not recommended for drinking, they can be used in medical purposes i.e., in balneology. This paradoxical situation of deliberate human exposure to a hazardous carcinogen is largely caused by the unresolved issue of the shape of the dose-effect relationship at low level doses. For workers and residents working or living under poor ventilation conditions in radon spas health effects have been noted, such as increased frequency of chromosome aberrations and atypical cells on sputum samples [33]. There are evidences that the risk to the patient undergoing from radon therapy in the form of thermal baths is smaller than that resulting from his annual exposure to the natural radiation environment. In the case of inhalation treatment or the drinking of thermal water the resulting doses can be considerably higher than for bathing, but the risk for the patient associated with this type of treatment is usually still relatively small in absolute terms [33]. Under unfavorable exposure conditions, for example poor ventilation, the resulting risk can be unacceptably high for some spa employees and inhabitants.

Exposure to radon 95% comes from indoor air; about one percent (1%) is from drinking water sources. Most of that 1% drinking water exposure is from inhalation of radon progeny (89% of risk estimates) associated primarily with increased risk of lung cancer, while ingestion exposure (11% of risk estimates) is associated primarily with elevated risk of stomach cancer. Since the effects of radon exposure are not fully understood, radon concentration of drinking water remains an important issue from the dosimetry aspect, because more attention is paid to the control of public natural radiation exposure.

FUNDING

The authors acknowledge the financial support of the Ministry of Education and Science of Serbia, within the projects No.171002, No.43002, No. 43007 and the Provincial Secretariat for Science and Technology Development within the project Development and application low-background alpha, beta spectroscopy for investigating of radionuclides in the nature.

REFERENCES

[1] P. K. Hopke, T.B. Borak, J. Doull, J. E. Cleaver, K. F. Eckeran, L. C. S. Gundersen, N. H. Harley, C. T. Hess, N. E. Kinner, K. J. Kopecky, T. E. Mckone, R. G. Sextro, S. L. Simone, Health Risks Due to Radon in Drinking Water. *ENVIRONMENTAL SCIENCE & TECHNOLOGY*, VOL. 34, NO. 6, 2000, pp. 921-926.

[2] US Environmental Protection Agency. Radon in drinking water health risk reduction and cost analysis. EPA Federal Register 64 (USEPA, Office of Radiation Programs, Washington, DC) (1999).

[3] Thomas, 1987; Lowry, Hoxie & Moreau, 1987; Dillon, Carter, Arora & Kahn, 1991; Zelensky, Buzing & Los, 1993; Otwoma & Mustapha, 1998.

[4] D. Amrani, D. E. Cherouati, M. E. H. Cherchali, Groundwater radon measurements in Algeria, *Journal of Environmental Radioactivity* 51 (2000) 173-180.

[5] Senior, A. 1998. Radon-222 in the Ground Water of Chester County, Pennsylvania, Water-Resources Investigations Report 98-4169, U.S. *Department of the Interior* U.S. Geological Survey, p.79.

[6] Appleton, J.D. 2005. Radon in air and water. In: Essentials of Medical Geology: Impacts of the Natural Environment on Public Health. Selinus, O. (ed). Elsevier Amsterdam, 227-262.

[7] UNSCEAR (The United Nations Scientific Committee on the Effect of Atomic Radiation) 2008.

[8] Antonović, A., 1990. Radioactivity in natural - important for investigation in geology. Special issue of Geoinstitute, Belgrade, No. 12, p. 191 (in Serbian).

[9] Durrance E.M., 1986. Radioactivity in geology: principles and applications. Ellis Horwood Ltd., New York, Halsted Press (John Wiley & Sons), p 441.

[10] Garver, E., Baskaran, M. 2004. Effect of heating on the emanation rates of radon-222 from a suite of natural minerals. *Applied Radiation and Isotopes* 61, 1477-1485.

[11] Fesenko, S., Carvalho, F., Martin, P., Moore, W.S., Yankovich, T. 2014. Chapter 3. Radium in the environment, in IAEA: Technical reports Series No. 476 *The Environmental Behaviour of Radium:* Revised Edition, 33-106.

[12] Kovačević J., Nikić Z., Papić P., 2009. Genetic model of uranium mineralization in the Permo-Triassic sedimentary rocks of the Stara Planina eastern Serbia. *Sedimentary Geology.* Elsevier 219, 252-261.

[13] Kovačević J. 2006. Metalogennic of mt. Stara planina. PhD dissertation, *Faculty of Mining and Geology*, Belgrade, p. 246 (in Serbian).

[14] Przylibski, T. A., Zebrowski, A. 1999. Origin of radon in medicinal waters of Lądek Zdrój (Sudety Mountains, SW Poland). *Journal of Environmental Radioactivity*, 46 (1), 121-129.

[15] Onishchenko, A., Zhukovsky, M., Veselinović, N., Zunić, Z. S. 2010. Radium-226 Concentrations in Spring Water Sampled in High Radon Regions, *Appl. Radiat. Isotopes* 68, 825-827.

[16] Surbeck, H., 2005. Dissolved gases as natural tracers in karst hydrogeology; radon and beyond, UNESCO Chair "Erdélyi Mihály" School of Advanced Hydrogeology, Budapest, Hungary.

[17] WHO (World Health Organization), 2008. Guideline for Drinking-water Quality, Third edition, Vol. 1, Geneva, p. 515.

[18] Medley, P. 2010. Barium sulphate method for consecutive determination of radium-226 and radium-228 on the same source, Internal Report 544, Supervising Scientist, Darwin.

[19] Šarin, A, 1988. Instructions for the preparation of the Basic Hydrogeological Map of SFR Yugoslavia,1: 100000. *Federal Geological Survey*, Belgrade. p. 124 (in Serbian).

[20] Ródenas, C., Gómez, J., Soto, J., Maraver, F., 2008. Natural radioactivity of spring water used as spas in Spain. *Journal of Radioanalytical and Nuclear Chemistry* 277 (3), 625–630.

[21] Przylibski, T., 2000. ^{222}Rn concentration changes in medicinal groundwaters of Lądek Zdrój. *Journal of Environmental Radioactivity* 48, 327-347.

[22] Deetjen P. 1997. Scientific Principles of the Health Treatments in Bad Gastein and Bad Hofgastein. Sem Reports. Salzburg, Austria: University of Innsbrook. (ISSN 0256-4173).

[23] Sarić, V. 1989. Radioactive water occurrences in Kitka mountain region, *Proceedings of Geointitute*, 23, 215-221.

[24] Protić, D., Antonović, A. 1988. The Majur-Slatina radioactive spring and uranium exploration in the Cer-Iverak area in west Serbia, *Proceedings of Geoinstitute*, 22, 133-141.

[25] Protić, D., 1995. Mineral and thermal water of Serbia, *Special issue of Geoinstitute*, No. 17 (in Serbian).

[26] Protić, D., 1993. A review of hydrogeochemical investigation in Bukulja granitoids in respect to uranium potentiality, *Proceedings of Geointitute*, 28, 213-219.

[27] Nikolov, J., Todorović, N., Petrović Pantić, T., Forkapić, S., Mrđa, D, Bikit, I., Krmar, M., Vesković, M. 2012. Exposure to radon in the radon spa Niška Banja, Serbia, *Radiation Measurements* 47, 443-450.

[28] Petrović Pantić, T., 2014. Hydrogeothermal resource of Serbian crystalline core, PhD dissertation, *Faculty of Mining and Geology*, Belgrade, p. 198 (in Serbian).

[29] Nikolov, J. Todorović, N., Bikit, I., Petrović Pantić, T., Forkapić, S., Mrđa, D, Bikit, K. 2014. Radon in thermal waters in south-east part of Serbia, *Radiation Protection Dosimetry* 160(1-3):239-43.

[30] Igarashi, G., Saeki, S., Takahata, N., Sumikawa, K., Tasaka, S., Sasaki, Y., Takahashi, M., Sano, Y. 1995. Groundwater Radon Anomaly Before the Kobe Earthquake in Japan, Science, New Series, Vol. 269, No 5220, 60-61.

[31] Protić, D. 1994. The indicators of hydrochemical anomalies of radioactivity, *Proceedings of Geointitute*, 30, 299-314.

[32] Spizzico, M., 2005. Radium and radon content in the carbonate-rock aquifer of the southern Italian region of Apulia. *Hydrogeology journal* 13, 493-505.

[33] Nikolov, J., Todorovic, N., Petrovic Pantic, T., Forkapic, S., Mrdja, D., Bikit, I., Krmar, M. and Veskovic, M. Exposure to radon in the radon-spa Niska Banja, *Serbia. Radiat. Meas.* 47, 443–450 (2012).

[34] Zunic, Z.S., Janik, M., Tokonami, S., Veselinovic, N., Yarmoshenko, I.V., Zhukovsky, M., Ishikawa, T., Ramola, R.C., Ciotoli, G., Jovanovic, P., Kozak, K., Mazur, J., Celikovic, I., Ujic, P., Onishchenko, A., Sahoo, S.K., Bochicchio, F., 2009. Field experience with soil gas mapping using Japanese passive Radon/Thoron discriminative detectors for comparing high and low radiation areas in Serbia (Balkan region). *Journal of Radiation Research* 50, 355-361.

[35] Zunic, Z.S., Kobal, I., Vaupotic, J., Kozak, K., Mazur, J., Birovljev, A., Janik, M., Bochicchio, F., 2006. High natural radiation exposure in radon spa areas: a detailed field investigation in Niska Banja (Balkan region). *Journal of Environmental Radioactivity* 89, 249-260.

[36] Zunic, Z.S., Yarmoshenko, I.V., Birovljev, A., Bochicchio, F., Quarto, M., Obryk, B., Paszkowski, M., Celikovic, I., Demajo, A., Ujic, P., Budzanowski, M., Olko, P., McLaughlin, J.P., Waligorski, M.P.R., 2007a. Radon survey in the high natural radiation region of Ni_ska Banja, Serbia. *Journal of Environmental Radioactivity* 92, 165-174.

[37] Zunic, Z.S., Kozak, K., Ciotoli, G., Ramola, R.C., Kochowska, E., Ujic, P., Celikovic, I., Mazur, J., Janik, M., Demajo, A., Birovljev, A., Bochicchio, F., Yarmoshenko, I.V., Kryeziu, D., Olko, P., 2007b. A campaign of discrete radon concentration measurements in soil in Niska Banja town, Serbia. *Radiation Measurements* 42, 1696-1702.

[38] Pecinar, M., 1961. Hidrogeologija termalnih vrela Niske banje i njihova zastita od rashladjivanja i mucenja. SAN, Glas CCXLVII, knjiga 5, Beograd.

[39] Natasa Todorovic, Ivana Jakonic, Jovana Nikolov, Jan Hansman and Miroslav Veskovic Establishment of a method for ^{222}Rn determination in water by low-level liquid scinntilation counter, *Radiation Protection Dosimetry* (2014) doi:10.1093/rpd/ncu240.

[40] Nataša Todorović, Jovana Nikolov, Branislava Tenjović, Ištvan Bikit, Miroslav Veskovic, Establishment of a method for measurement of gross alpha/beta activities in water from Vojvodina region, *Radiation Measurements* 47 (2012) 1053-1059.

[41] R. William Field, Radon Occurrence and Health Risk, Department of Occupational and Environmental Health Department of Epidemiology, College of Public Health, 104 IREH, University of Iowa, Iowa City, IA 52242.

[42] Natasa Todorovic, Jovana Nikolov, Sofija Forkapic, Istvan Bikit, Dusan Mrdja, Miodrag Krmar, Miroslav Veskovic, Public exposure to radon in drinking water in SERBIA, *Applied Radiation and Isotopes* 70 (2012) 543–549.

[43] WORLD HEALTH ORGANISATION (WHO), 2004. third ed. Guidelines for Drinking Water Quality, vol. 1. World Health Organisation, Geneva.

[44] Somlai K., Tokonami S., Ishikawa T., Vancsura P., Gaspar M., Jobbagy V., Somlai J., Kovacs T., 2007. ^{222}Rn concentrations of water in the Balaton Highland and in the southern part of Hungary, and the assessment of the resulting dose. *Radiat. Meas.* 42, 491–495.

[45] UNSCEAR, 1993. REPORT, Sources and Effects of Ionizing Radiation, United Nations Scientific Committee on the Effects of Atomic Radiation, Annex A: Exposures from Natural Sources of Radiation.

[46] Galan Lopez, M. et al.,2004.Estimatesofthedosedueto ^{222}Rn concentrations in water. *Rad. Prot. Dosimetry* 111(1), 3–7.

[47] International Atomic Energy Agency (IAEA), 1996. International Basic Safety Standards for Protection against Ionizing Radiation and the Safety Radiation Sources. Safety Report Series no. 115, Vienna.

[48] EPA Method 913.0. Determination of radon in drinking water by liquid scintillation counting. Radioanalysis Branch, Nuclear Radiation Assessment Division, Environmental Monitoring Systems Laboratory, U.S. *Environmental Protection Agency.* 89119.

[49] EC (European Commission). *Proposal for council directive* COM 147, 2012/0074 (NLE) (2012).

[50] EUROPEAN COMMISSION. Commission recommendation of 20th December 2001 on the protection of the public against exposure to radon in drinking water. 2001/982/Euratom, L344/85 (2001).

[51] Kendall, G.M., Smith, T.J., 2002. Doses to organs and tissues from radon and its decay products. *J. Radiol.* Prot. 22, 389–406.

In: Radon
Editor: Audrey M. Stacks

ISBN: 978-1-63463-742-8
© 2015 Nova Science Publishers, Inc.

Chapter 9

INDOOR RADON ACTIVITY CONCENTRATION MEASUREMENT USING CHARCOAL CANISTER

G. Pantelić[1], M. Živanović[1], J. Krneta Nikolić[1], M. Eremić Savković[2], M. Rajačić[1] and D. Todorović[1]

[1]University of Belgrade, Vinča Institute of Nuclear Sciences,
Belgrade, Serbia
[2]Serbian Agency for Radiation Protection and Nuclear Safety,
Belgrade, Serbia

ABSTRACT

Active charcoal detectors are used for testing the concentration of radon in dwellings. The method of measurement is based on radon adsorption on coal and measurement of gamma radiation of radon daughters. Detectors used for the measurement were calibrated by ^{226}Ra standard of known activity in the same geometry. The contributions to the final measurement uncertainty are identified, based on the equation for radon activity concentration calculation. The quantities that contribute to the combined measurement uncertainty in charcoal canister method for radon concentration screening were identified as uncertainties of: counting statistics, efficiency, calibration factor for radon adsorption rate, decay factor, time of exposure and measurement. Different methods for setting the region of interest for gamma spectrometry of canisters were discussed and evaluated. The obtained radon activity concentration and uncertainties do not depend on peak area determination method.

Standard and background canisters are used for QA&QC, as well as for the calibration of the measurement equipment. Standard canister is a sealed canister with the same matrix and geometry as the canisters used for measurements, but with the known activity of radon. Background canister is a regular radon measurement canister, which has never been exposed.

Carbon filters were unsealed and exposed in closed rooms for 2 to 3 days. Detectors were placed at distance of 1 m from the floor and the walls. Upon closing the detectors, the measurement was carried out after achieving the equilibrium between radon and its daughters (at least 3 hours) using NaI or HP Ge detector. Radon concentration as well as measurement uncertainty was calculated according to US EPA protocol 520/5-87-005.

Considering the measured concentration values of ^{222}Rn in dwelling units in Belgrade, as well as flaws of randomized sampling methods, the situation is not upsetting. Radon concentration in more than 80 % of apartments was lower than 200 Bq/m^3, which is within normal limits for apartments. Radon concentration in 6 % of apartments and in 4 % of schools was higher than 400 Bq/m^3 and intensive airing was recommended. For these dwellings additional measurements are required, followed by reparation of the facilities.

1. INTRODUCTION

Radon is an odorless and colorless noble gas with atomic number 86 and no stable isotopes. Radon isotopes that are occurring in nature are all members of radioactive decay chains – uranium, thorium and actinium series. Although there are several natural isotopes, the most important are ^{222}Rn and ^{220}Rn (also referred to as thoron), due to the long half-lives compared to other isotopes – 3.82 days and 55.6 seconds, respectively (BIPM, 2004).

Both radon and thoron undergo α-decay, which is followed by very weak photon emissions. However, their progenies include radioisotopes with intensive gamma radiation, such as ^{214}Pb and ^{214}Bi (radon progeny) and ^{212}Pb (thoron progeny). Finally, after several radioactive decays, both radon isotopes yield stable lead isotopes (BIPM, 2004).

The importance of radon from the radiation protection point of view lies in the fact that it is the only gas in natural radioactive decay chains. It easily escapes into the atmosphere not only from soil, but also from building materials and it is also readily dissolved in water.

Uranium and thorium are present in varying concentrations everywhere in soil and water, and so is radon. Radon exposure is universal for all human beings. However, since the uranium and thorium distribution is not uniform, radon exposure is also different for different populations. UNSCEAR report (UNSCEAR, 2000) states that the mean annual dose due to radon is 1.25 mSv, which constitutes roughly 50% of total annual dose for general population.

Detrimental health effects are most easily noticed in areas with high radon concentration. First observations were made in underground mines, long before the discovery of radioactivity and radon itself. Georgius Agricola (German physician and geologist, 1494 to 1555) wrote about the "consumption that eats miners' lungs" almost 5 centuries ago (Hoover and Hoover, 1950). Underground mines, especially uranium mines, provided an opportunity for systematic research of radon health effects and first epidemiological studies performed in the twentieth century (Holaday, 1956; Archer and Lundin, 1967; Walsh, 1970) and based on this research, IARC recognized radon as a cause of lung cancer in 1988 (IARC, 1988). The importance of influence of low radon concentrations on health has been recognized only recently. The research focus has shifted to radon in regular living and working quarters, which created a need for measurement methods that are fast, cheap and can be used on a large scale, preferably by untrained staff.

1.1. Radon Measurement

Radon concentration measurement is a problem that can be approached in many different ways. Measurements can be performed by active or passive methods, by measuring radon directly or via its progeny, by alpha or gamma spectrometry. Measurements can be performed almost instantaneously, or they can take months (EPA, 1996).

Radon measurement techniques that are in widespread use are quite diverse, and include techniques based on track detectors, active and continuous detectors, activated charcoal detectors, electrets (EPA, 1996).

Track detectors are usually made of plastics but can also be made of glass or inorganic crystals. Heavy charged particle incident on detector surface leaves a narrow track with the diameter of several tens of angstroms. These tracks are called latent tracks and they are "developed" by treating the detectors with appropriate aggressive chemicals. This process is termed etching. After etching, the tracks can be counted under optic microscope, and the density of tracks can be related to radon concentration. Track detectors exposure usually lasts for several months, but can be made shorter because track detectors are true integrating devices. Track detectors can be filtered or unfiltered, depending on whether alpha emitters other than radon are filtered (EPA, 1996; ECA, 1995).

Active and continuous detectors are based on scintillation detectors, ionization chambers or different types of semiconductor detectors. The air either diffuses through the measurement chamber or is pumped through it. Various filters can be used to discriminate between radon and other alpha emitting radionuclides. In some devices, radon concentration is estimated based on gross alpha activity, but alpha spectrometry can also be performed real time, which increases the method cost significantly (EPA, 1996; ECA, 1995).

Electrets are electrostatically charged disks that can retain the charge for a long period of time, on the scale of years. The disks are usually mounted in plastic casing, and the air containing radon is allowed to circulate. When radon and its progeny are undergoing radioactive decay, alpha particles are causing the electrets to be discharged at the rate proportional to radon concentration in air. Depending on the design, electret detectors can be used for long time or short time measurements, but typical exposure times are between 1 and 12 months. Electrets are true integrating devices (EPA, 1996; ECA, 1995).

Grab methods are modifications of several passive methods. These methods entail using pumps as a means of increasing the volume of air that comes into contact with sensitive detector volume. As a consequence, exposure times are significantly lowered. An example is the use of pumps with activated charcoal detectors (EPA, 1996; ECA, 1995). Activated charcoal method will be described in detail in the rest of the text. Other radon measurement techniques are also available, as well as the combinations and variations of the mentioned techniques.

Joint Research Centre published report "An overview of radon surveys in Europe" where the variety of methods used in the European countries for radon levels determination were presented (Dubois, 2005). The report emphasized the diversity of methods using various types of detectors and different time intervals.

1.2. Screening Measurements

The EPA recommends a two-step measurement strategy for assessing radon levels in homes (Ronca-Battista et al., 1987). The first step is an inexpensive screening measurement to determine whether a house has a potential for causing high exposures to its occupants. If the obtained result is bellow the screening level (150 Bq/m^3), follow-up measurements are probably not needed. If the results of screening measurements are between 150 Bq/m^3 and 740 Bq/m^{3}EPA recommends the follow-up integrated measurements in several areas of the house. If the screening measurement results are higher than 740 Bq/m^3 then short term follow-up measurements should be performed within several months and if the screening measurement results are higher than 7400 Bq/m^3 then short-term actions to reduce the radon levels should be done as soon as possible.

Short-term measurements of radon in homes and schools with charcoal canister are easy to perform, but need to be based on standardized protocols to ensure accurate and consistent measurements. We used EPA protocols for measurement (EPA, 1987; EPA, 1993a; EPA, 1993b).

1.3. Radon Regulation

Radon has been a subject of regulatory control in the European Union and in other European countries following the ICRP 50 and ICRP 60 (ICRP, 1987; ICRP, 1991) and in the EC Recommendation on the protection of the public against indoor exposure to radon (EC, 1990). They recommended to define a reference level for indoor exposure in existing buildings of 400 Bq/m³ and a design level of 200 Bq/m³ for future constructions, implying a protection of new buildings against radon.

In the new ICRP publications (ICRP, 1993; ICRP, 2007) the radon concentration of 600 *Bq/m³* was endorsed as a reference level expressed in radon gas concentration.

National radon programmes should aim to reduce the overall population risk and the individual risk for people living in object with high radon concentrations. The International Commission on Radiological Protection (ICRP), the World Health Organization (WHO) and the International Atomic Energy Agency (IAEA) have been encouraging countries to create programs on radon and to establish advisory levels for radon in homes and workplaces, issue guidelines for locating buildings with radon, check for radon risks prior to starting new constructions, and introduce limits for concentrations of natural radioactive elements in building materials.

The Republic of Serbia will start with national program for radon very soon, while intervention levels for chronic exposure to ^{222}Rn in homes are already defined in our regulation (RS Official Gazette, 2011). The intervention levels for chronic exposure to ^{222}Rn in homes are equal to the annual average concentration of 200 Bq/m^3 in the air in newly built housing object, and 400 Bq/m^3 in the air for existing housing objects.

2. MEASUREMENT METHOD

Activated charcoal detectors have been used for radon measurements for over 30 years. The method was described in papers by Cohen & Cohen in 1983 (Cohen and Cohen, 1983) and George in 1984 (George, 1984) and three years later US EPA published standard operating procedures. This method is now widely used for radon screening in working and living quarters, due to its advantages over the other measurement methods. Detectors can be used for other applications in places with medium to high radon concentrations, but are not well suited for outdoor measurements. The detectors are simply deployed on-site, and there is no requirement for special operating skills, electricity, temperature or pressure measurements etc. The detectors are unobtrusive and do not produce noise. Exposure times are short – typically 2 to 7 days. Detectors can be sent by mail, and then measured at a remote location. The shortcomings of the method are mostly related to many influence quantities (humidity, temperature, exposure time etc.), most of which are not taken into account of during the calculation of the radon concentration. Measurement uncertainty is thus increased, and it is not always easy to estimate it (Ronca-Battista and Gray, 1988; Jenkins, 2002).

Activated charcoal detectors are most commonly used for radon concentration screening in working and living quarters. Detectors can be used for other applications in places with medium to high radon concentrations, but are not well suited for outdoor measurements.

The measurement method is based on high affinity of activated charcoal for radon. In principle, radon is being adsorbed to charcoal during the exposure and it is undergoing radioactive decay. Approximately 3 hours after the detectors are closed, radioactive equilibrium between ^{222}Rn and its daughters is reached. At this point, it is possible to determine ^{222}Rn activity by measuring the activity of any of its daughters. Due to the intensive gamma lines, ^{214}Bi and ^{214}Pb are measured by gamma spectrometry (Grey and Windham, 1987).

2.1. Detector Construction

According to EPA protocol, the detectors are enclosed in metal cans 2.9 cm deep, with the diameter of 10.2 cm. 70 ± 1 grams of 6×16 mesh of activated charcoal is placed within the can, and covered with metal net in order to prevent spilling. Can lids are padded and a strip of vinyl tape is placed at the can – lid border. Activated charcoal should be of low activity – less than 0.04 Bq/l. This requirement doesn't apply to ^{40}K, because its gamma line lies outside this method's spectral region of interest (Grey and Windham, 1987).

2.2. Detector Exposure

There are a number of problems related to detector exposure that must be addressed before a reliable measurement result is obtained. First, radon adsorption is competitive with the adsorption of other chemical species present in air, most importantly water vapor. Therefore the rate of radon adsorption will be dependent on air humidity in particular and air composition in general. According to EPA method, radon detectors are exposed in radon chambers with controlled humidity at relative humidity of 20 %, 50 % and 80 %. The differences between measurement results with different humidity are used to obtain calibration factors. However, the relative humidity is typically not known at the place of exposure, so the humidity must be related to some measurable quantity. In this application, the detector mass change is easy to measure and it is dependent on the quantity of adsorbed water, which is in turn dependent on air humidity (Grey and Windham, 1987; Jenkins, 2002).

Measurement results are also dependent on the exposure time, but the relationship is complex. During the exposure, desorption is taking place at the same time as adsorption. Desorption rate is increasing with time, until at some point equilibrium is reached.

Radioactive decay should also be taken into account, as well as the influence of other competitive species. In order to avoid complicated theoretical predictions, radon detectors are exposed in radon chambers for different time intervals, and thus the calibration curves are obtained. It is necessary to construct separate calibration curves for different air humidity. Recommended exposure times are between 2 and 7 days (Grey and Windham, 1987; Ronca-Battista et al., 1987). In order to obtain average daily radon concentration, it is necessary that the exposure time is integer multiple of 24 hours. The problem arises when the radon concentration is fluctuating during the exposure. The method is not an integrating method, and the equilibrium between radon adsorption and desorption is shifting when the radon concentration is changing. This is especially important when radon concentration is lower at the end of the exposure, because the total activity of adsorbed radon may actually decline with time in this case. This can lead to serious underestimation of average radon concentration (Ronca-Battista and Gray, 1988).

Detector preparation before measurement and the detector construction itself can introduce new sources of measurement uncertainty, due to the fact that many laboratories do not perform detector calibrations themselves but instead use calibration factors provided by US EPA. These calibration factors are obtained in stationary atmosphere (Grey and Windham, 1987), or in the active environmental chamber (EERF, 1989). These factors are obtained for specific detector batches. Even though detector construction is specified in US EPA protocol, it is impossible to produce two identical batches, especially having in mind that the detectors are produced by several manufacturers and by using different activated coal. Also, the factors pertain to new detectors, which have never been exposed. These detectors contain some quantity of water and other chemical species, adsorbed during packaging, transport or storage. Some laboratories, however, recycle the canisters by heating them. In this case, the relationship between detector mass change during the exposure and air humidity is different than that specified in US EPA protocol (Jenkins, 2002).

Air temperature is an important factor that influences the quantity of adsorbed radon. If the temperature changes from 10 ^{0}C to 27 ^{0}C, radon collection efficiency will drop by as much as 40 % (depending on air humidity and other factors). Air temperature, however, is not taken into account by correction factors or in any other ways (Ronca-Battista and Gray, 1988).

2.3. Gamma Spectrometry Measurement

After the exposure, the canisters are sealed and usually analyzed by gamma spectrometry. Principal gamma peaks present in the spectrum originate from ^{214}Pb and ^{214}Bi decays. For this application, three energies are commonly used: 295 keV and 352 keV (^{214}Pb photons) and 609 keV (^{214}Bi photon). Radon detector spectra are usually very simple and free of interferences, especially in this spectral region. Due to this fact, gamma spectrometer resolution is of secondary importance, so NaI detectors can be used as well as high purity germanium detectors. With appropriate calibration, even single channel analyzers can be used. However, activated charcoal and other detector parts must have low radioactivity content and 239 keV peak of ^{212}Pb must be excluded from the spectral region of interest (Grey and Windham, 1987).

Detector efficiency is easily determined by using the standards with the same matrix and geometry as radon canisters, but with known ^{226}Ra activity (Radium nitrate uniformly distributed in charcoal, Isotope Products Laboratories, Los Angeles, USA). Radioactive equilibrium between ^{226}Ra and ^{222}Rn is achieved after approximately 30 days and radon concentration remains practically the same afterwards, due to the long half-life of radium (1600 years). ^{226}Ra does not have interfering gamma peaks in the spectral region of interest for this method. The result of the efficiency calibration is efficiency factor E_f, calculated by dividing counts per minute in the spectral region of interest with the radon activity in standard (Grey and Windham, 1987).

According to US EPA protocol (Grey and Windham, 1987), for radon concentration calculation, the following equation is used:

$$A_{Rn} = \frac{G - B}{t C_f D_f E_f} \tag{1}$$

In this equation, A_{Rn} is radon activity (usually expressed in Bq/m^3), G is the total number of counts in detector spectrum in the region of interest (for example between 270 keV and 720 keV, expressed in counts per minute or counts per second – CPM or CPS), B is the total number of counts in the same spectral region of the background spectrum, t spectrum collection time (in minute or second), C_f calibration factor specific for radon detector (expressed in l/min), D_f decay factor, used to correct for radon decay and E_f gamma spectrometer efficiency (expressed in CPM/Bq).

The relation between ^{214}Pb and ^{214}Bi activity in radon detector on one hand, and radon concentration in air on the other hand is not straightforward. It is most easily determined empirically, by exposing the detectors to known radon concentrations in known conditions and then collecting gamma spectrum. If the equation 1 is solved for C_f, it can be easily calculated. If exposure time and humidity are changed, calibration curves can be obtained.

Decay factor is calculated according to equation 2, where t_D is time from the exposure midpoint to the start of measurement and $t_{1/2}$ ^{222}Rn half-life.

$$D_f = e^{-\frac{\ln(2) t_D}{t_{1/2}}} \tag{2}$$

2.4. Methodology Used in the Vinča Institute

All laboratories in Serbia that are using activated charcoal detectors for radon measurements are accredited according to US EPA protocol 520/5-87-005 (Grey and Windham, 1987).

Based on the conclusions presented in previous text, EPA procedure and national and international intercomparison, our laboratory developed a set of procedures for charcoal detector exposure and measurement.

Charcoal detectors are dried for 3 hours at 105 ^{0}C in order to prepare them for the next exposure. After the detectors cool down, they are closed and sealed with vinyl tape and then the mass is measured. Exposures are performed by laboratory personnel, but the detectors can also be issued to users by mail or in person. Exposure guide and data sheet are always issued together with the detector. Data sheet is filled in by the user and it contains exposure beginning and end time, address, room characteristics (wall materials, floor, area, ceiling height, etc) and other necessary data. Users are told that the detectors should be positioned approximately 1 m away from the floor and walls as possible. The room door and windows should be closed during the exposure in order to examine the worst case scenario. Detectors should be opened, closed and handled in such a way that no mass is lost, including the mass of the vinyl tape; vinyl tape should not be discarded or replaced, even if it is damaged. This is necessary to obtain the correct estimate of air humidity. The exposure time should not be shorter than 2 days and no longer than 6 days.

After the exposure, the detectors should be returned to the laboratory as soon as possible. Ideally, if more than a week has passed from the exposure end, the exposure should be repeated due to ^{222}Rn decay. Another mass measurement is performed in laboratory and the humidity is estimated (20 %, 50 % or 80 %). Based on the exposure time and humidity level, appropriate calibration factor is selected. The gamma spectrometric measurement is performed no less than three hours after exposure. NaI detector is the first choice, but HPGe detectors are occasionally used. Region of interest is set in such way that all three peaks (295, 352 and 609 keV) are contained, while, at the same time, the region is kept as narrow as possible. The gross count will depend on the selection of Region of Interest (ROI). The different methods for ROI selection were used, and the results obtained by different methods were compared. One ROI which includes all three peaks of interest (ROI covers the whole spectral area between 270 keV and 720 keV for NaI detector, figure 1) or sum of three separate ROI for the three peaks selected (270-318 keV, 320-415 keV, 510-720 keV, figure 2) (Pantelić et al., 2014). Radon activity concentration calculated using equation (1) does not depend on ROI selection method when the calibration was performed using the same ROIs.

The set of calibration factors from EPA procedure addendum is used. According to the literature, these calibration factors are better suited for the application in residential and working areas, than stationary atmosphere calibration factors (Jenkins, 2002; EERF, 1989). The validity of using these calibration factors was checked in international intercomparison (Pantelić et al., 2013).

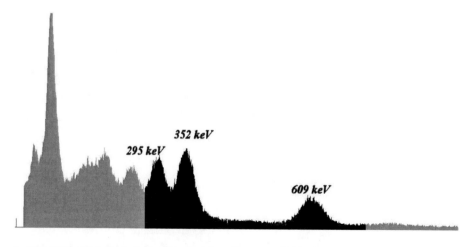

Figure 1. One ROI which include the three peaks of interest (ROI covers the whole spectral area between 270 keV and 720 keV) for NaI detector.

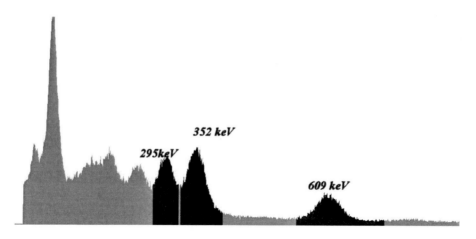

Figure 2. Three separate ROI for the three peaks selected (270-318 keV, 320-415 keV, 510-720 keV) for NaI detector.

3. UNCERTAINTY EVALUATION IN RADON CONCENTRATION MEASUREMENT USING CHARCOAL CANISTER

The contributions to the combined measurement uncertainty are identified, based on the equation (1) for radon activity concentration calculation and guide to the expression of uncertainty in measurement (ISO, 1995). The components which were considered were uncertainties of: counting statistics, efficiency, calibration factor for radon adsorption rate, which includes uncertainty of adjustment factor, decay factor, exposure time and measurement time. Each of these components contribute to the measurement uncertainty, and it was necessary to evaluate their magnitude.

In the case when the calculation is done with one ROI, which includes the three peaks of interest, uncertainty of radon concentration is calculated using following equation (Pantelić et al., 2014):

$$\left(\frac{\Delta A_{Rn}}{A_{Rn}}\right)^2 = \frac{(\Delta G)^2 + (\Delta B)^2}{(G-B)^2} + \left(\frac{\Delta t}{t}\right)^2 + \left(\frac{\Delta E_f}{E_f}\right)^2 + \left(\frac{\Delta C_f}{C_f}\right)^2 + \left(\frac{\Delta D_f}{D_f}\right)^2 \qquad (3)$$

where ΔX represents uncertainty of quantity X. If three separate peaks are used instead (three ROIs), the first term on the right hand side of equation (3) becomes:

$$\frac{(\Delta G_1)^2 + (\Delta G_2)^2 + (\Delta G_3)^2 + (\Delta B_1)^2 + (\Delta B_2)^2 + (\Delta B_3)^2}{(G_1 + G_2 + G_3 - B_1 - B_2 - B_3)^2}$$

while the other terms remain the same.

Uncertainties of measurement time and decay factor are neglected, because they are below 0.1 %. Relative uncertainty of detector efficiency and calibration factor for radon adsorption rate are estimated to be 2 % and 5.8 % respectively. Uncertainties of peak areas ΔG_i and ΔB_i are equal to the square root of appropriate peak area.

A typical uncertainty budget which gives all of the individual components of uncertainty and some typical values for one radon concentration measurement is presented in table 1.

Table 1. A typical uncertainty budget for one radon concentration measurement

Symbol	Quantity	Type	Probability distribution	Relative Standard Uncertainty (%)	
				one ROI	three ROIs
G_i - B_i	Net counts	A	Poisson	2.6	3**
T	Time	B	Rectangular	< 0.1	< 0.1
Ef	Efficiency	B	Normal	2	2
Cf	Calibration factor	B	Rectangular	5.8	5.8
Df	Decay factor	B	Rectangular	< 0.1	< 0.1
Combined standard uncertainty			(k=1)	6.7	6.8
			(k=2)	13	14

**Uncertainty from all G_i and B_i.

Only the first component of the equations (3) depend on ROI settings and counting statistics. Because of the poor counting statistics, the uncertainties are higher at low radon concentrations for all ROI settings. At the higher radon concentrations, the contribution of the peak area uncertainties are small. In these cases, the radon concentration uncertainty depends only on detector efficiency uncertainty and calibration factor for radon adsorption rate uncertainty and becomes the same for all ROI settings. It was shown that the different methods that were used for ROI settings do not influence the results for radon concentration and appropriate combined measurement uncertainty (Pantelić et al., 2014). Because of that, we recommend to use the method with one continuous ROI which is less sensitive to gamma spectrometry system instabilities, manifested by energy calibration shifts.

4. QUALITY ASSURANCE AND QUALITY CONTROL

Objective evidence of precision and accuracy is a key component of data defensibility, and critical to the success of any environmental program which relies on analytical data for decision making (Betti and Aldave de las Heras, 2004). For that purpose, an effective quality control (QC) and quality assurance (QA) program is necessary to maintain high quality of results. QC ought to be planned, described in the QC documentation, performed in a systematic manner, recorded and reviewed. To reduce the fraction of results that has to be rejected, QC must be embedded into an overall systematic approach to avoid mistakes *before* they are made, and this is commonly referred to as 'quality assurance' (IAEA, 2004; ISO, 2006). Planning should identify and define type and frequency of QC, acceptance limits, actions if those limits are exceeded and periodic review of results. Also, an external control can be implemented in a form of various interlaboratory proficiency tests and intercomparisons.

A QA program includes several steps: (a) the selection and validation of analytical methodology; (b) the resources used for the analysis (qualified personnel, work area, instrumentation and equipment, consumables, supplies, etc.); (c) laboratory operations for sample handling and analysis (reception, recording samples, data handling, reporting, archiving of results, etc.); and (d) QC, monitoring and auditing. All of these proposed measures for QC/QA are implemented in the Radiation and Environment Protection Department of the Institute for Nuclear Sciences in Vinča, Belgrade. Since the operation of this department includes daily measurements of a large number of samples it is essential to have a stable and accurate measuring system, so that the results are accurate, precise and reproducible. Especially when measuring radon, due to the influence of background, the stability of the system is checked immediately before measurement.

The accuracy and reproducibility of gamma spectrometry systems are verified on a weekly basis. Total background count rate, using an unexposed standard radon canister, is monitored to verify that the detector and shielding have not been contaminated by radioactive materials. Energy calibration is checked in a whole region of energies before applying usual QC procedure for radon measurement. The total activity of calibration source is used to check the efficiency calibration and the general operating parameters of the gamma spectrometry system (source positioning, contamination, library values, and energy calibration). For that purpose, standard canister (a sealed canister with the same matrix and geometry as the canisters used for measurements, but with the known activity of radon) is used. The detector-shield background, detector efficiency, peak shape, and peak drift are measured to ascertain whether they are within the warning and acceptance limits.

For the purpose of checking the accuracy of activity measurement of radon, a standard charcoal source containing ^{226}Ra (Certificate No. 236-4-2-1-2 issued by Isotope Product Laboratories, with 913.8 Bq activity on 01.01.1989.) is used. Since ^{222}Rn is detected using its daughters ^{214}Pb and ^{214}Bi with appropriate energies, 295keV, 352keV and 609keV, the efficiency calibration can be derived directly from the measurement. The source was placed directly on the detector cap for the measuring time of 1800s. The Shewhart Charts for NaI detector are presented in figure 3.

Also, background measurements are routinely performed. Since radon progeny is present in the background, this measurement constitutes important part of the quality control. The total background count over 8192 channels for NaI detector is depicted in figure 4 showing that background count is stable and within the acceptance limits. The acceptance limits are set to be $\pm 2\sigma$ and $\pm 3\sigma$ around the mean value taken over a chosen period of time.

For the NaI detector checking the efficiency and the background is enough for the quality control of the detector. Points falling between $\pm 2\sigma$ are considered to be satisfactory, the ones between $\pm 3\sigma$ are warning and those exceeding $\pm 3\sigma$ indicate that a problem in measurement of the background has occurred.

For the HP Ge detector which are used for the radon and other gammaspectrometry measurements, several more parameters were checked with radioactive standard material (Pantelić et al., 2009). Additional parameters are full width at half maximum (FWHM) ratio (FWHM spectrum/FWHM calibration) for a list of energies emitted by a calibration source and average full width at tenth maximum (FWTM) ratio (FWTM spectrum/FWTM calibration) for a list of energies emitted by a calibration source. As it can be seen, all the values over the time period are within the acceptance limit, so no correction measures were necessary (Krneta Nikolic et al., 2014).

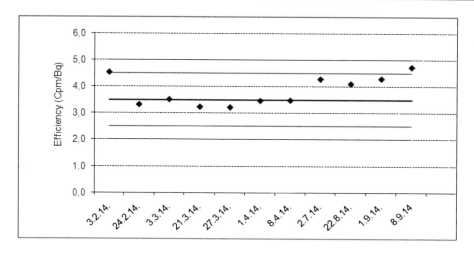

Figure 3. The Shewhart Charts for efficiency of NaI detector.

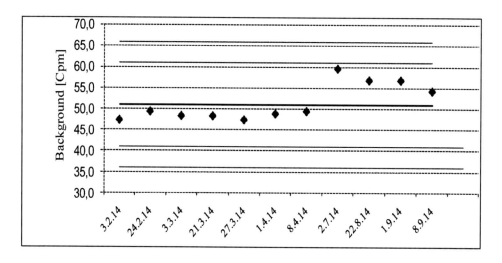

Figure 4. The Shewhart Charts for the backround measurement with NaI detector.

Besides weekly QA/QC procedures, the Laboratory takes part in interlaboratory proficiency tests on a regular basis (Pantelic et al., 2010). The last radon intercomparison test was organized by Federal Office for Radiation Protection of Germany 2012 (Foerster et al., 2012). The performance evaluation of the proficiency tests showed that the laboratory results were in a good agreement with the target value, which served as a confirmation of the reliability and traceability of all measurements conducted in this laboratory.

Also, as a part of QA/QC procedures, control measurements are periodically performed. Namely, to check the stability of measurements, the same exposed charcoal canister was measured on 3 different HpGe detectors (Canberra models GC 2018-7500, 7229N-7500-1818 and GC 5019) and one NaI detector. Obtained results showed agreement within the measurement uncertainty proving the reproducibility of the measurement. The results are presented in table 2.

Control measurements were performed also by different analysts to prove the competence of all staff that deals with the measurement. For that purpose, one exposed charcoal canister was measured and the spectrum was analyzed by three different analysts (table 3). The results proved competence for all three of the staff.

Table 2. The control measurements on different detectors

Detector	Date of measurement	Sample number	Result ± measurement uncertainty (Bq/m^3)
GC 2018-7500	26.09.2013.	1	1750 ± 90
7229N-7500-1818	26.09.2013.	1	1750 ± 90
GC 5019	26.09.2013.	1	1780 ± 90
NaI	31.03.2014.	2	1062 ± 57
GC 5019	01.04.2014.	2	1105 ± 57
GC 2018-7500	01.04.2014.	2	1097 ± 59

Table 3. The control measurements of one exposed charcoal canister analyzed by three different analysts

Detector	Date of measurement	Sample number	Result ± measurement uncertainty (Bq/m^3)
GC 5019	13.02.2013.	3	519 ± 30
GC 5019	14.02.2013.	3	527 ± 31
GC 5019	19.02.2013.	3	521 ± 40

5. RESULTS OF SCREENING MEASUREMENTS IN SERBIA

Following EPA's recommendations, we performed short-term test with charcoal canisters for a period of 2-3 days during the weekend in closed-building conditions (windows and doors kept closed during exposure). Obtained results were compared with Serbian regulations (RS Official Gazette, 2011).

5.1. Radon in Dwelings

Screening measurements are very useful for homeowners and quickly determine whether their homes contain high radon concentrations and to decide whether additional measurements are needed.

Systematic survey of radon concentration in dwellings was performed in Belgrade during the period 1991-2010 (Petrović and Pantelić, 1999; Eremić-Savković et al., 2012). The measurement were performed in a livable room on a level closest to the underlaying soil, susch as basement (EPA, 1987; EPA, 1993a). In the dwellings with multiple floors the measurement were performed in the basement and in the first and second floor. The charcoal canister was placed in the room expected to have the lowest ventilation rate and with closed windows during the exposure.

Figure 5 illustrates the number of examined dwellings (shown in %) in Belgrade divided in three intrevals with radon concentration: below 200 Bq/m^3, between 200 Bq/m^3 and 400 Bq/m^3 and above 400 Bq/m^3. In the twenty years period radon concentration in dwellings in Belgrade is estimated to be 89 Bq/m^3. This result falls into the interval from 20 Bq/m^3 to 144 Bq/m^3 in the other countries in Europe (Dubois, 2005). The latest research in nearby countries showed that the average radon concentration in dwellings in Romania is 126 Bq/m^3 (Cosma et al., 2013) and from 89 Bq/m^3 to 127 Bq/m^3 in Macedonia (Stojanovska et al., 2012).

In 84.3 % examined dwellings radon concentration was lower than 200 Bq/m^3 and in 9.8 % dwellings radon concentration was between 200 Bq/m^3 and 400 Bq/m^3. In total 94.1 % homes meet the requirement for radon concentration lower that 400 Bq/m^3. The highest radon concentration was found in homes that used coal for heating and homes with bad floor insulation. We suggested the owners to do a long-term test to get a measurement that will help them to decide if it is worthwhile to take action to reduce their radon exposure.

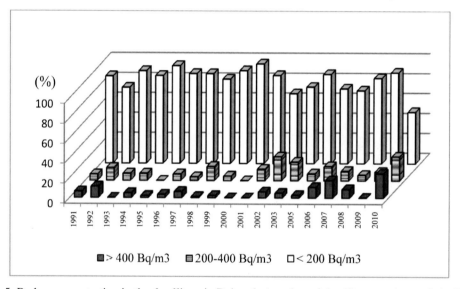

Figure 5. Radon concentration in the dwellings in Belgrade (number of dwellings per intervals in %).

5.2. Radon in Kindergartens and Schools

EPA's research in schools has shown that radon levels often vary greatly from room to room in the same building (EPA, 1993b). A known radon measurement result for a given classroom cannot be used as an indicator of the radon level in adjacent rooms. Therefore we recommend that schools conduct initial measurements in 4-5 classrooms on the ground level and the first floor. Sometimes measurement was performed in the basement if the school had clasrooms there.

Radon surveys in the kindergarten and schools in Belgrade were initiated by city government. Radon measurements were performed in 40 kindergarten, 30 elementary schools and 22 high schools during the winter in 2010 and 2011 (Eremić-Savković et al., 2012). The range (minimum and maximum) and average radon concentration with standard deviation in the kindergartens and schools are presented in tables 4-6.

Table 4. Radon concentration in kindergartens in Belgrade

	Min (Bq/m^3)	Max (Bq/m^3)	Average (Bq/m^3)	Standard deviation (Bq/m^3)
1	49	74	66	12
2	5	44	30	15
3	5	35	18	13
4	12	15	14	2
5	13	29	19	8
6	23	33	29	5
7	36	89	62	22
8	39	347	142	143
9	49	88	69	28
10	51	83	62	18
11	60	465	218	181
12	107	429	262	161
13	117	419	292	156
14	132	166	148	17
15	67	106	83	16
16	115	691	375	300
17	27	37	32	5
18	25	144	80	48
19	179	507	300	133
20	40	56	48	6
21	10	17	13	3
22	11	27	20	7
23	25	32	29	3
24	28	34	30	3
25	33	49	42	7
26	40	43	42	2
27	42	80	65	16
28	48	85	70	16
29	47	75	56	14
30	30	123	54	39
31	61	173	105	44
32	37	375	181	128
33	103	305	191	72
34	107	305	172	78
35	177	481	323	143
36	55	266	129	81
37	74	421	205	141
38	64	212	132	71
39	13	43	29	12
40	27	138	64	48

Table 5. Radon concentration in the elementary schools in Belgrade

	Min (Bq/m^3)	Max (Bq/m^3)	Average (Bq/m^3)	Standard deviation (Bq/m^3)
1	16	142	73	53
2	5	9	7	2
3	10	67	28	26
4	17	37	26	10
5	17	26	21	4
6	68	311	178	106
7	58	280	116	94
8	38	138	101	41
9	25	25	25	0
10	28	208	101	87
11	28	48	37	10
12	19	152	66	51
13	14	27	23	5
14	23	293	119	119
15	56	333	195	114
16	62	198	112	55
17	41	75	60	15
18	185	646	354	204
19	13	204	107	100
20	14	362	144	140
21	19	232	120	97
22	46	90	70	17
23	5	16	10	6
24	14	23	19	4
25	5	80	27	30
26	21	309	97	142
27	31	125	73	37
28	30	78	46	20
29	35	95	68	23
30	47	645	217	288

The obtained results suggested that elevated radon levels (levels > 400 Bq/m^3) may exist in some rooms in kindergartens (6 %) and some classrooms in the schools in Belgrade (4.5 %). We have repeated all measurements in schools and kindergartens with the results higher than 400 Bq/m^3 in the same classrooms and in the other classrooms at the same level and in the other floors. The repeated measurement showed similar results. We suggested to directors of schools to do a long-term test to get reliable results for average annual radon concentrations in their objects.

It can be seen from the tables, when the average radon concentrations per object was calculated, that the radon concentrations above recommended level of 400 Bq/m^3 were obtained only in 2 high schools (4 % of all schools).

Table 6. Radon concentration in the high schools in Belgrade

	Min (Bq/m^3)	Max (Bq/m^3)	Average (Bq/m^3)	Standard deviation (Bq/m^3)
1	9	148	50	56
2	5	20	12	8
3	57	496	214	245
4	44	456	188	193
5	16	56	31	18
6	39	74	55	18
7	55	378	124	142
8	66	979	557	438
9	15	129	65	52
10	21	297	174	119
11	5	95	44	38
12	61	214	117	59
13	62	204	114	79
14	310	588	430	143
15	99	402	213	117
16	16	191	83	77
17	14	30	20	9
18	23	53	38	11
19	28	59	41	14
20	16	89	39	30
21	12	357	134	155
22	16	257	115	113

The arithmetic mean values of indoor radon concentrations in our kindergartens (108 Bq/m^3) and schools (88 Bq/m^3 in elementary schools and 130 Bq/m^3 in high schools) based on our short term measurement with charcoal are comparable with the results from the other countries. In Romania radon concentration was measured in schools with nuclear track detectors one month, during the winter, and obtained arithmetic mean was 146 Bq/m^3 (Dumitrescu et al., 2004). The indoor radon concentration obtained by short term measurements in Bulgarian kindergartens was 411 Bq/m^3 and schools 287 Bq/m^3 (Vuchkov et al., 2013).

CONCLUSION

Indoor radon concentration was measured with charcoal canisters under closed-building conditions (windows and doors kept closed during exposure). The charcoal canisters were exposed in kindergartens, schools and dwellings following EPA's recommendations.

The gamma spectrometric measurement is performed no less than three hours after exposure on NaI or HPGe detectors. Detector efficiency is determined by using the [226]Ra standards with the same matrix and geometry as radon canisters. The accuracy and reproducibility of gamma spectrometry systems are verified on a weekly basis measuring the total background count rate, using an unexposed standard radon canister, and checking the stability of detector efficiency. The quantities that contribute to the combined measurement uncertainty were identified as uncertainties of counting statistics, efficiency, calibration factor for radon adsorption rate, decay factor, time of exposure and measurement. Part of quality assurance programme is participating in domestic and international intercomparison with successful results.

According to 20 years investigation in Belgrade, only 6 % of homes have elevated levels of radon. The radon concentration measurements in the kindergartens and schools are still at an early stage. Based on 2 years measurements we can conclude that situation is not alarming because only 4 % schools have elevated radon concentrations. Our results are similar to results obtained in the other Balkan's country.

We recommended intensive airing in objects with elevated radon concentration and suggested owners of dwellings to perform additional measurements, followed by reparation of the facilities if it is necessary.

The obtained results merely show a preliminary picture of indoor radon in dwellings, schools and kindergartens in Belgrade and it is a good base for a national radon monitoring program, which is being established in this year.

REFERENCES

Archer Victor E., Lundin Frank, Jr., 1967. Radiogenic Lung Cancer in Man: Exposure-Effect Relationship *ENVIRONMENTAL RESEARCH* 1, 370-383.

BIPM, 2008. Monographie BIPM-5; BIPM: Sevres, France; Vol 4.

Betti M. and Aldave de las Heras L., 2004. Quality assurance for the measurements and monitoring of radioactivity in the environment, *Journal of Environmental Radioactivity* 72, 233-243.

Cohen B.L., Cohen E.S., 1983. Theory and practice of radon monitoring with charcoal adsorption. *Health Physics*, 45 (2), pp. 501–508.

Cosma C., Cucos A.D., Dicu T., 2013. Preliminary results regarding the first map of residential radon in some regions of Romania. *Radiation Protection Dosimetry* Vol. 155, No. 3, 343-350.

EC, 1990. Commission recommendation of 21 February 1990 on the protection of the public against indoor exposure to radon. (90/143/Euratom). *Official Journal of the European Commission* 1996 39 L80, 26-27.

EERF, 1989. Addendum to the EERF Standard Operating Procedures for Radon-222 Measurement Using Charcoal Canisters. United States Environmnetal Protection Agency.

ECA, 1995. European Collaborative Action "Indoor Air Quality & It's Impact on Man", Report No. 15: *Radon in Indoor Air*, EUR 16123, EN 1995.

EPA, 1987. Interim Protocols for Screening and Followup Radon and Radon Decay Product Measurements. United States Environmnetal Protection Agency. EPA 520/1-86-014.

EPA, 1993a. Protocols for Radon and Radon Decay Product Measurements in Homes. *United States Environmnetal Protection Agency*. EPA 402-R-93-003.

EPA, 1993b. Radon Measurement In Schools. *United States Environmnetal Protection Agency*. EPA 402-R-92-014.

EPA, 1996. Radon Proficiency Program (RPP), *United States Environmnetal Protection Agency*. EPA 402-R-95-013 Handbook.

Eremić-Savković M., Javorina Lj., Tanasković I., Pantelić G., 2012. Indoor radon Concentration in Schools and Kindergartens in Belgrade in 2010. *13th International Congress of the International Radiation Protection Association*, 13-18 May 2012, Glasgow, IRPA13 Abstracts, P10.60.

Eremić Savković M., Pantelić G., Kolarž P., Javorina Lj., Arsić V., 2012. Indoor radon concentration in buildings and public institutions and recommendations for radon reduction, Counseling Masonry building and technical regulations, Modern building practices in Serbia and Europe, Belgrade, 16th May 2012, 113-120, in Serbian.

Foerster E., Beck T., Buchröder H., Döring J., Schmidt V., 2012. Instruments to Measure Radon Activity Concentration or Exposure to Radon – Interlaboratory Comparison 2012, Federal Office for Radiation protection, Salzgitter.

Dumitrescu A., Milu C., Vaupotič J., R. Gheorghe R., Stegnar P., 2004. Preliminary indoor radon and gamma measurements in kindergartens and schools in Bucharest. 11th International Congress of the International Radiation Protection Association, 23rd-28th May 2004, Madrid, IRPA11, 6a50, 1-6.

Dubois G., 2005. An overview of radon surveys in Europe. EUR21892 EN EC.

George, A. C., 1984. Passive, Integrated Measurement of Indoor Air Using Activated Carbon. Health Physics, 46:867.

Grey, D.J., Windham, S.T., 1987. EERF Standard Operating Procedures for Radon-222 Measurement Using Charcoal Canisters. *United States Environmnetal Protection Agency*. EPA 520/5-87-005.

Holaday Duncan A., 1956. Radiation hazards of uranium mining. *The International Journal of Applied Radiation and Isotopes*, Volume 1, Issues 1–2, July 1956, Pages 130-131.

Hoover H.C., Hoover L.H., 1950. Geogrius Agricola De Re Metallica, translated from the first Latin edition of 1556. Dover Publications, Inc. New York.

IAEA, 2004. Quality system implementation for nuclear analytical techniques. Training course series 24. IAEA Vienna, Austria.

IARC, 1988. Man-made mineral fibres and radon; IARC Monographs on Evaluation of Carcinogenic Risks to Humans; IARC: Lyon, France; Vol. 43, pp 1-300.

ICRP, 1987. Lung cancer risks from indoor exposures to radon daughters. Publication 50, *Annals of the ICRP* Vol. 17, No 1, Pergamon Press.

ICRP, 1991. 1990 Recommendations of the International Commission on Radiological Protection. Publication 60, *Annals of the ICRP* Vol. 21, No 1-3, Pergamon Press.

ICRP, 1993. Protection Against Radon-222 at Home and at Work. Publication 65, Annals of the ICRP Vol. 23, No 2, Pergamon Press.

ICRP, 2007. The 2007 Recommendations of the International Commission on Radiological Protection. Publication 103, Annals of the ICRP Vol. 37, No 2-4, Pergamon Press.

ISO, 1995. Guide to the expression of uncertainty in measurement. International Organization for Standardization, Geneva, Switzerland, ISBN 92-67-10188-9.

ISO, 2006. General requirements for the competence of testing and calibration laboratories, ISO/IEC 17025 ISO, Geneva, Swizerland.

Jenkins P.H., 2002. A Critique of the "EPA Method" for Analyzing and Calibrating Charcoal Canisters for Radon Measurements. 2002 International Radon Symposium Proceedings, *American Association of Radon Scientists and Technologists*, Inc.

Krneta Nikolic J.D., Todorović D.J., Janković M.M., Pantelić G.K., Rajačić M.M. , 2014. Quality assurance and quality control in environmental radioactivity monitoring, *Qality assurance and safety of crops and foods*, 6(4) 403-409.

Pantelić G., Vuletić V., Eremić Savković M., Tanasković I., Javorina Lj., 2009. QA /QC in gammaspectrometry laboratory. Report on Technical Meeting on Quality Assurance for Nuclear Spectrometry Techniques (G4-TM-36923), IAEA, Vienna, Austria, 12th–16th October 2009.

Pantelic G., Vuletic V., Mitrovic R., 2010. Proficiency test of gamma spectrometry laboratories in Serbia. *Applied Radiation and Isotopes*, 68, 1270-1272.

Pantelić G, Živanović M, Eremić Savković M, Forkapić S, 2013. Radon concentration intercomparison in Serbia, Proceedings of the ninth symposium of the Croatian Radiation Protection Association; Eds. Ž. Knežević, M. Majer, I. Krajcar-Bronić, 10.-12. April 2013. Krk, Croatia, 193-198, ISBN 978-953-96133-8-7, CRPA Zagreb.

Pantelić G., Eremić Savković M., Živanović M., Nikolić J., Rajačić M., Todorović D., 2014. Uncertainty evaluation in radon concentration measurement using charcoal canister, *Applied Radiation and Isotopes*, 87, 452-455.

Petrović I., Pantelić G., 1999. Analysis of the radon concentration measurements in the Republic of Serbia. Proceedings of the XX Yugoslav Symposium for Radiation Protection, Tara, 3rd -5th November 1999; 135-139, in Serbian.

Ronca-Battista M., Magno P., Nyberg P., 1987. Interim Protocol For Screening And Followup Radon and Radon decay Product Measurements. *United States Environmnetal Protection Agency*. EPA 520/1-86-014.

Ronca-Battista M., Gray D., 1988. The influence of changing exposure conditions on measurements of radon concentrations with the charcoal adsorption technique; *Radiation protection dosimetry*, Vol. 24, no 1-4, pp 361 – 365.

RS Official Gazette, 2011. *Rulebook on limits of exposure to ionizing radiation and measurements for assessment the exposure levels*. Official Gazette of the Republic of Serbia No 86. National Assembly of the Republic of Serbia, Belgrade.

Stojanovska Z., Januseski J., Boev B., Ristova M., 2012. Indoor exposure of population to radon in the FYR of Macedonia. *Radiation Protection Dosimetry* Vol. 148, No. 2, 162-167.

UNSCEAR, 2000. United Nations Scientific Committee on the Effects of atomic Radiation. Sources and effects of ionizing rdiation. Report to the General Assembley. *Annex B: Exposures from natural radiation sources*, New York.

Vuchkov D., Ivanova K., Stojanovska Z., Kunovska B., Badulin V., 2013. Radon measurements in schools and kindergartens (Kremikovtsi municpality, Bulgaria). *Rom.Journ.Phys.*, Vol.58, Supplement, S328-S335.

Walsh P.J., 1970. Radiation dose to the respiratory tract of uranium miners—A review of the literature, *Environmental Research*, Volume 3, Issue 1, January 1970, Pages 14–36.

In: Radon
Editor: Audrey M. Stacks

ISBN: 978-1-63463-742-8
© 2015 Nova Science Publishers, Inc.

Chapter 10

METHODS OF RADON MEASUREMENT

J. Nikolov, N. Todorović, S. Forkapić, I. Bikit, M. Vesković, M. Krmar, D. Mrđa and K. Bikit*

University of Novi Sad, Faculty of Sciences,
Department of Physics, Novi Sad, Serbia

ABSTRACT

Radon-222 is a radioactive, noble gas. As a chemically inert gas, it is easily released from soil, building materials, and water, to emanate to the atmosphere. Research carried out in recent decades has shown that, under normal conditions, more than 70% of a total annual radioactive dose received by people originates from natural sources of ionizing radiation, whereby 40% is due to inhalation and ingestion of natural radioactive gas radon ^{222}Rn and its decay products. Many techniques have been developed over the years for measuring radon and radon progenies, because of its hazard effects on human health. Conceptually, measurement techniques can be divided into three board broad categories: (1) grab sampling, (2) continuous and active sampling, and (3) integrative sampling. In this section different techniques for radon measurement and comparison of those techniques will be presented. For radon in air measurement: RAD7, Alpha Guard, passive detectors and charcoal canisters. For activity concentration of radon in water: RAD7-H_2O, liquid scintillation counting. In general, the following generic guidelines should be followed when performing radon measurements during site investigations:

- The radon measurement method used should be well understood and documented;
- Long term measurements are used to determine the true mean radon concentration;
- The impact of variable environmental conditions (e.g., humidity, temperature, dust loading, and atmospheric pressure) on the measurement process should be accounted for when necessary. Consideration should be given to effects on both the sample collection process and the counting system;

---

*Corresponding author's email: jovana.nikolov@df.uns.ac.rs.

- The background response of the detection system should be accounted for;
- If the quantity of interest is the working level, then the radon progeny concentrations should be evaluated. If this is not practical, then the progeny activities can be estimated by assuming they are 50% of the measured radon activity.

INTRODUCTION

Radon is a radioactive noble gas produced by the decay of radium isotopes in both the uranium and thorium radioactive decay series. ^{222}Rn is a gas and it may release from the origin material to indoor air and accumulate to high enough concentrations to provide a significant dose. Radon is a natural inert radioactive tasteless and odorless gas, whose density is 7.5 times higher than that of air. It dissolves in water and can readily diffuse with gases and water vapor, thus building up significant concentrations. The physical half-life of radon is 3.825 days and half-elimination time from lungs 30 min. Radon decays to a number of short lived progenies (^{218}Po, ^{214}Pb, ^{214}Bi, ^{214}Po) that are themselves radioactive. It is soluble in water, and its origin in water is the radium from water surrounding soil and bedrock [1].

Many techniques have been developed over the years for measuring radon and radon progenies, because of its hazard effects on human health. One of the possible ways to divide measurement techniques is into three board broad categories: (1) grab sampling, (2) continuous and active sampling, and (3) integrative sampling. The choice between these categories will depend on the costs involved, the time over which an instrument can be devoted to measurements at a single location, the kind of information required, and the desired accuracy with which measurements can be related to an estimate of risk [2].

Some of the mostly used techniques for radon in air measurement are: RAD7, Alpha Guard, passive detectors, charcoal canisters. For radon in water, in this section, only two techniques will be explained and compared: RAD7-H$_2$O and liquid scintillation counting (LSC). LSC method for radon measurement is very rarely used but gives very good results which will be explained in detail.

MONITORING RADON-222 IN THE ENVIRONMENT

In outdoor air concentrations of ^{222}Rn vary with surrounding soil type, meteorological conditions, etc. Outdoor radon is a small health hazard, for example in US there are around 700 lung cancer deaths per year [3].

^{222}Rn can also be dissolved in water, it is logical and also can be concluded from published results that there are highest concentrations of radon in ground water compared to surface water. Radon gas which emanates from a residential water source produces radon progeny. Radon from water can affect humans by ingestion or inhalation. Inhaling or ingesting waterborne radon progeny is usually a very small health hazard.

Concerning indoor air, activity concentrations of ^{222}Rn vary with underlying surrounding rocks or soil type. There are residential and occupational exposures. Occupational exposure to radon levels is highest for miners (uranium, iron and fluorospar) [4]. Residential exposure to radon levels is highest in basement and ground floor rooms. Both residential and occupational levels depend from building type, construction, level of repair and the most important ventilation of the indoor. For example, in radon spas there are elevated levels of radon in homes, but also in medical centers [1]. Indoor radon is the most significant health hazard.

It is necessary to monitor the ^{222}Rn in the environment, especially in indoor air, but also in outdoor air, as well as in drinking water in order to protect inhabitants. According to the geological composition of the examined area there are different expectations for radon activity concentrations.

MEASUREMENT TECHNIQUES IN GENERAL

If we use the above mention way to divide radon measurement techniques into three broad categories, than we have to explain the difference between those categories.

Grab sampling technique is used if we have essentially instantaneous measurements of radon or radon progeny in air over time intervals that are short (on the order of minutes) compared to the time scale of fluctuations in activity concentrations. The air is usually collected in container and after that analyzed in the laboratory conditions. Typical containers include plastic bags, metal cans and glass containers. The volumes of the containers are usually between 5 and 20 liters.

Continuous sampling means that there is an automatic taking of measurements at closely spaced time intervals over a long period of time. The results of such measurements are series of measurements. The advantage of those techniques is that they include the information on the pattern in which the activity concentration varied throughout the whole measuring interval.

Integrating sampling uses devices that collect information on the total number of radiation events in a long period of time (several days to months). The results are an estimation of the approximate average concentration during the whole measuring interval.

Numerous radon measuring devices are commercially available - differing in the kind of the measuring system, in sensitivity and in handling. Each device is characterized by advantages and disadvantages. These have to be evaluated depending on the purpose, for which the instrument will be used.

The following table gives an overview of common radon measuring devices, distributed by different companies [5].

Table 1. Overview of common radon measuring devices, distributed by different companies [5]

Device	manufacturer	measuring method
E-PERM Introductory Kit	Rad Elec Inc. (USA)	electret
AlphaGUARD-Radon Monitor PQ 2000 Pro	Genitron (Germany)	Ionisation chamber
Continous Radon Monitor	RTCA (USA)	ionisation chamber
CRM-510 Continous Radon Monitor	F & J Specialty Products, Inc. (USA)	ionisation chamber
CRM-510LP Continous Radon Monitor	F & J Specialty Products, Inc. (USA)	ionisation chamber
ERS/RM Radon Monitor	Tracerlab GmbH (Germany)	silicon semiconductor

Table 1. (Continued)

Device	manufacturer	measuring method
LCD-ARDM-Plus	Tracerlab GmbH (Germany)	silicon semiconductor
LCD-BWLM-Plus	Tracerlab GmbH (Germany)	silicon semiconductor
RAD7 Professional Radon Detector	Durridge Company (USA)	silicon semiconductor
AB-4 Radon Monitor	Pylon (USA)	Lucas cell
AB-5 Radon Monitor	Pylon (USA)	Lucas cell
CRM-1	Pylon (USA)	Lucas cell
CRM-2	Pylon (USA)	Lucas cell
Extraction System	Storm King Associates (USA)	Lucas cell
Radon Monitor MR 1	Tesys (Italy)	Lucas cell
TM372 Sample Counter	Env. Instr.Inc. (Canada)	Lucas cell

RADON IN AIR – METHODS OF MEASUREMENT

According to the recommendations of the International Commission on Radiological Protection (ICRP) from 1994, the civil engineering standard for building new apartments and houses is 100 Bq m^{-3} radon as an average annual level, 200 Bq m^{-3} for the recommendation of inexpensive remediation measures, and 600 Bq m^{-3} for the recommendation of expensive remediation measures. Namely, even in the case of observing these recommendations and norms, ALARA (*As Low As Reasonably Achievable*) principle applies [6].

Whichever technique is used for measuring of radon-222 in air, there are two possibilities for testing period: *short-term* measurements that lasts 2 days to 3 months, and *long-term* measurements that lasts more than 3 months. If the short-term measurement is used than it is important to choose appropriate location – lowest level of dwelling that is commonly used, at least 50 cm above the floor. It is also important not to measure in the kitchen or bathroom because of the higher humidity and ventilation systems. The best conditions which should be provided for short-term measurements of radon in air are: doors and windows closed 12 hours before and during test period. Almost all types of devices, radon monitors could be used for short-term measurements. The main limitation for this method is the fact that radon concentration in air vary day to day and also seasonally. The main advantage of short-term measurements is obtaining quick results, but if high level of activity concentration of ^{222}Rn is registered than the "follow-up" testing is required [7].

All the measurements of radon which lasts more than three months continually are registered as long-term measurements. The best location for those measurements is the lowest level of dwelling that is commonly used, at least 50 cm above the floor, also as in the case of short-term measurements, avoiding kitchen and bathroom. Devices usually used for long-term measurement of radon in air are alpha track detectors, electric ion detectors and continuous monitors. The main disadvantage of this method is that the results are delayed up to 1 year, but it is more likely to indicate accurate average radiation level and public exposure [7].

Methods for radon measurement in air can be divided in these three categories [2, 7]:

- alpha-particle scintillation counting with ZnS (known as Lucas Cell)
- internal ionization chamber counters
- two-filter methods

Using a scintillation cell is one of the earliest methods for measurement of radon activity concentration; it is usually set up together with grab sampling method. Historically, this cell has been known as Lucas cell. The inside wall of the cell is coasted with zinc sulfide (ZnS), except one end which is covered with a transparent window for coupling to a photomultiplier tube. When an alpha particle strikes the wall of the cell, a flash of light is emitted from the ZnS coating. The light is detected by the photomultiplier tube and translated into an electrical signal. The efficiency of these cells is typically 70 to 80%. Background rates in typical Lucas cells are low, about 0.1 or 0.2 counts per minute (cpm).

Ionization chambers are also used for detection of alpha particles from the radon decay series. In these counters, contrary to previous ones, an electrical signal is produced without the intermediary of scintillation counting. Ionization counters can be used either to count electrical pulses from individual decay events or to measure currents resulting from the integrated effect of all decays. In general, ionization chambers are not as widely used as scintillation counters, since ionization chambers are more expensive to construct than Lucas cells and for radon measurements they do not appear to have a major advantage over Lucas cells.

If there is a need to measure both radon and radon daughter concentrations, than the two filter method is the best choice. In this method, air is passed through the first filter where daughter products are removed. Then the air is passed through a long decay chamber, where daughter products are allowed to grow in and are collected on a second filter. The filters can be counted separately to determine the concentration of radon (from the second filter) and daughter products (from the first filter).

Unlike radon, radon daughters deposit readily on dust particles or other surfaces. By drawing air through a filter, the radon daughters can be collected with high efficiency. In all cases, counting must begin shortly after the sample is collected because the daughters all have short half-lives (the longest is 27 minutes). In order to determine the daughter concentrations, expressed as the number of working levels, it is necessary to know the individual concentrations of polonium-218, lead-214 and bismuth-214. The concentration of polonium-214 is not relevant, because it has really short half-life (< 0.0002 sec).

In this section a couple of commercially available methods for radon in air measurement will be mentioned.

First one is AlphaGUARD (Figure 1). AlphaGUARD is portable equipment designed for instantaneous or continuous measurement of radon gas activity (^{222}Rn). AlphaGUARD is used for the measurement of Radon radon in the environment, mines and laboratories and also for complementary investigations in buildings. Air, water, soil, exhalation measurements are performed thanks to a large range of accessories and external probes.

The portable radon monitor AlphaGUARD (PQ-2000 Genitron Instruments GmbH) is easy to use for continuous monitoring of radon, and it is capable of circulation measuring. It has two selectable modes, the diffusion mode and the flow mode. The disadvantage is that the AlphaGUARD does not distinguish between radon and thoron. The sensitivity to thoron of the AlphaGUARD is not clear [8]. When the radon concentration in air is measured with the AlphaGUARD, air circulation needs to be stopped until the thoron in the chamber decays. Then the raw data need to be reread [9]. In the mentioned reference, it was shown that radon and thoron concentrations exhaled from soil could be separately measured using the AlphaGUARD. First, the sensitivity to thoron concentration in air with the AlphaGUARD was determined. The thoron concentration from the RAD7 was used to get a conversion equation to calculate thoron levels with the AlphaGUARD. However the conversion factor was found to be affected by the air flow rate.

Figure 1. AlphaGUARD - radon in air measurement.

The DURRIDGE RAD7 (Figure 2) uses a solid state alpha detector. A solid state detector is a semiconductor material (usually silicon) that converts alpha radiation directly to an electrical signal. The RAD7 draws samples of air through a fine inlet filter, which excludes the progenies, into a chamber for analysis [10]. The radon in the RAD7 chamber decays, producing detectable alpha emitting progenies, particularly the polonium isotopes. Though the RAD7 detects progeny radiation internally, the only measurement it makes is of radon gas concentration.

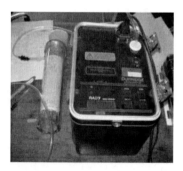

Figure 2. RAD 7, radon in air measurement.

Radon in the air can be measured by the RAD7 detector; usually all indoor air samples were taken according to the EPA test protocol [11]. Air sucked by the built in pump is passed through the drierite to the solid state detector for the measurement of radon concentration. The number of cycles and recycles depends of the used preferences and expected accuracy in activity concentration determination. The radon concentration in surrounding soil gas is one of many parameters that impact the radon health risk. The porosity of the soil, the height of the water table and several other factors are important. In radon spas, for example, it is of great interest to determine the radon concentration in soil gas [1].

One more method for radon in air measurement is measuring by passive detectors, for example, in the reference [12] radon and thoron concentrations in soil gas were measured by using two types of passive radon-thoron discriminative detectors [13, 14]. These detectors were developed and evaluated at the National Institute of Radiological Sciences (NIRS) in Japan. They can measure both radon and thoron concentrations. One type of the detector is called RADOPOT, and the other one, new type of the detector is called RADUET. This is a remodeled version of the RADOPOT monitor, which is now made in Hungary and is commercially available. A comparison of these detectors showed that the differences between the results do not exceed 10% [12]. Construction of these low cost detectors is simple, so the monitors have been widely used for radon and thoron surveys throughout the world [12]. In the investigation presented in the reference [12] RADOPOT and RADUET detectors were used for the first time to measure radon and thoron concentrations in soil gas. The influence of humidity on radon/thoron measurement was tested at the NIRS and no significant influence (about 10%) was found. The typical error on the radon/thoron concentration measurement was assumed to be 20–25% [12] and therefore the humidity has only a small effect on the results. One way for the installation of the mentioned passive detectors at the survey areas is to drill holes in the soil to a depth of about 80 cm depending on soil hardness. A detector can be hung in the hole on a 70-cm stick and then the hole has to be covered by a polyethylene bag, which was pressed with an appropriate tile [12].

Charcoal canisters (Figure 3) represent the simplest device for measuring radon concentration. Accuracy is fair (less than 20%) and is greatly affected by humidity and air movement. The accuracy of the charcoal canister decreases the longer the test evaluation is delayed. Radon is adsorbed onto the charcoal grains and decays to several particulate decay products: ^{218}Po, ^{214}Pb, ^{214}Bi, ^{214}Po and ^{210}Pb. Radon concentration can be determined by counting the gamma-ray emissions of both ^{214}Pb (295 and 352 keV) and ^{214}Bi (609 keV). This is possible due to the relatively short half life of these progenies. Within 3 hours, the progenies are in equilibrium with ^{222}Rn. The efficiency of the gamma detectors can be determined using the Environmental Protection Agency (EPA) ^{226}Ra reference source [15]. The passive nature of activated charcoal allows both adsorption and desorption; in addition, the adsorbed radon undergoes radioactive decay during the exposure period. Therefore, the canister cannot uniformly integrate over the entire exposure period. However, the canister can be calibrated to yield precise result for radon concentrations in structures during the deployment period of 48 hours used by EPA [1, 16]. The exposure time for the canister must not be less than 24 hours (1 day) noror longer than 144 hours (6 days). The ideal exposure time is 48 hours (2 days). During the sampling by active charcoal, the room to be surveyed has to be closed for 48 hours. The typical time between the end of the exposition and the beginning of the measurement has to be minimal.

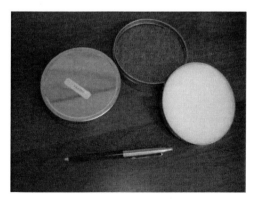

Figure 3. Charcoal canisters, radon in air measurement.

CR39 radon detectors represent plastic films with an area of 1 cm^2, 1 cm thick, and sensitive to traces of ionizing alpha particles [17, 18]. During the exposure, these detectors were glued to the cover of a closed diffusion plastic chamber 5 cm high. The detector is sensitive to alpha radiation only, and its sensitivity equals 2.9 traces/(cm^3 kBq h /m^{-3}). The detectors were etched in 25% solution of NaOH at a constant temperature of 900°C during 4 hours. The traces were read and treated by RADOSYS 2000 electronic equipment (Figure 4) in the Radosys Company, Hungary. This sophisticated equipment (Figure 4) includes: RADOBATH 2000 (thermostated bath for chemical etching of traces on the detectors) and RADOMETER 2000 equipment for reading traces, with a B&W CCD camera and a compatible PC. Average exposure time of CR39 detectors was 90 days.

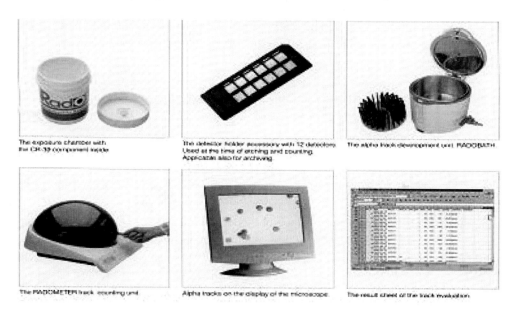

Figure 4. CR39 radon detectors placed in plastic diffusion chambers with displayed procedures of developing and etching traces and RADOSYS 2000 electronic equipment [17].

There are also reported investigations [19, 20] in which radon concentration measurements were performed using 1 cm^2 CR-39 detectors, enclosed in small cylindrical (5 cm height, 3 cm diameter) diffusion chambers. Detectors were etched in 6.2 M NaOH at 90°C for 4.5 h, washed in clean water for 10 min and dried overnight. Tracks on the CR-39 were counted with an automatic setup consisting of an optical microscope connected to a CCD camera controlled by a personal computer. Durations of exposure (typically 3 - 5 months) were chosen to reduce the probability of saturating the detector response. Nevertheless, in some cases, especially for indoor winter measurements, detector saturation did occur at exposures exceeding 10 MBq m^{-3} h^{-1}.

RADON IN WATER – METHODS OF MEASUREMENT

Radon concentration in water is due to ^{226}Ra decay associated with the rock and soil [21]. ^{222}Rn is picked up by the groundwater passing through rocks and soil containing such radioactive substances. When the groundwater reaches the surface (spas, wells and springs), ^{222}Rn concentration decreases sharply with the water movement and with purification treatment. If the water is consumed directly from the point of emergence, as is habitual in rural sites, the time is often not long enough to prevent the health risks associated with its short-lived daughters [22]; therefore, determination of ^{222}Rn activity concentration in waters is important. Inhalation of ^{222}Rn dissolved in and released from water for human consumption accounts for 89 % on radon-related cancers [23].

Radon is soluble in water and this route of exposure is important if high concentrations are found in drinking water [24]. On some occasions, water is consumed immediately after leaving the faucet before its radon is released into the air. This water goes directly to the stomach. Before the ingested water leaves the stomach, some of the dissolved radon can diffuse into and through the stomach wall. During that process, the radon passes next to stem or progenitor cells that are radiosensitive. These cells can receive a radiation dose from alpha particles emitted by radon and radon decay products that are created in the stomach wall. After passing through the wall, radon and decay products are absorbed in blood and transported throughout the body, where they can deliver a dose to other organs [25, 26].

The contribution of drinking-water to total exposure is typically very small and is due largely to naturally occurring radionuclides in the uranium and thorium decay series [25]. For drinking water radioactivity total indicative dose TID (effective dose from radionuclides in drinking water except ^{3}H, ^{40}K, radon and radon progenies) is 0.1 mSv year^{-1} [27]. The guidance levels for radionuclides in drinking-water are presented in [28] for radionuclides originating from natural sources or discharged into the environment as the result of current or past activities.

The measurement of radon in water is relatively simple compared to air measurement. The rate and magnitude of variations are much lower, and there are fewer sampling problems. But the measurement of radon in water has its own set of sampling issues and analytical difficulties. All methods require correction for decay of radon during the decay between sampling and analysis and this time should be the smallest possible.

Methods for radon measurement in water can be divided in these three categories [2, 7]:

- liquid scintillation counters
- gas extraction
- direct gamma counting

Usually in water there are higher concentrations of radon, comparing to the concentrations of radon in air. If it is expected to have radon concentration in water more than 1000 pCi l^{-1} (around 37 Bq l^{-1}), which is often the case for well water, than direct liquid scintillation counting should be applied as a rapid and practical method. The water sample can be mixed with the counting material - scintillator and counted by the conventional liquid scintillation counters used for radon in air samples.

A more sensitive method for detecting radon in water, which is suitable for lower concentrations, is to extract the radon as a gas and count the emitted alpha particles in a ZnS scintillation cell. Helium is bubbled through the water, striping the radon. The mixture of gases is then passed through a cold trap, for example activated charcoal at liquid nitrogen temperature that traps radon while the helium passes through. The trap is then warmed and radon is transferred into a Lucas counting cell by stripping with a small amount of helium.

For relatively high activity concentrations of radon in water (more than 500 pCi l⁻¹), there is a possibility to measure radon by counting gamma rays from radon daughter decay by using a standard gamma spectroscopy techniques. The original radon concentration can be distinguished from the radium-226 concentration by repeating the count after 30 days, at which time the original radon will have virtually all decayed and the only remaining radon is that in secular equilibrium with radium-226.

A number of factors affect the accuracy and precision of a radon in water measurement. Most critical among these factors is the sampling technique. Other factors include the sample concentration, sample size, counting time, temperature, relative humidity and background effects [26].

The sampling method is the main source of error in measurements of radon in water. Sampled water must be representative of the water being tested and it should be provided that this water has not been in contact with air at all. This task is not easy to fulfill. To provide this one should put up a bowl to the faucet so that the water overflowing the bowl prevents the water to be in touch with air when leaving the faucet. The vial should be filled with water at the bottom of the bowl. The minimal contact of water with air, the better and more accurate results can be obtained.

In this section two possible techniques (instruments) for measurement of activity concentration of radon in water will be explained.

One of the mostly used methods for radon in water measurement is RAD7 instrument with special RAD-H$_2$O accessory suitable for these measurements (Figure 5). The lower limit of detection (LLD), proposed by the manufacturer is less than 0.37 Bq l⁻¹ [29]. In order to determine the background the best is to use distilled water sample which stands closed and undisturbed for more than four weeks, this sample can be assumed as a radon-free sample. Usually background sample has to be measured several times using the same protocol in order to determine minimal detectable activity (MDA). In some papers [1, 26] this value is determined to be 0.1 Bq l⁻¹. According to the manufacturer [29], it is possible to achieve this low activity value of background sample if completely eliminate all radon and its progeny before measurement. This is very important fact in order to have reliable measurements. The best way for collecting the samples is the one proposed by the manufacturer [29]. A 250 ml vial and Wat-250 protocol was used if it is expected that radon concentrations in water are less than 100 Bq l⁻¹. If the radon concentration is higher than 100 Bq l⁻¹, a 40 ml vial and Wat-40 protocol was used. For radon in water measurements by RAD7 instrument it is important to choose appropriate protocol (according to the used vial) and Grab mode. It is also very important that before measurement RAD7 must be free of radon and dry. In order to obtain this condition it is necessary to provide around 10 min of purging with adequate drying unit. It is important to reach relative humidity be under the 6% in the instrument and that assures that it will stay under 10% during whole measurement process. This is the only way to deal with humidity which greatly affects accuracy of the measurement results. The RAD-H$_2$O requires that desiccant be used all the time to dry the air stream before enters the instrument. When measuring water small desicant is usually used, as it is shown on Figure 5. Another very important thing is that a measuring vial containing a water sample is set up in a closed air loop with the RAD7. The RAD-H$_2$O method employs a closed loop aeration scheme whereby the air volume and water volume are constant and independent of the flow rate. The air circulates again and again through the water and continuously extracts the radon until a state of equilibrium develops. This system reaches the state of equilibrium within about 5 min, after which no more radon can be extracted from the water. The RAD7 pump operates automatically for 5 min to aerate the sample, distributing the radon that was in the water throughout the loop. During the 5 min of aeration more than 95% of the available radon is removed from the water. The RAD7 instrument waits a further 5 min while the ^{218}Po count rate approaches equilibrium and then counts for four 5 min cycles. The number and duration of cycles can be changed according to measurement conditions, those mentioned are the ones proposed by manufacturer of RAD7 instrument. The radon concentration in the water is calculated directly and displayed on the monitor of the instrument.

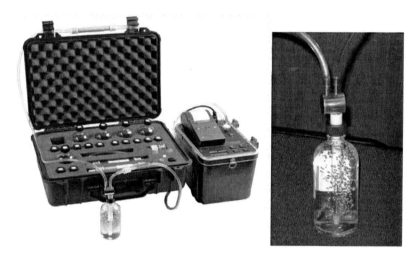

Figure 5. RAD7 – H$_2$O, radon in water measurement.

A relative humidity showed the greatest impact on measurement error in the measurement results. For accurate readings, the RAD7 should be dried out thoroughly before making the measurement. If the RAD7 is thoroughly dried out before use, the relative humidity inside the instrument will stay below 10% for the entire 30 min of the measurement. If not, then the humidity will rise during the 25 min that the RAD7 is counting and the pump is stopped, and may rise above 10% before the end of the measurement period. High humidity reduces the efficiency of collection of the polonium-218 atoms, formed when radon decays inside the chamber. However, the 3.05 min half-life of ^{218}Po means that almost all the decays that are actually counted come from atoms deposited in the first 20 min of measurement. So a rise in humidity above 10% over the last 10 min of the counting period will not have a significant effect on the accuracy of the result. On the other hand, if the humidity rises above 10% before the end of the first counting cycle, there will be an error whose size is indeterminate [26, 29]. The most significant background effects in the RAD-H$_2$O are counts from radon daughters and traces of radon left from previous measurements. The RAD7 has the unusual ability to tell the difference between the "new" radon daughters and the "old" radon daughters left from previous tests. Even so, a very high activity of radon sample can cause daughter activity that can affect the next measurement [29].

The liquid scintillation counting (LSC) technique (Figure 6) is also suitable for radon in water measurements. This technique allows the processing of a large number of environmental samples with minimal handling and has the advantage that samples are collected in situ, thereby avoiding radon loss by emanation [21, 22]. Liquid scintillation counter equipped with an alpha–beta discriminator allows a rapid and simple determination of ^{222}Rn [30].

Figure 6. LSC Quantulus, radon in water measurement.

According to Environmental Protection Agency (EPA) [31], samples should be collected from a non-aerated faucet of spigot to minimize desorption of radon: a 1 l glass beaker have to be filled with water from the source, carefully allowing to gently overflow, and the sample container has to be opened and submerged inverted into a beaker. After that, the sample container should be rotated so that it was filled with water. The Teflon-lined cap should be applied with the sample container still under water to eliminate headspace. Analysis should begin within 3 days after receipt of sample in the laboratory.

For the determination of radon in drinking water from groundwater and surface water sources by Liquid Scintillation Counting, EPA Method 913.0 [31] can be used. For calibration and standardization, the Radium Solution Method could be applied, where standard of 100 ml of ^{226}Ra in water should be prepared such that we get the final activity to be ~1.3 kBq l^{-1}. According to this single-phase method, 10 ml of the diluted standard should be transferred into a 20-ml scintillation vial to which had been added 10 ml of miscible scintillation cocktail OptiPhaseHiSafe 3 [21]. There are some other scintillation cocktails also applicable for this method; this one is taken for example. The background samples should also be prepared using 10-ml distilled water. The standards and background samples should be prepared and set aside for 30 days to allow radon to attain secular equilibrium. Radon will diffuse from the sample into scintillaton cocktail, for which it has a much greater affinity than for water. After this long period of time, the samples can be counted in LS counter (in this case Perkin Elmer LSC Quantulus 1220, Figure 6, was used) for 50 min using energy discrimination for alpha/beta particles. In the study presented in reference [21] a standard radioactive source ^{226}Ra activity produced from Czech Metrology Institute, Inspectorate for Ionizing Radiation, was used for the calibration of the instrument, $c(^{226}Ra) = 1.0844$ ng ml^{-1}. All LSC samples (10 ml of sample + 10 ml of OptiPhaseHiSafe 3 scintillation cocktail) were prepared in 20-ml high-performance glass vials (Perkin Elmer) and measured on the LS counter Quantulus. The spectra were acquired by WinQ and analysed by Easy View software by Perkin Elmer. LSC Quantulus has its own background reduction system around the vial chamber, which consists of both passive shield (lead, copper and cadmium) and active shield (mineral oil scintillator). Low-activity materials were used in the construction of the Quantulus, and they are useful for measuring low-level radiation activity. The Quantulus 1220 has two multichannel analysers (MCA), one is used for active shield and the second one is used for spectra record [33]. The system is provided with two pulse analysis circuits that are accessible for the users: a pulse shape analysis (PSA) and pulse amplitude comparator circuit. PSA discriminates alpha- from beta-radiations and directs them separately into alpha-MCA or beta-MCA. In the mentioned study [22] the PSA parameter influence was investigated with the set of ^{226}Ra standards left for 1 month after preparation to reach radioactive equilibrium. It was determined that PSA parameter is not necessary to set at the crossover point (e.g., the least alpha/beta misclassification) [33] before the sample measurement; PSA variation impacts the CF (Calibration Factor) value which then impacts calculated ^{222}Rn activity of sample. The most important is to keep PSA parameter fixed during CF determination and sample counting, in which case PSA value itself does not influence ^{222}Rn determination. In this investigation it was noticed that ^{226}Ra standard activity does not change significantly CF determination during calibration. So, it can be concluded that the most important thing for radon measuring by LSC method is to do the best possible calibration which among others means to determine calibration factor correctly and to keep PSA value constant, as it is explained in detail in reference [21]. Another, also very important part of LSC measurements is setting an optimal ROI (region of interest) or window selection. For this purpose, standard source should be used, in the case of radon measurement it should be radium standard which is measured for a short time (5 minutes is enough if the standard source is with high enough activity concentration). In the mentioned study, the region of greatest alpha activity defined by two large peaks in the energy spectrum, extended by 10 % on each side, forms the optimal window [31], and it was fixed between channels 420 and 780. The ^{226}Ra content of the water samples has to be determined from a cumulative sample by some other additional method, for example, gammaspectrometry measurements.

Ultra-low-level LS counting coupled to alpha–beta discrimination allows rapid and simple simultaneous determination of gross alpha- and beta activities, and therefore, it can be applied to the determination of ^{222}Rn activity in waters. Radon diffusion from the sample into scintillator cocktail for which it has a much greater affinity than for water ensures good efficiency collection. With this method an MDA of 0.029 Bq l^{-1} in 20 ml of sample has been achieved in glass vials during measurement time of 300 min [21].

CONCLUSION

Radon has been recognized as a radiation hazard causing excess lung cancer among underground miners [34]. Since the 1970s evidence has been increasing that radon can also represent a health hazard in non-mining environments [35]. In Europe the annual effective dose from all sources of radiation in the environment is estimated to 3.3 mSv with indoor doses from radon, thoron and their short lived progenies accounting to 1.6 mSv at this value [36]. The studies on indoor levels in the EU indicated that radon is a cause of about 20,000 lung cancer deaths each year; that is 9% of the total lung cancer deaths and about 2% of cancer deaths in the EU overall [37].

Determination of radon-222 activity concentration in air and water samples is really important. Many techniques have been developed over the years for measuring radon and radon progenies, because of its hazard effects on human health.

For indoor radon exposure, according to recommendations from the International Commission on Radiological Protection Statement on Radon, from November 2009, the upper level for radon gas in dwellings is 300 Bq m^{-3}. Taking account of differences in the lengths of time spent in homes and workplaces of about a factor of three, a level of radon gas of around 1000 Bq m^{-3} defines the entry point for applying occupational protection for existing exposure situations at workplaces [38].

Radon is soluble in water, and its origin in water is the radium from water surrounding soil and bedrock. The release of radon from rocks and soils is controlled largely by the types of mineral in which uranium and radium occur [39]. The special importance is to measure radon in drinking water [25], because the humans are exposed to radon in drinking water both by ingestion and inhalation. Special intention should be devoted to the control of the deep wells which are used as drinking water supplies. The European Commission recommends for ^{222}Rn in drinking water, that a reference level should be appointed above an activity concentration of 100 Bq l^{-1}, and with radon activity concentrations above 1000 Bq l^{-1} measures are justified [27].

There is a wide spectrum of techniques which could be used for radon measurement, both in air and in water samples. This section presents the general methods and principals used in detection of radon and its progeny in measured samples. In this chapter just a few techniques were explained in detail. For radon in air measurement five different commercially available instruments were presented (AlphaGUARD, RAD7, passive detectors, CR-39 and charcoal canisters). All of the mentioned methods have its advantages and disadvantages, some of them are fast and can be easily be distributed on a wide territory in order to do the best possible radon mapping (passive detectors, charcoal canisters....), and others are more accurate and precise.

For radon in water, two methods were presented, RAD7 H$_2$O and LSC method. RAD7 instrument is portable and can be easily used on the field, while Ultra-low-level LS counting coupled to alpha–beta discrimination allows rapid and simple simultaneous determination of gross alpha- and beta activities, and therefore, it can be applied to the determination of ^{222}Rn activity in waters [21], but the instrument calibration and sample preparation is more complex.

FUNDING

The authors acknowledge the financial support of the Ministry of Education and Science of Serbia, within the projects No.171002, No.43002, No. 43007 and the Provincial Secretariat for Science and Technology Development within the project Development and application low-background alpha, beta spectroscopy for investigating of radionuclides in the nature.

REFERENCES

[1] Nikolov, J., Todorovic, N., PetrovicPantic, T., Forkapic, S., Mrdja, D., Bikit, I., Krmar, M. and Veskovic, M. Exposure to radon in the radon-spa NiskaBanja, Serbia. *Radiat. Meas.* 47, 443–450 (2012).

[2] Bodansky D., Robkin MA., Stadler DR., Indoor Radon and Its Hazards, University of Washington Press, USA (1987).

[3] BEIR VI Report, published in 1999.

[4] Nikolov J., Forkapic S., Hansman J., Bikit I., Veskovic M., Todorovic N., Mrdja D., Bikit K., Natural radioactivity around former uranium mine, Gabrovnica in Eastern Serbia, *J. RadioanalNucl. Chem.* (2014) 302:477–482.

[5] http://www.radon-analytics.com/index_e.php?show=measuring_devices_overview.

[6] EC, 1997: Radiation Protection 88. Recommendations for implementation of Title VII of the European Basic Safety Standards concerning significant increase in exposure due to natural radiation sources. European Commission. Office for Official Publications of the European Commission. Radiation Protection Series, 4. Radon Legislation and National Guidelines. SSI report. Swedish Radiation Protection Institute. No 99: July 1999. ISSN 0282-44344.

[7] C. Richard Cothern. James E. Smith, Jr "Environmental Radon." Plenum Press, New York (1987).

[8] Vaupotic, J., Gregoric, A., Kobal, I.,Zvab, P., Kozak, K., Mazur, J., Kochowska, E. and Grzadziel, D. Radon concentration in soil gas and radon exhalation rate at the Ravne Fault in NW Slovenia. *Nat. Hazards Earth Syst. Sci.* 10, 895–899 (2010).

[9] Y. Yasuoka, A. Sorimachi, T. Ishikawa, M. Hosoda, S. Tokonami, N. Fukuhori and M. Janik, SEPARATELY MEASURING RADON AND THORON CONCENTRATIONS EXHALED FROM SOIL USING ALPHAGUARD AND LIQUID SCINTILLATION COUNTER METHODS, *Radiation Protection Dosimetry* (2010), pp. 1–4.

[10] RAD7 RADON DETECTOR, User Manual, Revision 7.2.8. © 2014 DURRIDGE Company.

[11] EPA, 1992. Indoor Radon and Radon Decay Product Measurement Device Protocols, Office of Air and Radiation. *U.S. EPA Publication.* 402-R-92-004.

[12] Zunic, Z.S., Janik, M., Tokonami, S., Veselinovic, N., Yarmoshenko, I.V., Zhukovsky, M., Ishikawa, T., Ramola, R.C., Ciotoli, G., Jovanovic, P., Kozak, K., Mazur, J., Celikovic, I., Ujic, P., Onishchenko, A., Sahoo, S.K., Bochicchio, F., 2009. Field experience with soil gas mapping using Japanese passive Radon/Thoron discriminative detectors for comparing high and low radiation areas in Serbia (Balkan region). *Journal of Radiation Research* 50, 355-361.

[13] Zhuo, W., Tokonami, S., Yonehara, H. and Yamada, Y. (2002) A simple passive monitor for integrating measurements of indoor thoron concentrations. *Rev. Sci. Instrum.* 73: 2877–2881.

[14] Tokonami, S., Takahashi, H., Kobayashi, Y. and Zhuo, W. (2005) Up-to-date radon-thoron discriminative detector for a large scale survey. Rev. Sci. Intrum. 76: 113505.

[15] EPA (1987) EERF standard operating procedures for Rn-222 measurement using charcoal canisters, U.S. EPA Publication, 520/5-87-005.

[16] Todorovic, N., Forkapic, S., Bikit, I., Mrdja, D., Veskovic, M., Todorovic, S., 2011., Monitoring for exposures to Tenorm sources in Vojvodina region. *Radiation Protection Dosimetry* 144 (1-4), 655-658.

[17] S. Forkapić, I. Bikit, Lj. Čonkić, M. Vesković, J. Slivka, M. Krmar, N. Žikić-Todorović, E. Varga, D. Mrđa, METHODS OF RADON MEASUREMENT, FACTA UNIVERSITATIS Series: *Physics, Chemistry and Technology* Vol. 4, No 1, 2006, pp. 1 – 10.

[18] I.S.Bikit, D.S.Mrdja, I.V.Anicin , J.M.Slivka , J.J.Hansman , N.M.Zikic-Todorovic, E.Z.Varga , S.M.Curcic , J.M.Puzovic: THE FIRST RADON MAP OF VOJVODINA, IRPA 11, Madrid, Spanija: 23-28 Maj, 2004, str. 6a11.

[19] Zunic, Z.S., Yarmoshenko, I.V., Birovljev, A., Bochicchio, F., Quarto, M., Obryk, B., Paszkowski, M., Celikovic, I., Demajo, A., Ujic, P., Budzanowski, M., Olko, P., McLaughlin, J.P., Waligorski, M.P.R., 2007a. Radon survey in the high natural radiation region of NiskaBanja, Serbia. *Journal of Environmental Radioactivity* 92, 165-174.

[20] Zunic, Z.S., Kozak, K., Ciotoli, G., Ramola, R.C., Kochowska, E., Ujic, P., Celikovic, I., Mazur, J., Janik, M., Demajo, A., Birovljev, A., Bochicchio, F., Yarmoshenko, I.V., Kryeziu, D., Olko, P., 2007b. A campaign of discrete radon concentration measurements in soil in NiskaBanja town, Serbia. *Radiation Measurements* 42, 1696-1702.

[21] Natasa Todorovic, IvanaJakonic, JovanaNikolov, Jan Hansman and Miroslav VeskovicEstablishment of a method for [222]Rn determination in water by low-level liquid scinntilation counter, *Radiation Protection Dosimetry* (2014) doi:10.1093/rpd/ncu240.

[22] Galan Lopez, M., Martin Sanchez, A. and Gomez Escobar, V., Application of ultra-low level liquid scintillation to the determination of [222]Rn in groundwater, *J. Radioanal. Nucl. Ch.* 261(3), 631–636 (2004).

[23] Khalid Abdulaziz, A., SayeedAlghamdi, A., Fahad Ibrahim, A. and Md. Shafiqul, I., Measurement of radon levels in groundwater supplies of Riyad with liquid scintillation counter and the associated radiation dose, Radiat. Prot. Dosim. 154(1), 95–103 (2013).

[24] Kendall, G.M., Smith, T.J., 2002., Doses to organs and tissues from radon and its decay products, *Journal of Radiological Protection* 22, 389-406.

[25] WHO (World Health Organization), 2008. *Guideline for Drinking-water Quality*, Third edition, Vol. 1, Geneva, p. 515.

[26] Todorovic N., Nikolov J., Forkapic S., Bikit I., Mrdja D., Krmar M., Veskovic M., Public exposure to radon in drinking water in SERBIA, *Applied Radiation and Isotopes* 70 (2012) 543–549.

[27] EUROPEAN COMMISSION, 2001. Commission recommendation of 20th December 2001 on the protection of the public against exposure to radon in drinking water, 2001/982/Euratom, L344/85.

[28] WORLD HEALTH ORGANISATION (WHO), 2004. third ed. Guidelines for Drinking Water Quality, vol. 1. *World Health Organisation*, Geneva.

[29] RAD H_2O User Manual, © 2014 DURRIDGE Company, Revision 2014-02-19.

[30] Jowzaee, S. Determination of selected natural radionuclide concentration in the southwestern Caspian groundwater using liquid scintillation counting. *Radiat. Prot. Dosim.* 1–8 (2013).

[31] EPA Method 913.0, Determination of radon in drinking water by liquid scintillation counting, Radioanalysis Branch, Nuclear Radiation Assessment Division, Environmental Monitoring Systems Laboratory, U.S. Environmental Protection Agency, 89119.

[32] Quantulus 1220, 2002, Instrument Manual, Ultra Low Level Liquid Scintillation Spectrometer, PerkinElmer, 1220-931-06.

[33] Todorovic, N., Nikolov, J., Tenjovic, B., Bikit, I. and Veskovic, M, Establishment of a method for measurement of gross alpha/beta activities in water from Vojvodina region, *Radiat. Meas.* 47, 1053–1059 (2012).

[34] ICRP, 1981. International Commission on Radiological Protection, Limits on Inhalation of Radon Daughters by Workers. Publication 32. Pergamon Press, Oxford.

[35] IAEA, 2003. Radiation Protection against Radon in Workplaces Other than Mines, Jointly Sponsored by IAEA, ILO. International Atomic Energy Agency, Vienna. Safety Reports Series No.33.

[36] BEIR IV, 1988. Health Risks of Radon and Other Internally Deposited Alpha-Emitters. National Academy Press, Washington, DC. 20055.

[37] Manic, G., Petrovic, S., Manic, V., Popovic, D., Todorovic, D., 2006, Radon concentrations in a spa in Serbia, *Environment International* 32, 533-537.

[38] ICRP (2007) The 2007 recommendations of the international commission on radiological protection, vol. 37, ICRP Publication No. 103, Ann. ICRP, pp 2–4.

[39] Appleton, J.D. 2005, Radon in air and water. In: Essentials of Medical Geology: *Impacts of the Natural Environment on Public Health*. Selinus, O. (ed). Elsevier Amsterdam, 227-262.

In: Radon
Editor: Audrey M. Stacks

ISBN: 978-1-63463-742-8
© 2015 Nova Science Publishers, Inc.

Chapter 11

RADON BUILDUP IN DWELLINGS, SPAS AND CAVES: FACTS AND INTERPRETATIONS

I. Bikit, S. Forkapic, D. Mrdja, K. Bikit[*],
N. Todorovic and J. Nikolov

University of Novi Sad, Faculty of Sciences, Department of Physics, Novi Sad, Serbia

ABSTRACT

Radon as a natural radioactive gas, the daughter nucleus of the long lived ^{238}U is present in all parts of nature. Although its concentration in open air is very low and contributes negligibly to health risk, underground or in closed spaces radon might be a serious health risk problem.

This was emphasized a long time ago, and a series of radon mapping measurements and legislatives have been established in the meantime.

Despite the fact that radon levels in dwellings are usually limited to about 300 Bq/m^3, lot of spas use much higher radon levels (about 10,000 Bq/m^3) for medical purposes. A lot of experimental techniques and methods have been adopted for radon measurements. For long time measurements (about 6 months) usually solid-state track detectors are used. Alpha spectroscopy combined with radon samplers is a method for determination of temporary radon concentrations in air, water and soil gas. Frequently, activated charcoal can be exploited for radon sampling from the air, followed by gamma spectroscopy determination of radon concentration.

In the chapter the experimental results of the Novi Sad Nuclear Physics Group are presented and compared with worldwide published results. The associated health effects are estimated and discussed.

1. INTRODUCTION

Radon is naturally occuring radioactive gas. It is colorless and odorless noble gas, 7,5 times weighted than air. Uranium ^{238}U decays through several different isotopes. When it

---

[*] Corresponding author: Istvan Bikit, e-mail: bikit@df.uns.ac.rs.

reaches radium ^{226}Ra, it decays to radon ^{222}Rn and undergoes a change to state to a gas. Radon partially decays in material where it has been generated and partially moves rapidly by concentration-driven diffusion into the open air and houses. It dissolves in water and can readily diffuse with gases and water vapor, and thus build up significant concentrations. Physical half-life of radon is 3.825 days and half-elimination time from lungs 30 min. Radon ^{222}Rn, which is the daughter of uranium ^{238}U, represents most important radon isotope. Decay of the radon nuclei ^{222}Rn yields short-living daughters: polonium ^{218}Po, lead ^{214}Pb and bismuth ^{214}Bi.

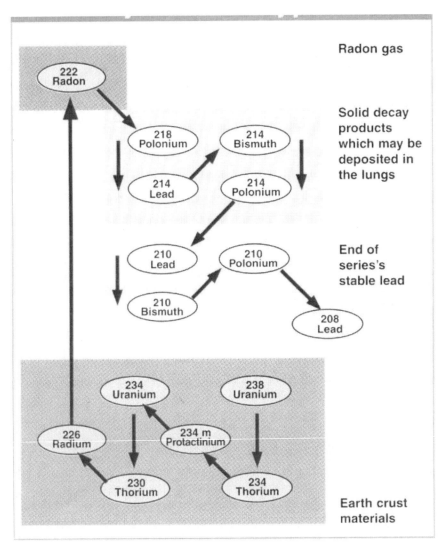

Figure 1. Uranium ^{238}U decay chain [1].

Sources of indoor radon are: soil under the object, building's materials (BM) and water and gas used in household. Indoor radon concentration depends on various meteorological and geological factors and also surface area and isolation quality of structures in the contact with soil, building floor, air ventilation and indoor-outdoor temperature and pressure differences. Meteorological factors modifying radon concentrations are: atmospheric pressure, temperature, humidity, precipitation, wind speed and direction. Geological factors that have an impact on radon emanation from soil are: radium and radon content of the soil, soil characteristics (density, porosity and soil granulation), tectonic constraints, earth-quakes, land-slides, volcanoes and water table level.

BM causing high concentration has been used in several countries. In some cases these are materials of natural origin (i.e., granite or alum shale concrete), and in other cases they are by-products from different industries (by-product gypsum, waste rock from mining etc.). Republic of Serbia has advisory reference levels for internal BM [2]: ^{226}Ra< 200 Bq/kg; ^{232}Th<300 Bq/kg; ^{40}K<3000 Bq/kg; all anthropogenic radionuclides< 4000 Bq/kg. Maximum permitted annual indoor dose due to gamma radiation from BM is 1 mSv, which corresponds to the following activity index:

$$^{226}Ra/200 + {}^{232}Th/300 + {}^{40}K/3000 + \text{all anthropogenic radionuclides}/4000 < 1 \qquad (1)$$

Researches carried out in the recent decades have shown that, under normal conditions, more than 70% of total annual radioactive dose received by people originated from natural sources of ionizing radiation, whereby 40% is due to inhalation and ingestion of natural radioactive gas radon ^{222}Rn and its decay products [3] (Fig. 2). Exposure to radon via inhalation in closed rooms is the cause of about 10% of all death cases from lung cancer. Changes at the cellular and molecular levels are significantly more pronounced in early stages of life [4].

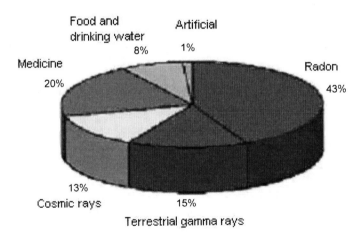

Figure 2. UNSCEAR Report 1982: Sources of radiation exposure [3].

The base of developing the public indoor radon politics is a definition of risks connected with the exposition of population to radon and its daughters in the living rooms. The actions that are conducted in order to reduce radon activity concentrations in flats and offices suppose the establishment of national strategies and recommendations or directives about maximal permitted radon concentrations.

2. Radiation Health Risk and Alara Principle

According to the recommendations of the International Commission on Radiological Protection (ICRP) from 1994, the civil engineering standard for building new apartments and houses is 100 Bq/m^3 of radon as an average annual level, 200 Bq/m^3 for the recommendation of inexpensive remediation measures, and 600 Bq/m^3 for the recommendation of expensive remediation measures. Namely, even in the case of observing these recommendations and norms, ALARA (As Low As Reasonably Achievable) principle applies [5, 6]. Hence, if the absolute abundance in parts of the lithosphere or geochemical processes, regionally or locally, concentrates natural radionuclides, then higher ionization levels are to be unavoidably tolerated.

According to the Rulebook of the ionising radiation exposure limits and measurement to assess the level of exposure to ionizing radiation ("Official Gazette of Republic of Serbia'', No.86, 2011) [7], intervention levels for exposure to radon in apartments are equal to an annual average concentration of 200 Bq/m^3 ^{222}Rn-222 in the air in newly built apartments and 400 Bq/m^3 ^{222}Rn in the air in existing residential properties. Intervention level for exposure to radon in a working place are equal to an annual average concentration of 1000 Bq/m^3 ^{222}Rn in the air.

The national radon program should aim to reduce both the risk for the overall population exposed to an average radon concentration and the risk of individuals living with high radon concentrations. The development of a national radon program involves the setting-up of a clear organizational structure and a range of components in order to monitor radon levels, facilitate prevention and migration, and provide radon risk communication services to the public. The crucial stage in the every national radon program is production of the national radon survey and radon risk map. There are two key objectives in the design of a national radon survey:

- To estimate the average exposure of the population to indoor radon and the distribution of the exposures occurring. This can be done through a population-weighted survey by measuring indoor radon levels in randomly selected homes,
- To identify those areas within the country where high indoor radon concentrations are more likely to be found. This can be achieved with a geographically based survey [8].

Based on the data obtained from the national radon surveys, it is possible to set a national reference level as low as reasonably achievable (the chosen reference level should not exceed 300 Bq/m^3 which represents approximately 10 mSv per year according to recent calculations by the International Commission on Radiological Protection, ICRP) and Council Directive 2013/59/Euratom of 5 December 2013 laying down basic safety standards for protection[9].

3. Measurement Method

Radon, the natural radioactive gas which can be accumulated in indoor air to significant concentrations, gives the great contribution to annual effective dose of population in the world. The internal exposure is caused by the inhalation of radon (^{222}Rn), thoron (^{220}Rn) and their short lived decay products. The most important source of indoor radon is the underlying soil, so the enhanced levels of radon are usually expected in mountain regions where the rocks contain high concentrations of uranium ^{238}U radon radioactive parent. However, radon levels in plain areas might be also elevated due to soil porosity.

The results of indoor radon survey in Serbian Northern Province named Vojvodina are presented in this chapter. Vojvodina lies on the south part of Panonian low-land (Fig.3) without elevate uranium levels in the surface soil. However the presence of numerous underground hot spring and sources of natural gas, as well as some crude oil reservoirs, point to the possibility of elevated radon levels. In order to examine this problems the series of measurements on building materials, soil, radon emanation from soil and radon buildup in houses and flats in Novi Sad, the capital city of Vojvodina Province, were performed.

Figure 3. The map of Pannonian Basin with country borders and marked position of Vojvodina – northern province of Serbia.

Since 1992, Nuclear Physics Group of Faculty of Sciences in Novi Sad carried out radon activity concentration measurements in closed room as well as radon emanated from soil, using the method of adsorption on activated carbon. To the present, almost 400 measurements have been carried out with charcoal canisters (Fig. 4). Calibration of detection efficiency was performed using EPA radium standard. Concentrations of radon activity were determined on the basis of the intensity of short-living radon daughters ^{214}Bi and ^{214}Pb gamma-lines. Gamma-spectrometric measurements were carried out using a high-resolution HPGe detector nominal efficiency of 22%, placed in a shield chamber with 25-cm thick iron walls and NaI(Tl) scintillation spectrometer. The canisters were exposed for two days. The typical time between the end of the exposition and the beginning of the measurement was about two hours. In order to achieve 5% statistical accuracy at 100 Bq/m^3 the time of measurement was usually 1 hour. In our measuring chamber the radon levels are very low (less than 5Bq/m^3) so radon fluctuations could not affect the results of most measurements.

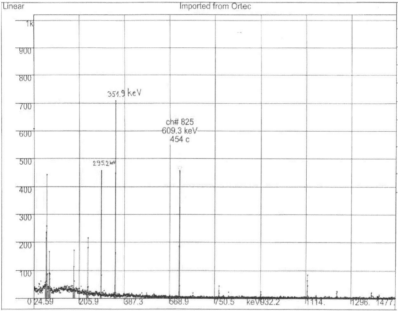

Figure 4. Picture of charcoal canisters F&J Speciality Products, USA according to EPA standards – left; gama spectrum of charcoal canister with marked post radon gamma lines – right.

In real conditions the charcoal power of adsorption depends on the time of exposure and must be corrected to moisture. For each series of charcoal canisters manufacturer calibrates a number of canisters (5 canisters for different exposure times 1, 2, 3, 4, 5, and 6 days) in the radon chamber on stabilized conditions. After the time of exposure the adsorbed radon concentrations in sealed canisters are determined by gamma-spectrometric method and calibration factors CF [l/min] (power of adsorption) calculate according the formula:

$$CF = \frac{I - I_F}{T_S \, E \, RN \, DF} \qquad (2)$$

where CF is calibration factor (power of adsorption in [l/min] or [m^3/ks]), I is total intensity in post radon gamma lines [C/ks], IF is background spectral intensity [C/ks], Ts is time of exposure [ks], E is detection efficiency [C/(ks Bq)], RN is radon concentration [Bq/m^3] and DF is decay correction ($DF = e^{-\frac{0.693\,t}{T_{1/2}(Rn)}}$; t is time in days from the midpoint of exposure to the time of measurement and T$_{1/2(Rn)}$ – radon half-life (3.824 days)).

Based on data obtained from manufacturer for the used series of canisters curves of humidity for different times of exposures and different water gain are plotted in Table Curve software packages and Mathematica in order to determine the best mathematical fit of these functions (Fig. 5).

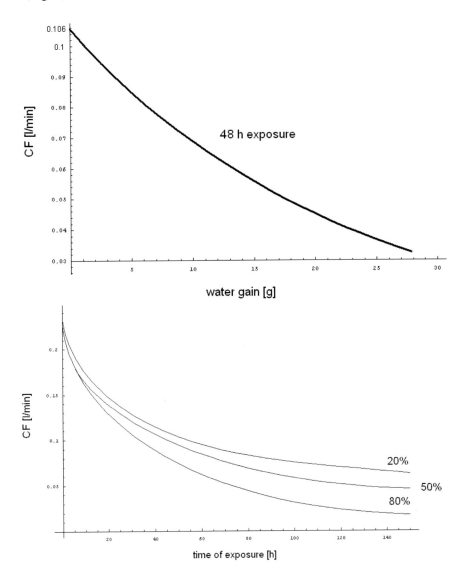

Figure 5. Humidity curves for: 48 h of exposure – up; and different times of exposure and differrent water gains – down.

Radon emanation from soil was measured using steel probe dug in the earth to a depth of 70 cm, with the upper end closed and a metal grid on which charcoal canister was placed at the bottom end (Fig. 6). Charcoal canisters exposure lasted in average two days. Since soil represents the dominant radon source, soil samples taken from the location of radon emanation, after the appropriate preparation, were subjected to measurement of radionuclides concentration by gamma-spectrometric method.

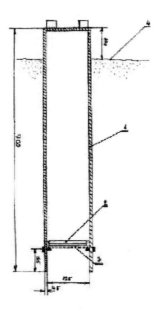

Figure 6. Schema of the probe for radon emanation measurement: 1 - steel tube, 2 – charcoal canister, 3 - metal net, 4 – soil surface layer.

Soil samples with a mass of about 300 g were dried up at 105°C, homogenized and sealed in cylindrical containers. The measurements where performed after more than 40 days after the sealing, in order to bring the ^{222}Rn daughters in equilibrium. Activity concentrations of radionuclides were determined by gamma-spectrometry method [10, 11] on two ultra low level gamma spectrometers. The first detector is HpGe gamma-spectrometer GMX type manufactured by Canberra model GC2520-7600. Detector operated in a 25 cm thick iron shield and has relative efficiency of 25% with resolution of 1.8 keV. The second detector is HPGe coaxial end cup detector produced by CANBERRA has nominal efficiency of 36% and resolution of 1.79 keV. The detector was operated inside the 12 cm thick lead shield with 3 mm Cu inner layer. Data acquisition and on-line spectral analysis in the form Canberra *.CNF were performed by Canberra Genie2000, version 2.1 licensed software.

The second method applied in radon emanation investigations is RAD7 alpha-spectrometer (DURRIDGE Company) grab protocol with stainless soil gas probe (Fig.7). While pumping, the air flow rate is about 0.7 l/min and therefore 3.5 l of soil gas is extracted from the soil from the depth of 70 cm.

Figure 7. RAD7 (Durridge Co.) continuous monitor equipment for radon emanation measurements.

In order to construct the first radon map of Vojvodina, radon activity concentrations have been measured using CR39 solid track detectors on about 1000 different locations in all 45 municipalities in the period December 2002 - March 2003. The main aim of this radon survey was to determine the maximal annual effective dose due to exposure to radon and to highlight the regions with elevated levels of indoor radon. Based on previous indoor radon measurements, target sampling locations for the first year of measurement were chosen to be old adobe houses with no concrete construction of the floor or poorly isolated concrete constructions. These dwellings are also representative for villages and suburban districts in the province. The time of exposure was 90 days during the winter season, from December 2002 to March 2003. The number of detectors per municipality was determined proportional to the number of citizens in the municipality. To cover all municipalities, the detectors were distributed to local physics teachers in high schools, who further distributed them among their pupils. The teachers were trained in our laboratory about sampling procedure in order to explain their pupils how to place and expose the diffusion chambers with detectors and to supervise them. The questionnaires were given to pupils to fulfill in order to obtain the following data for each site:

- Site: district, municipality, village, address;
- Type of room, floor level, location of detector in the room;
- Date of detector deployment and collection;
- Detector codes;
- Construction conditions, age of construction,
- Heating mode and air ventilation,
- Presence or absence of a basement and
- Other parameters that influence the radon levels (observations, remarks).

Teachers recommended pupils to place CR39 diffusion chambers in living rooms and bedrooms where they spend a lot of time, on the ground floor, 1 m distance from the walls, not near to sources of water, windows or heat sources. Measurements within the same scope were repeated, also in the winter season of 2004 and 2005.

CR39 radon detectors represent plastic films with an area of 1 cm^2 and 1 cm thick, sensitive to traces of ionizing alpha particles [12]. During the exposure, these detectors were glued to the cover of a closed diffusion plastic chamber, with 5 cm in height (Fig. 8). The detector is sensitive to alpha radiation only, and its sensitivity equals to 2.9 traces/(cm^3 kBqh/m^3). After the exposure detectors were etched in 25% solution of NaOH at a constant temperature of 90^{0}C during 4 hours. The traces were read and treated by RADOSYS 2000 electronic equipment (RADOBATH 2000 (thermo stated bath for chemical etching) and RADOMETER 2000 with a B&W CCD camera and a compatible PC) in the Radosys Company, Hungary.

The exposure chamber with the CR-39 component inside

The detector holder accessory with 12 detectors

The alpha track development unit RADOBATH

The RADOMETER track counting unit

Alpha tracks on the display of the microscope

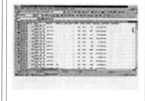
The result sheet of the track evaluation

Figure 8. CR39 detectors: a) plastic diffusion chamber with detector pasted on the inner side of the cover; b) holder frame with 12 CR39 detectors; c) bath for simultaneous etching of 432 detectors; d) RadoMeter 2000 unit; e) visualization of tracks and f) software for radon activity concentrations assessment.

4. RESULTS

Results of measuring activity concentration of radon in closed rooms that have been carried out on about 400 different loacations in the area of Novi Sad municipality in the period from 1992 to the present are shown in a tabular form (mean, minimal, and maximal measured radon concentration) in Table 1 and as the ln-normal distribution in Fig. 9. Only 1% of measurements is above the intervention level for existing dwellings of 400 Bq/m^3 and 5 % of indoor radon measurements is in the range of (200-400) Bq/m^3 [13, 14].

Table 1. Results of radon measurements in Novi Sad apartments by charcoal canister method

Number of measurements	A_{AV} [Bq/m^3]	σ [Bq/m^3]	Geo. A_{AV} [Bq/m^3]	Geo. σ [Bq/m^3]	A_{min} [Bq/m^3]	A_{max} [Bq/m^3]
381	52	75	28	3	2	544

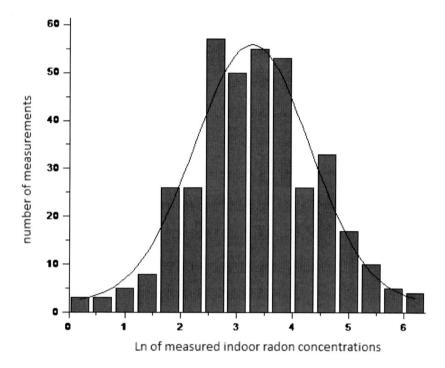

Figure 9. Ln-normal distribution of radon concentration in Novi Sad apartments measured by charcoal canister method.

These values can be compared with the values of radon concentration obtained in the long-term measurements by the method with CR39 detector for Novi Sad (Table 2). The values measured by this method are somewhat higher compared with those in Table 3, which can be explained in terms of weather conditions (mainly winter ones) and the location of the detectors mainly in suburban settlements in old ground-floor houses with poor floor isolation.

The obtained activity concentrations of natural radionuclides ^{238}U, ^{232}Th, ^{40}K and ^{226}Ra and artificial radionuclide ^{137}Cs for 91 soil samples are listed in Table 3. The frequency distribution of the emanated radon concentrations for all these locations in Novi Sad municiplity is presented on Figure 10., yielding an arithmetic mean value of 1361 Bq/m^3, with the standard deviation of 1007 Bq/m^3 from the minimum value of 141 Bq/m^3 to the maximum of 6022 Bq/m^3. The distribution falls almost ln-normal, which indicates that radon emanation from soil is a natural process. The distribution is shifted towards lower concentrations of radon activity, and measured values are common for this region of Europe [15].

Table 2. Results of measuring radon concentration in Novi Sad apartments by CR39 method

Municipality	A_{AV} [Bq/m^3]	$\sigma(A_{AV})$ [Bq/m^3]	n No. of meas.	A_{min} [Bq/m^3]	A_{max} [Bq/m^3]
Novi Sad	133	115	86	10	445

Table 3. Arithmetic means, standard deviations and range of measured radionuclide activity concentrations in soil

radionuclide	A_{AV} [Bq/kg]	$\sigma(A_{AV})$ [Bq/kg]	Range [Bq/kg]
^{40}K	509	93	317-718
^{226}Ra	34	8	15.9 - 66
^{232}Th	39	10	17.8 – 61.6
^{238}U	46	17	20 - 130
^{137}Cs	6	4	0.6 – 16.5

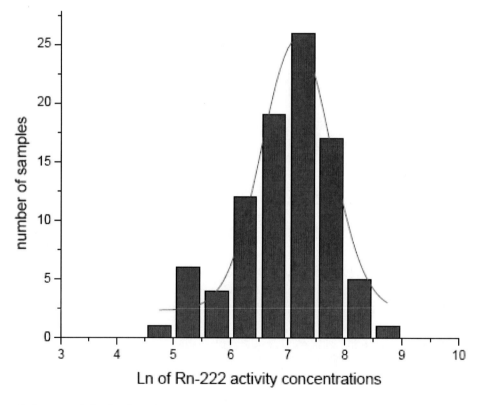

Figure 10. Ln normal distribution of emanated radon concentrations.

For all investigated samples a basic statistical analysis included calculation of correlation factor between radon emanation from soil and radium content was performed (Fig.11).

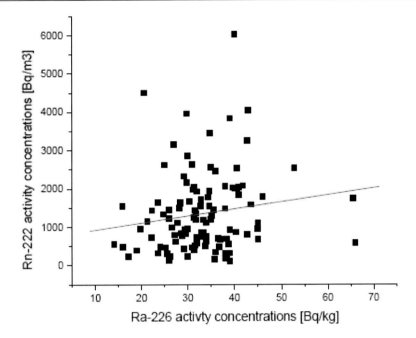

Figure 11. Correlation between radium content in the soil and emanated radon concentration from the soil.

$$r = \frac{\sum_i x_i y_i - \frac{1}{n}\sum_i x_i \sum_i y_i}{\sqrt{\left[\sum_i y_i^2 - \left(\frac{1}{n}\right)(\sum_i y_i)^2\right]\left[\sum_i x_i^2 - \left(\frac{1}{n}\right)(\sum_i x_i)^2\right]}} \tag{3}$$

Correlation factor value of 0.149 indicates that no correlation between ^{226}Ra concentration in surface soil and radon emanation intensity has been established. This means that the radon emanation from soil is mainly determined by **rock disposition in the soil**, deep soil cracks, **soil porosity and groundwater flows** [16]. The highest radon concentrations were detected in chernozem type soil which keeps humidity and radon. Despite that the sand type soil is more porous and doesn't keep radon gas. The relationship of radon emanation and the granulation effects of soil could be investigated in further research.

In Table 4 the statistic of the first radon map of Vojvodina for the period December 2002 to March 2003 is presented.

The radon map was prepared by calculations of the local averages on the basis of administrative boundaries of municipalities and published in [17] (Fig.12). Based on the log-normal distribution of measured radon activity concentrations (Fig.13) the geometric averages were chosen.

Almost 20% of the measurements are in the range 200 Bq/m^3 - 400 Bq/m^3, and 4% of the measurements are elevated indoor radon concentrations above 400 Bq/m^3 according to radiation protection legislative in Serbia. The geometric mean of indoor radon concentration measurements by CR-39 in the winter season December 2002 – March 2003 in the Province Vojvodina is 104.2 Bq/m^3 with standard deviation of 2.3 Bq/m^3.

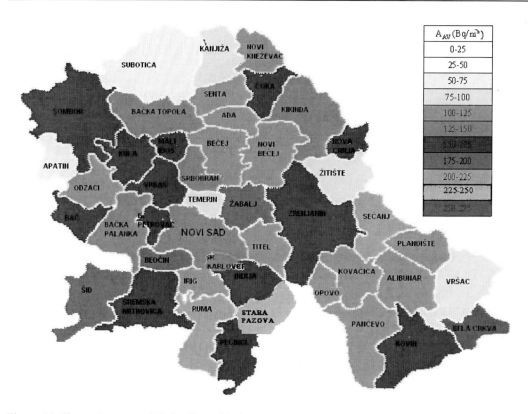

Figure 12. First radon map of Vojvodina with the measurement results for winter season 2002/2003.

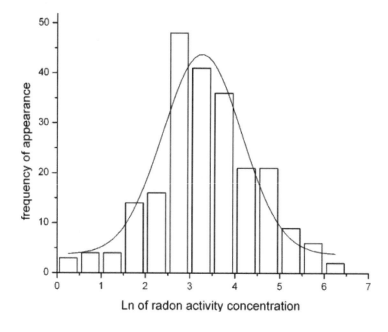

Figure 13. Ln–normal distribution of indoor radon measurements by CR-39 during the period December 2002 – March 2003.

Table 4. The number of measurements, arithmetic mean (A_{AV}) with standard deviation, geometric mean (Geo.A_{AV}) with standard deviation and locations with minimum and maximum radon levels for 2003/2004

Number of measurements	A_{AV} [Bq/m^3]	σ [Bq/m^3]	Geo. A_{AV} [Bq/m^3]	Geo. σ [Bq/m^3]	Location with A_{min}	Location with A_{max}
986	144	120	104.2	2.3	Ada	Zrenjanin

The dose from radon can be estimated from the radon gas concentration (Bq/m^3), or from the progeny energy concentration (J/m^3). On grounds of cost benefit analysis, it is recommended that [222]Rn concentration be the preferred measurement method for screening measurements where dose estimates are less than 5 mSv/year. According to [18] an estimated mean annual radon concentration is:

$$C_0 = 0.73 \times AML \tag{4}$$

where 0.73 is the seasonal correction factor for 3 months duration of the measurement started in December.

The effective dose [nSv] can be calculated from the following formula [19]:

$$Dose = C_0(\varepsilon_0 + \varepsilon_d F)\, O \tag{5}$$

where C_0 is the mean annual radon activity concentration in Bq/m^3, ε_r (0.17 nSvh^{-1} per Bqm^{-3}) and ε_d (9 nSvh^{-1} per Bqm^{-3}) are dose conversion factors for radon and its short-lived progeny respectively, F is the equilibrium factor between radon and its short-lived progeny (F=0.4 assumed) [20], and O is the occupational factor (time spent indoors by average European, O= 0.7x8.76x10^{3}h).

Using these relations, with the mean annual radon activity concentration of 76.1 Bq/m^3, an annual effective dose for the whole body of 1.76 mSv/year was estimated. The result is in good agreement with the WHO data [21] for the effective annual radon dose of 2,5 mSv/year.

For the next sequence of 1000 indoor radon gas measurements for the period from December 2003 to March 2004 sampling strategy were no more adobe old houses than no-basement single-stored houses and ground floor living rooms and bedrooms. The number of measurement locations in municipalities with elevated mean radon concentrations was enlarged and better statistics obtained. The results of measurement and radon map are given in Table 2 and Figure 5. The estimated effective annual radon dose of 1.57 mSv/a was slightly decreased but still in good agreement. The elevated radon levels were observed in all municipalities in the mountain region of Fruška gora and probably were caused by the geology under the houses.

Table 5. The number of measurements, arithmetic mean (A_{AV}) with standard deviation, geometric mean (Geo.A_{AV}) with standard deviation, minimum and maximum radon levels for 2003/2004

Number of measurements	A_{AV} [Bq/m^3]	σ [Bq/m^3]	Geo. A_{AV} [Bq/m^3]	Geo. σ [Bq/m^3]	Location with A_{min}	Location with A_{max}
941	102		68.1	2.6	Plandište	Inđija

The last 1000 measurements of indoor radon concentrations in Vojvodina were done in winter 2004/2005 during the same three months of exposure. The results (Table 6) were lower maybe due to the weather conditions – relatively high air temperatures and pressures for winter season. The calculated effective annual dose of 1.23 mSv/year confirms that assumption.

Table 6. The number of measurements, arithmetic mean (A_{AV}) with standard deviation, geometric mean (Geo.A_{AV}) with standard deviation, minimum and maximum radon levels for 2004/2005

Number of measurements	A_{AV} [Bq/m^3]	σ [Bq/m^3]	Geo. A_{AV} [Bq/m^3]	Geo. σ [Bq/m^3]	Location with A_{min}	Location with A_{max}
926	103	92	73.3	2.3	Bečej	Kula

Based on the results of indoor radon mapping of Vojvodina in 2005 the Radioactivity Environmental Monitoring (REM) group, Institute for Environmental and Sustainability (IES) from the Joint Research Centre Directorate (JRC) of the European Commission (EC) identified our department as the national contact point for further collaboration on the European atlas of natural radiation (http://radonmapping.jrc.it). By compiling the results of radon surveys of three years in the single report the useful database for further developments and investigations in the fields obtained.

On the end the influence of house parameters to radon levels was determined concerning the database of 3000 measurements. In Table 4 the statistics of radon potential are showed. According to [18] radon potential is the number of dwellings exceeding a given radon level (200 Bq/m^3) among 1000 dwellings.

It was concluded that the construction parameters and life habits of population have dominant influence to radon levels. New buildings with poor isolation concrete floors also have elevated radon levels like the old ones with adobe walls and without floor construction. Geological aspects effect the indoor radon concentration if the houses are single-storied with no basement. Heating modes don't effect too much to radon accumulation in rooms, but if the ventilation is poor it may result with significant elevated radon levels.

Table 7. The analysis of radon potentials for all radon maps and percentage abundance of house parameters in dwellings exceeding 200 Bq/m³

Winter period	2002/2003	2003/2004	2004/2005
Radon potential	22%	13%	11%
old house	46%	45%	42%
new house	54%	55%	58%
basement	9%	8%	13%
no-basement	91%	94%	87%
concrete floor	64%	66%	71%
soil / wood	36%	34%	29%
single storied	94%	99%	99%
more-storied	6%	1%	1%
electric heating	35%	31%	38%
solid fuel	38%	25%	44%
gas	21%	42%	18%

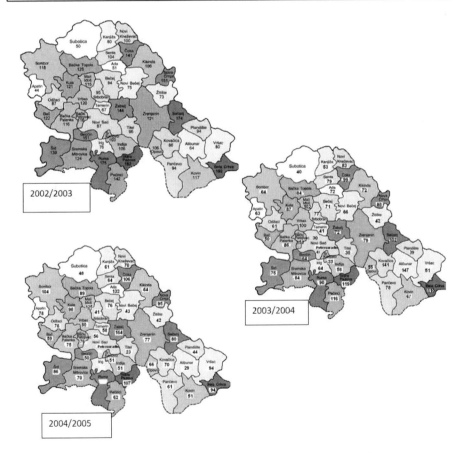

Figure 14. Radon maps of Vojvodina during 3 winter seasons from 2002 to 2005. Numbers given along the names of municipalities indicate geometric mean radon concentrations in Bq/m³.

Beside the survey of radon levels in dwellings in Novi Sad and Vojvodina, Novi Sad Nucler Physics Group applied a lot of experimental techniques and methods to investigate radon spa in Niška Banja on the South-East of Serbia [22, 23]. Radon in the air of pool rooms and room for preparing of therapy mud was measured by the RAD7 detector and charcoal canisters also. The measurement of water samples from pools and main springs in spa was made using the RAD7 accessory RAD-H₂O. In the measurement, the samples were collected according to the technique proposed by the manufacturer (RAD7 RAD H₂O). A 250 ml vial and Wat-250 protocol was used for radon concentrations less than 100 Bq/l. If the radon concentration is higher than 100 Bq/l , a 40 ml vial and Wat-40 protocol was used. The activity of radionuclides in four solid samples was determined using two HPGe detectors, by standard gamma spectroscopy. The measured samples were: soil powder which is the main component for the preparation of therapy mud (used in the peloidotherapy) and two samples of soil taken from the pool "number 5" source and from the public park in Niška Banja. The samples were taken from the surface (0-2 cm). We measured radon concentration in air in one hole in the basement of hotel – dispensary "Radon" which is just below the thermal pool. We inserted the 1,5 m long tube in the soil hole in the floor and sucked the air by automatic pump of RAD7 monitor. The measurement lasted for 30 minutes in two cycles.

The results of gamma-spectrometry measurements of soil and peloid samples from Niška Banja spa are presented in Table 8.

Table 8. Activity concentrations of radionuclides in samples from Niška Banja spa

	Activity concentration of radionuclides [Bq/kg]				
sample	^{238}U	^{226}Ra	^{232}Th	^{40}K	^{137}Cs
peloid	83±10	60±3	104±13	1210±40	< 2,3
soil from thermal source	20±5	6,8±2,5	20,3±1,4	361±16	2,0±0,4
soil from park	18±5	91±6	37,1±2,4	537±21	21,2±1,0

The results of measurements with alpha spectrometer RAD7 are presented in Table 9. The measurements were performed on the following locations in Niška Banja spa: air: around pool "Zelengora", pool in the hotel "Radon", pool "number 5" and "Glavno Vrelo" the main source in Niška Banja and radon exhalation from soil in the basement of hotel – dispensary "Radon".

Table 9. Activity concentrations of ^{222}Rn in air and water samples in Niška Banja spa measured by RAD7 monitor

	Activity concentrations of ^{222}Rn in air [kBq/m³]	Activity concentrations of ^{222}Rn water [Bq/l]
pool "Zelengora"	2,81± 0,15	28± 5
pool "Radon"	0,44± 0,16	24,5± 2,4
pool "number 5"	0,21± 0,05	34±3
Spring "Glavno Vrelo"	0,17± 0,04	61± 5
Spring "Staro kupatilo"	-	69± 8

Very high radon levels have been measured beneath hotel "Radon". In there is a high level of radon concentration in soil gas emanated from a hole in the floor of the hotel-dispensary "Radon" basement. The measured activity concentration of ^{222}Rn was (22.90 ± 0.57) kBq·m^{-3}.

Table 10. Activity concentrations of ^{222}Rn in air samples by charcoal canisters

Location	Activity concentration of ^{222}Rn A [kBq/m^3]
basement of hotel "Radon"	4.03 ± 0.06
room for preparing of therapy mud	0.584 ± 0.023
pool "Zelengora"	3.81 ± 0.07
pool "number 5"	0.259 ± 0.017

According to formula (5) with dose conversion factors introduced by UNSCEAR2000 Report [18] and ICRP recommendation for equilibrium factor value of 0.4 the annual effective dose from inhalation of radon and radon daughters around the thermal pools in Niška Banja spa were estimated to very high values: 18 mSv/year for Zelengora pool and 3 mSv/year for Radon pool which are in good agreement with worldwide published results [24] and [25]. Maybe the problem of high levels could be solved by improving the ventilation system in indoor spaces especially around pool "Zelengora".

Equilibrium factor between radon and short lived progenies is very important for dose assessment from inhalation of radon and it must be determined in each radon monitoring, especially in spas and caves this factor depends largely on environmental conditions (AMD distribution, big humidity etc. [26]). Because that it is necessary to develop a method of measuring the progeny concentrations in order to calculate equilibrium factor according to the formula:

$$F = \frac{EEC_{Rn}}{C_{Rn}} \tag{6}$$

$$EEC_{Rn} = 0.105C_1 + 0.515C_2 + 0.380C_3 \tag{7}$$

Where EEC_{Rn} is radon equilibrium equivalent concentration; C_{Rn}, C_1, C_2 and C_3 are the activity concentrations (in Bqm^{-3}) for ^{222}Rn, ^{218}Po, ^{214}Pb and ^{214}Bi, respectively.

In the literature, most often methods for measuring the equilibrium factor are based on the detection of gross alpha activities that bring a great deal of uncertainty and error. In published paper [27], we propose a method for simultaneous alpha spectrometry measurement of the activity concentration of radon in the air and determining of radon progeny activity concentrations by gamma spectrometry measurement of the filter paper. The advantage of gamma spectroscopy method is accurate and fast determination of radionuclide activities and problem of long cooling time and transport to the laboratory could be avoided by portable HPGe spectrometry systems.

Radon activity concentration was measured by RAD-7 monitor which enables continual and direct reading of radon concentration in air and therefore it is suitable for such simultaneous measuring. High volume air sampler - F&J model DHFV-1SE with fiber glass filter paper of high collection efficiency of ε=98% were used for aerosol sampling. The pump flow velocity was adjusted to value of v=0.0303 m³/s. During the experiment radon short lived daughters attached to aerosols were collected on the fixed filter paper. After the suction the filter paper was approximately homogenous packed to the cylindrical geometry and gamma spectrometry measured in ten successive measurements in duration of 1000 s, with time of 120 s elapsed to first measurement. Gamma spectrometry measurement were performed by HPGe low level detector, Canberra manufacturer, type GX10021 with extended range from 6 keV to 3 MeV in original lead shield with wall thickness of 15 cm. Relative detector efficiency is 100% (equivalent to absolute efficiency of 3"x 3" NaI(Tl) detector on 1332 keV gamma line).

In order to connect the results of gamma spectrometry measurements of the filter paper and the radon progeny concentrations in the air at the start of suction (C_1(^{218}Po), C_2(^{214}Pb), C_3(^{214}Bi)), it is necessary to take into account decay corrections during the time of suction t_U, cooling time t_H after the suction but before the measurements and during the measurements t_M, because these periods of time are not negligible relative to short life times of radon daughters (^{218}Po – 3 min, ^{214}Pb – 26.8 min and ^{214}Bi – 19.9 min).

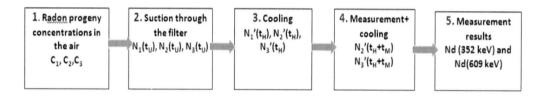

Figure 15. Method algorithm for radon progenies determination.

After the time t_U elapsed from the beginning to stopping of suction the number of radon progeny atoms ^{218}Po, ^{214}Pb and ^{214}Bi collected on the filter paper N_1, N_2 and N_3 change according to differential equations (4-6) if one consider the approximation that on the beginning of suction there were no radon progenies on the filter paper $N_1(0)=0$, $N_2(0)=0$, $N_3(0)=0$:

$$\frac{dN_1}{dt_U} = C_1 v \varepsilon - \lambda_1 N_1 \tag{8}$$

$$\frac{dN_2}{dt_U} = C_2 v \varepsilon + \lambda_1 N_1 - \lambda_2 N_2 \tag{9}$$

$$\frac{dN_3}{dt_U} = C_3 v \varepsilon + \lambda_2 N_2 - \lambda_3 N_3 \tag{10}$$

After the suction of the air through the filter paper the radon progeny atoms captured on the filter paper decay during the cooling and measuring time and the number of atoms have changed in accordance with the equations:

$$\frac{dN'_1}{dt} = -\lambda_1 N'_1 \tag{11}$$

$$\frac{dN'_2}{dt} = \lambda_1 N'_1 - \lambda_2 N'_2 \tag{12}$$

$$\frac{dN'_3}{dt} = \lambda_2 N'_2 - \lambda_3 N'_3 \tag{13}$$

The initial conditions for this system of equations were obtained by solving the first system of differential equations (8-10) in *Matematica* program using the values of constant parameters: $\lambda_1 = 0.003787$ s^{-1}, $\lambda_2 = 0.000431$ s^{-1}, $\lambda_3 = 0.00058$ s^{-1}, $v = 0.0303$ m^3/s, $\varepsilon = 0.98$ and $t_U = 1320$s:

$$N_1{}'(0) = 7.78814\,C_1 \tag{14}$$

$$N_2{}'(0) = -78.4011\,(-0.31813C_1 - 0.381261\,C_2) \tag{15}$$

$$N_3{}'(0) = -1.90951 \times 10^9\,(-2.72364C_1 - 3.79308 \times 10^{-9}C_2 - 1.43426 \times 10^{-8}C_3) \tag{16}$$

Now it could be solved the system of equations (11-13) and obtained the number of not decayed nuclei of radon progenies on the filter paper $N'_1(t)$, $N'_2(t)$, $N'_3(t)$.

The number of decayed nuclei of ^{214}Pb, and ^{214}Bi respectively during the measurement could be connected with the gamma spectrometry detected results for measured filter paper:

$$\frac{N_d}{\varepsilon_d p_\gamma} = N_r \tag{17}$$

where N_r is the number of decayed nuclei during the time of measurement t_M, N_d -area under the photopeak, ε_d -photopeak detection efficiency and p_γ - γ-ray emission probability. Detected decays are actually the difference between the not decayed nuclei after the cooling time $t = t_H$ and not decayed nuclei after the time of measurement $t = t_H + t_M$:

$$N_r = N'(t = t_H) - N'(t = t_H + t_M) \tag{18}$$

where $N'(t = t_H)$ is the number of not decayed nuclei after the cooling time, and $N'(t = t_H + t_M)$ - is the number of not decayed nuclei after the cooling time and measurement.

CONCLUSION

The gamma spectroscopy measurements on building materials and soil resulted in moderated natural radioactivity (^{238}U, ^{232}Th, ^{226}Ra) levels with order of magnitude 50 Bq/kg.

The activity concentration of the radon emanated from soil was about 1000 Bq/m³ with significant local variations. The main aim of the present study was to explore the critical group of population for radon exposure and to estimate maximal annual doses. The sampling strategy was oriented towards suburban and urban regions in the Province. Almost 16% of the results of the radon concentration measurement in apartments and houses of Vojvodina exceeded the reference value of 200 Bq/m³. This result leads to the annual dose estimate of 1.76 mSv/year which is under the recommended action limit of ICRP. For urban dwellings in Novi Sad (Province capital) the annual mean value of 54 Bq/m³ (220 measurements) is obtained. By comparison of these two results it is concluded that radon surveys based on measurements in urban environment may seriously underestimate the radon related health risk. The elevated radon levels could not be explained by elevated uranium levels of surface soil. The benefit of the indoor radon survey of Vojvodina is that the houses with elevated radon levels were found and not expensive reparation were recommended to owners, like plastic isolation, better ventilation etc. The results of repeated measurements confirm the decreasing of indoor radon concentrations to ordinary levels. It was concluded that the construction parameters and life habits of population have dominant influence to radon levels. Geological aspects effect the indoor radon concentration if the houses are single-storied with no basement. The future plans are to collaborate in regional and national projects in producing a European map of indoor radon levels, using a reference grid with resolution 10 km x 10 km in order to obtained randomly sampling [28]. Radon survey in spa resulted with very high estimated annual effective dose discussed in this chapter. The new exact method for radon equilibrium determination by gamma spectrometry measuring of radon progeny concentrations in the air is suggested in order to validate widely used portable alpha spectroscopy systems for radon measurements in spas and caves. Equilibrium factor between radon and short lived progenies is very important for dose assessment from inhalation of radon and it must be determined in each radon monitoring.

ACKNOWLEDGMENTS

The authors acknowledge the financial support of the Provincial secretariat for the environmental protection and Ministry of science and technology development of Serbia, within the projects 171002 and 43002.

REFERENCES

[1] WHO Regional Office for Europe: Radon - Booklet for Local authorities, health and environment, Scientific advisers: Dr Denis Bard and Margot Tirmarche (1996)

[2] Rulebook on limits of radionuclides content in drinking water, foodstuffs, feeding stuffs, medicines, general use products, construction materials and other goods that are put on market, ("Official Gazette of Republic of Serbia'', No.86/2011 and 97/13

[3] UNSCEAR 1982 REPORT: *Ionizing Radiation: Sources and Biological Effects*, United Nations Scientific Committee on the Effects of Atomic Radiation 1982 Report to the General Assembly, with annexes.

[4] NRC (National Research Council). Committee on Health Effects of Exposure to Radon (BEIR VI), and Commission on Life Sciences. Health Effects of Exposure to Radon in Mines and Homes. Washington, D.C. *National Academy Press.* 1994.

[5] EC, 1997: Radiation Protection 88. Recommendations for implementation of Title VII of the European Basic Safety Standards concerning significant increase in exposure due to natural radiation sources. European Commission. Office for Official Publications of the European Commission. Radiation Protection Series.

[6] Radon Legislation and National Guidelines. SSI report. Swedish Radiation Protection Institute. No 99: July 1999. ISSN 0282-4434.

[7] Rulebook of the ionising radiation exposure limits and measurement to assess the level of exposure to ionizing radiation ("Official Gazette of Republic of Serbia", No.86, 2011).

[8] WHO handbook on indoor radon-a public health perspective (2009).

[9] Council Directive 2013/59/Euratom of 5 December 2013 laying down basic safety. standards for protection against the dangers arising from exposure to ionizing radiation and repealing Directives 8/618/Euratom, 96/29/Euratom, 97/43/Euratom and 2003/122/Euratom, *Official Journal of the European Union* L13.

[10] I. Bikit , J. Slivka, Lj. Čonkić, M. Krmar, M. Vesković, N. Žikić-Todorović, E. Varga, S. Ćurčić, D. Mrdja: Radioactivity of the soil in Vojvodina (northern province of Serbia and Montenegro), *Journal of Environmental Radioactivity,* Volume 78, Issue 1, (2004), Pages 11–19.

[11] Bikit I., Mrđa D., Todorović N., Varga E., Forkapić S., Vesković M., Slivka J., Čonkić Lj. : Possibility of prompt U-238 activity concentration determination by gamma-ray spectroscopy, *Japanese Journal of Applied Physics Part 1-Regular Papers Short Notes & R Eview Papers,* 2005, Vol. 44, Broj 1A, str. 377-379.

[12] RADOSYS User Manual: www.radosys.com.

[13] Todorović N., Forkapić S., Bikit I., Mrđa D., Vesković M., Todorović S.: Monitoring for Exposure to TENORM Sources in Vojvodina Region, *Radiation Protection Dosimetry,* Volume 144, Numbers 1-4, 9 March 2011 , pp. 655-658(4).

[14] Todorović, N., Bikit, I., Vesković, M., Krmar, M., Mrda, D., Forkapić, S., Hansman, J., Nikolov, J., Bikit, K.: Radioactivity in the indoor building environment in Serbia, Radiation Protection Dosimetry, Volume 158, Issue 2, (2014), Pages 208-215.

[15] National Environmental Health Action Plans, Hungary, (1997), W.H.O. Reg. Office for Europe.

[16] King P.T, Michel J. and Moore W.S., Ground Water Geochemistry of ^{228}Rn, ^{226}Rn and ^{222}Rn, *Geochim. Cosmochim. Acta,* 46 (1982).

[17] S.Forkapić, I.Bikit, J.Slivka, Lj.Čonkić, M.Vesković, N.Todorović, E.Varga, D.Mrđa and E.Hulber: *Indoor radon in rural dwellings of the South-Pannonian region,* Radiation Protection Dosimetry, Vo 123, Issue 3, pp 378-383 (2006).

[18] Miles, J.C.H. and Howarth, C.B. *Validation scheme for laboratories making measurements of radon in dwellings: 2000 Revision NRPB-M1140.* (Chilton: NRPB) p.11 (2000).

[19] United Nations Scientific Committee on the Effects of Atomic Radiation, *Sources and effects of ionizing radiation.* (New York: United Nations) (2000).

[20] International Commission on Radiological Protection. *Recommendations of the International Commission on Radiological Protection,* ICRP Publication 60 (Anals of the ICRP 21 (1991)).

[21] Mettler, F; Upton, A.: Medical Effects of Ionizing Radiation, Second Edition, ISBN 0-7216-6646-9; 1995.

[22] J. Nikolov, N. Todorovic, T. Petrovic Pantic, S. Forkapic,D. Mrdja,I. Bikit, M. Krmar, M. Veskovic: Exposure to radon in the radon spa Niška Banja, Serbia, *Radiation Measurements* Volume 47, Issue 6, (2012), p. 443–450.

[23] Jovana Nikolov, Natasa Todorovic, Istvan Bikit, Tanja Petrovic, Pantic, Sofija Forkapic, Dusan Mrda and Kristina Bikit: *Radon in Thermalwaters in South-East Part of Serbia,* Radiation Protection Dosimetry (2014) Vol.160, No.1-3, pp.239-243.

[24] Lettner et al.: Occupational exposure to radon in treatment facilities of the radon-spa Badgastein, Austria, Environment International, 22 (Suppl.1) (1996), pp. S399–S407

[25] Somlai etal.: Contribution of 222Rn, 226Ra, 234U and 238U radionuclides to the occupational and patient exposure in Heviz-spas in Hungary, *Journal of Radioanalytical and Nuclear Chemistry,* 272 (1) (2007), pp. 101–106.

[26] Karel Jilek, Josef Thomas and Ladislav Tomašek: First results of measurement of equilibrium factors F and unattached fractons fp of radon progeny in Czech dwellings, *Nukleonika* 55(4), 2010,439-444.

[27] Forkapić, S., Mrda, D., Vesković, M., Todorović, N., Bikit, K., Nikolov, J., Hansman, J: Radon equilibrium measurement in the air Romanian Reports of Physics 58 (2013) , pp. S141-S147.

[28] G.Dubois, P.Bossew, T.Tollefsen, M.De cort. First steps towards a European atlas of natural radiation: status of the European indoor radon map, *Journal of Environmental Rdaioactivity* 101 (2010) 786-798.

In: Radon
Editor: Audrey M. Stacks

ISBN: 978-1-63463-742-8
© 2015 Nova Science Publishers, Inc.

Chapter 12

THE ANALYSIS OF RADON DIFFUSION, EMANATION AND ADSORPTION ON DIFFERENT TYPES OF MATERIALS

Sofija Forkapić[1,], Dušan Mrđa[1], Ištvan Bikit[1], Kristina Bikit[1], Selena Grujić[2] and Uranija Kozmidis-Luburić[2]*

[1]Department of Physics, Faculty of Sciences, University of Novi Sad, Novi Sad, Serbia
[2]Faculty of Technical Sciences, University of Novi Sad, Novi Sad, Serbia

ABSTRACT

Since people spend most of the time inside the buildings it is of great importance to analyze the radon diffusion through buildings materials in order to prevent indoor radon build-up. Insulating properties of different types of materials against radon were studied by means of radon diffusion coefficient. A method has been developed in our laboratory by using closed air circulation system which includes RAD 7 radon detector connected by tubes to tightly closed glass chamber with radon source materials covered by well known thickness of insulating materials. This experimental setup has been upgraded for radon emanation coefficient measurements. The granulation effects on the radon adsorption and radon emanation rate of several building materials (ceramic plates, sand, red brick and siporex brick) with different radium Ra-226 content were investigated and discussed. The possibility of using natural zeolite for radon concentration reduction was also considered.

On the other hand powder and liquid substances which are stored within closed rooms and exposed to higher radon concentrations can adsorb a certain radon amounts, depending on characteristics and granulation of powder, as well as radon solubility in liquids. This can lead to increase of dose received by general public, if such substances are used as human food, components for food or cosmetics. Research of radon adsorption by liquids and powders is also useful for correction of gamma spectrometry determination of Ra-226 concentrations in such samples mostly based on post-radon gamma lines.

---

[*] Corresponding author: dr Sofija Forkapić, Department of Physics, Faculty of Sciences, University of Novi Sad, Trg Dositeja Obradovica 4, Novi Sad, Serbia, tel: +381 21 459 368, e-mail: sofija@df.uns.ac.rs.

Radon adsorption by zeolite on various granulation was explored in this chapter. The radon adsorption coefficients were calculated based on gamma spectrometry measurements of materials and countinuous monitoring of radon inside the chamber with examined material and radon source.

1. INTRODUCTION

The indoor radiation exposure to humans is mainly due to radioactive radon gas and emissions from building construction materials. There are three main factors of radon entrance in to a building environment are: a) Emission of radon from building materials used in construction b) By convection through cracks and openings in the building and c) Diffusion from soil through pore space of construction materials. The transport phenomenon of radon through diffusion is a significant contributor to indoor radon entry.

The diffusion of radon in dwellings is a process mainly determined by the radon concentration gradient across the building material structure between the radon source and the surrounding air. Keeping this in mind the radon diffusion studies have been made through six types of building materials: tarkett linoleum, condor, Speer records, tarkett laminate, aluminum plate and rubber using RAD 7 alpha spectrometer.

Transport of radon takes place through two different processes: diffusion conditioned difference radon concentration in the system of pores, which is conducted by Fick's law:

$$J_d = -D_e \frac{dC}{dz} \tag{1}$$

where J_d – radon flux in Bq/(m^{2}s), D_e – diffusion coefficient (m^2/s) and dC/dz – concentration gradient along z – axis (Bq/m^4); and radon diffusion, conditioned by a difference in the pressures prevailing between the air in the pores and the outer air, which is carried by Darcy's law [1]:

$$v = -\frac{k}{\eta} \frac{dP}{dz} \tag{2}$$

where v – radon flow velocity (m/s), k - medium transmission (m^2), η - dynamic viscosity of air at 10 ^{0}C (Ns/m^2) and dP/dz - pressure gradient along the layer thickness z (Pa/m). Radon diffusion equation in porous environment for the stationary case has the form [1]:

$$\frac{D_e}{\varepsilon} \frac{d^2C}{dz^2} - \frac{1}{\varepsilon} \frac{d(vC)}{dz} - \lambda C + \Phi = 0 \tag{3}$$

In this case in material there is no source and the abyss of radon. A constant gradient of radon concentration is along z-axis, so the Fick's law is:

$$J_d = -D_e \frac{C_s - C_1}{d} \quad \left[\frac{Bq}{m^2 s} \right] \tag{4}$$

where d – material thickness (m), D_e – diffusion coefficient, C_s - radon concentration in reaching saturation after the diffusion, C_1 - concentration of radon below the investigated materials, while the radon flux calculated by simplified approximate formulae $J_d = \frac{C_s \cdot V}{t_s \cdot S}$, where V - volume of containers in which radon particles undergo by diffusion through the material, t_s - saturation reaching time, S - area through radon undergo by diffusion. Substituting the expression for the flux into the equation (4) gives an expression for calculation the diffusion coefficient of radon:

$$D_e = \frac{C_s \cdot V \cdot d}{t_s \cdot S \cdot (C_1 - C_s)} \quad \left[\frac{m^2}{s} \right] \tag{5}$$

The quantity that determines the radon exhalation rate from building materials is the effective specific activity of radium-226, $A_{s.eff}$ (^{226}Ra), and it is defined as a product of the specific activity of radium A_s (^{226}Ra) and the radon emanation coefficient η:

$$A_{seff} (^{226}Ra) = \eta \cdot As (^{226}Ra) \; [Bq/kg]. \tag{6}$$

The radioactivity of building products depends on the minerals that were used for its production (e.g., granite, aluminium shale and volcanic tuff have elevated contents of natural radionuclides). Sand and gravel usually have a specific activity of radium close to the average value for soil. The concentration of ^{226}Ra in building materials also depends on the place of production of the raw material. Radium is characteristic to Sweden and other Scandinavian countries that because of climatic and geological conditions, have an increased content of radium compared with other countries. An increased content of natural radionuclides may also appear as a result of using secondary products from industrial processes, such as electrofilter ash, which is produced by combusting coal in thermal power stations, slag from combusting bauxite ore, and waste from the uranium industry. The use of phosphate gypsum, for example, can significantly increase the radon concentration in closed spaces because of a high Ra-226 content and a high radon emanation factor [2].

The radon emanation coefficient is used to designate the fraction of radon generated that escapes from the source material [3]. The rate of emanation varies widely from 1% to 30% [4].

According to the value of the emanation rate, all building materials can be roughly divided into two groups [5]:

- Materials that are exposed to high-temperature processing during production (red brick, ash, cement, slag) whose emanation rate ranges from 1% to 2%.
- Materials that are not exposed to high-temperature processing (silicate brick, gravel, sand), which have an emanation rate around ten times higher (10%-20%).

High-temperature processing of materials decreases the radon exhalation rate from building materials (because of the change in the microstructure caused by high temperatures).

The powder and liquid substances which are stored within closed rooms and exposed to higher radon concentrations can absorb a certain radon amounts, depending on characteristics and granulation of powder, as well as radon solubility in liquids. Having in mind that determination of Ra-226 concentrations by gamma spectrometry is based on intensity measurements of post-radon lines (295 keV, 352 keV, 609 keV) [6, 7] from Pb-214 and Bi-214, these results can be incorrectly assigned to Ra-226 presence in sample, instead of absorbed radon. To investigate radon adsorption various zeolite granulation were used. Different granulations of zeolite in the range of μm-mm were achieved by ball mill, whereas the particle size distributions was determined by particle size analyzer, Master Sizer 2000. The zeolite samples were exposed to elevated radon concentrations up to 1800 Bq/m^3 inside a closed chamber (volume $\approx$ 5.4 x 10^{-3} m^3). The absorbed radon quantity was measured by high resolution gamma spectroscopy. The influence of particle size was measured and discussed.

2. EXPERIMENTAL DETERMINATION OF DIFFUSION COEFFICIENT, RADON EMANATION RATE AND RADON ADSORPTION

In order to accurately measure the coefficient of diffusion it is necessary to achieve the controlled conditions of radon passing through the set material. The aim is to prevent air flow in the measured area and provide air flow to the detector. The detector measured 20 cycles for 1 h. The ideal setting is glass container, made of fire-proof glass with a thickness of 1 cm. The bottom of the container is narrowed to the upper part, which provides a favorable setting for materials whose diffusion coefficient measured (Figure 1). The examined material is located on the extension between the lower and upper part of bowl. Container contains one outlet to vent through the valve RAD7 detector pump that draws air containing radon gas and makes it's detection. RAD7 detector is a type of alpha-spectrometer that allows constant monitoring of the concentration of radon in the air, and therefore monitoring of variations of radon levels with time (Figure 2). Used radioactive sample was white, powdery substance called zirconium, mass m = 873.9 g, and activity A = 3760±230 Bq/kg. In order to match the radium and radon activity it is necessary to approximate that the entire amount caused by radon remains in the "ideal closed container". Requirement also applies to a complete radon emanation from sources which accumulates above the source.

The data that are invariant of measured material are: the volume of container vessels of the lower V, cross section passing through the openings radon particles with the surface of the court between the lower and upper part of the vessel where the filter paper is placed, and C_1, which is saturation concentration obtained from measurements of radon concentration when there is no set of material for measurement.

In order to determine the exact value of C_1 first is necessary to perform the measurement of the concentration of diffusion without material, and to determine the saturation value C_1 for air from the histogram (Figure 3).

Each material has its own value of concentration and time required to reach saturation t (calculated from the time of closing vessels, to the time reach saturation point), under which the diffusion coefficient value calculate.

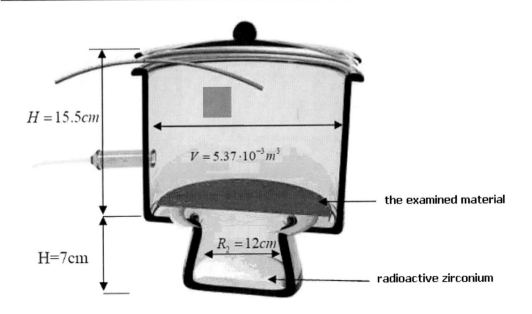

Figure 1. Schematic diagram of container.

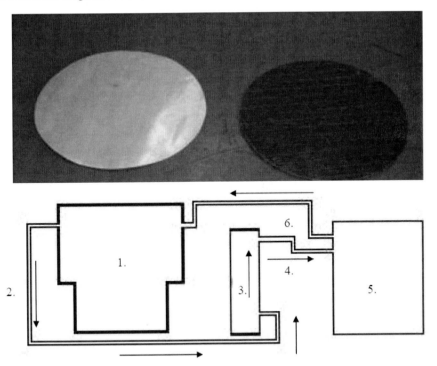

Figure 2. The discs of examined materials – up and the scheme of experimental setup: 1. Chamber with thick glass walls, 2. Plastic tube, 3. Humidity absorber (CaSO$_4$), 4. Input plastic tube, 5. Alpha spectrometer (RAD7), 6. Output plastic tube – down.

Based on Figure 3 the value of the concentration for the air is determined (C_1 = 973 Bq/m^3). The uncertainty of measurement obtained by numerical method and has the value δ = 77.5 Bq/m^3[8].

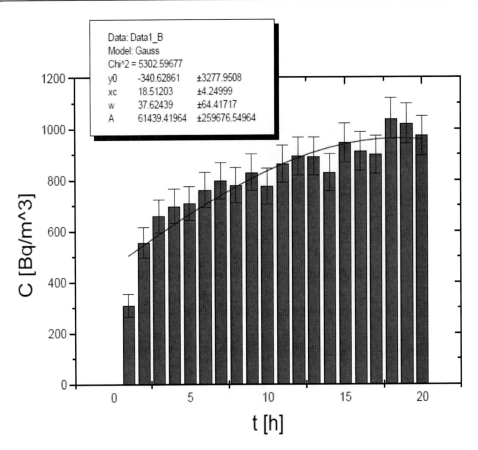

Figure 3. ^{222}Rn concentration in container during the time t.

The radon emanation and the granulation effect on the emanation rate of several building materials (ceramic plates, sand, red brick, and siporex brick) with different Ra-226 concentrations were investigated [9]. A ball mill was used to achieve different granulations of the materials. The particle size distributions were determined by a particle size analyser (Mastersizer 2000). The increase in the Rn-222 concentration inside a closed chamber (volume ≈ 5.4 x 10^{-3} m^3) due to emanation from each material with different granulations was measured by an alpha spectrometer (RAD7).

Thus, time-dependent curves for radon concentrations were obtained. The highest radon emanation coefficient (27%) was obtained for the siporex sample with the smallest grain size (0.34 μm). For the ceramic pads, the granulation effect was negligible and the emanation coefficient was very low (approximately 0.4%). The strongest influence of granulation on the radon emanation rate was found for the siporex brick sample.

The building materials that are most common on the market were chosen for testing the radon emanation rate: ceramic pads, sand, red brick, and siporex block. The emanation of radon from these materials was tested when they were in larger fragments as well as when they were in a powder form. Each sample was dried for a few hours at the constant temperature of 100 °C (water content less than 1%). To obtain samples of diverse grain size, different times of milling with different rotational speeds of the main panel were used. A particle size distribution analysis was conducted using a Malvern Mastersizer 2000 Particle Size Analyser capable of analyzing particles between 0.02 μm and 2000 μm.

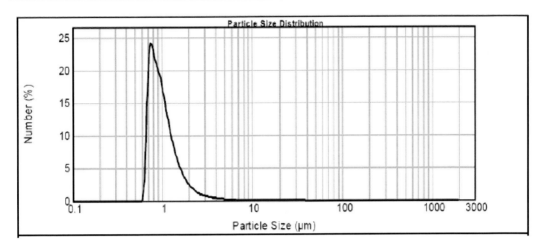

Figure 4. Distribution of particles of siporex based on their size after being milled for three minutes.

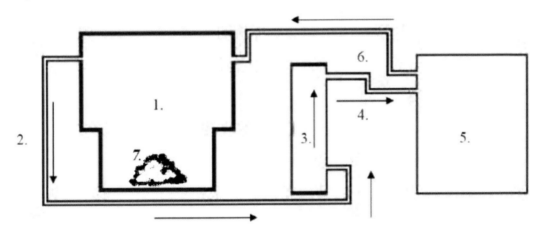

Figure 5. 1. Chamber with thick glass walls, 2. Plastic tube, 3. Humidity absorber (CaSO₄), 4. Input plastic tube, 5. Alpha spectrometer (RAD7), 6. Output plastic tube, 7. Sample of the material.

The Malvern Mastersizer 2000 records the light pattern scattered from a field of particles at different angles. It then uses an analytical procedure to determine the size distribution of spherically shaped particles that created the patterns.

The result of the analysis is the relative distribution of the volume (number) of particles in a range of size classes. The typical distribution of the size of the particles of the sample is shown in Figure 4.

The experimental setting used to test the radon emanation rate is presented in Figure 5. The samples of the material are placed inside a chamber with thick glass walls, whose volume is equal to 5.4×10^{-3} m^3. Circulation of air is provided by using proper tubes, without mixing the air inside the chamber with the outside air.

The direction in which the air moves is labeled by arrows. The air inside this closed system contained radon that emanated from the sample. Humidity absorption was performed by passing the air through a column of calcium sulphate (CaSO$_4$). Air flowed continuously at a rate of 0.5 l/min, thanks to the pump that is an integral component the alpha-spectrometer RAD7.

The measurement time for each sample was four days (96 h), except for zirconium silicate (20 h), which had a high content of Ra-226, and thus the measured radon concentrations for zirconium silicate were high and the corresponding measurement uncertainties were low.

The masses of the samples and the average size of the particles of each sample are given in Table 1.

For radon adsorption measurements, we used highly porous natural zeolite material, produced by FiMö-Aquaristik GmbH, Germany. Different granulations of this material are obtained by different times of milling (10, 20 and 40 minutes). Milling was performed in a planetary ball mill Fritsch Pulverisette 5. A hardened steel vial (250 cm^3 volume) and hardened steel balls (10 mm in diameter) were used. The mass of the milled sample was 50 g, and the angular velocity of the supporting disc and vial were 400 and - 800 rpm, respectively. Particle size distribution was determined using a Malvern Mastersizer 2000 particle size analyzer capable of analyzing particles between 0.01 and 2000 μm.

The experimental setup designed to test the radon adsorption by zeolite is presented in Figure 2. Zeolite samples were exposed to radon concentrations of about 1400-1800 Bq/m^3, during 48h for each sample, by placing them inside a chamber with thick glass walls, whose volume is 5.4 10^{-3} m^3.

The changes of radon concentration inside the chamber over time were continuously measured by alpha-spectrometer RAD7. Circulation of air at a rate of 0.5 l/min is provided thanks to the pump that is an integral component of the alpha-spectrometer, without mixing the air inside the chamber with the outside air. The air inside this closed system contained radon that emanated from the zirconium-oxide (m=0.9 kg), containing Ra-226 radionuclide (4100 ± 200 Bq/kg) which was placed at the chamber bottom.

The direction in which the air moves through the closed system is labeled by arrows on Figure 6. The temperature of air and relative humidity were measured by devices that are integral parts of RAD7 device. However, in this experimental setup the temperature of the air and its relative humidity were not adjustable parameters.

Humidity absorption was performed by passing the air through a column of calcium sulphate ($CaSO_4$). Each sample was also measured by low-level gamma spectrometry system before exposure to radon elevated concentrations and after exposure.

Table 1. Characteristics of the analysed samples

Sample	Mass of the sample [kg]	Average size of particles [μm]		
		Sample 1	Sample 2	Sample 3
Zirconium silicate	0.874	0.39	-	-
Siporex	0.584	0.34	0.75	bulk (about $1 \cdot 10^{-3}$ m^3)
Sand	0.548	0.87	3.39	3.89
Ceramic plates	0.520	0.34	0.75	pieces (~ 5 mm)
Red brick	0.505	0.75	1.18	pieces (~ 5 mm)

3. RESULTS AND DISCUSSION

Based on previous data and determining time of saturation, diffusion coefficient for each material were calculated using the formula (5). The results of the measurements are given in Table 2.

Gamma spectrometry analysis of the samples was carried out to check the content of Ra-226 (Rn-222 is produced as a result of Ra-226 disintegration) in the samples. The samples were packed in cylinder-shaped iron boxes (diameter = 10 cm, height = 3 cm) and were left hermetically closed for 30 days so that a radioactive equilibrium of Rn-222 and Ra-226 could be achieved. The measurements were carried out using a low-background HPGe detector with a relative efficiency of 35% placed inside a lead shield with a wall thickness of 12 cm. The samples were directly placed on the end-cap of the detector (contact geometry) so that the best geometrical efficiency was achieved. The measurement time was typically 18 h. The results are given in Table 3.

The sample of zirconium silicate was sealed inside a glass container for 24 h before measuring the radon emanation rate. The increase in the concentration of Rn-222 from the moment the measuring started and during the next 20 hours is shown in Figure 5.

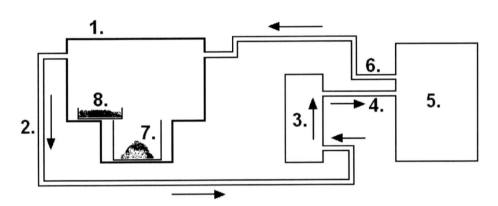

Figure 6. The experimental setup designed to test the radon adsorption by zeolite: 1. Chamber with thick glass walls, 2. Plastic tube, 3. Humidity absorber ($CaSO_4$), 4. Input plastic tube, 5. Alpha spectrometer (RAD7), 6. Output plastic tube, 7. Zirconium-oxide containing Ra-226, 8. Zeolite.

Table 2. Measured coefficient diffusion values for different building materials De relate to the value De_2 (from literature)

Material	$\rho [gcm^{-3}]$	$D_e(\delta)[m^2 s^{-1}]$	$D_{e_2}[m^2 s^{-1}]$	$\dfrac{D_e}{D_{e_2}}$
Tarkett linoleum	1.289	$22(12) \cdot 10^{-11}$	$1.3 \cdot 10^{-11}$[10]	19.62
Kondor	0.418	$18(14) \cdot 10^{-11}$	$2.3 \cdot 10^{-11}$[10]	7.83
Speer records	0.302	$13(17) \cdot 10^{-7}$	$2.35 \cdot 10^{-6}$ [10]	0.55
Tarkett laminate	0.86	$25(18) \cdot 10^{-10}$	$1.3 \cdot 10^{-11}$[10]	192.3
Aluminium plate	1.964	$9(4) \cdot 10^{-11}$	$<10^{-8}$	/
Rubber	1.16	$13(9) \cdot 10^{-11}$	$2.5 \cdot 10^{-10}$[11]	0.52

Table 3. Ra-226 content in the samples

Sample	Ra-226 [Bq/kg] Activity concentration
Zirconium silicate	3760 ± 230
Siporex	28.0 ± 0.7
Sand	9.3 ± 0.3
Ceramic plates	75.9 ± 0.9
Red brick	39.4 ± 0.7

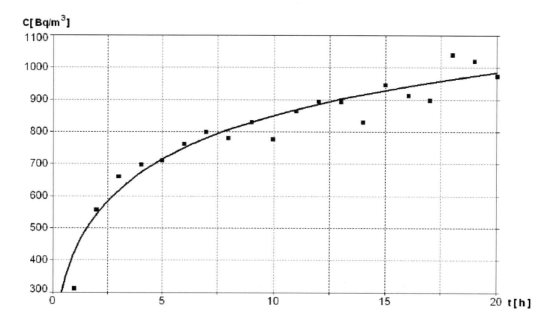

Figure 7. Increase in the concentration of Rn-222 that emanated from zirconium-silicate.

The data obtained in this experiment were fitted using the following formula:

$$C(t) = \frac{1}{a + \frac{b}{t^{0.5}}} \ [Bq/m^3], \qquad (7)$$

where t = time (h), a = 0.0006355, and b = 0.001704. The saturated activity concentration was determined with 1/a, while the speed of the increase in the concentration of Rn-222 was characterised by $b/t^{0.5}$. Taking into account the fact that the zirconium-silicate was already in the form of a fine powder with an average particle size of 0.39 μm, it was not possible to test the emanation rate of Rn-222 for larger granulations.

Knowing the concentration of Ra-226 in the sample of zirconium silicate, it is possible to estimate the amount of Rn-222 that emanates from the sample, i.e., Rn-222 that did not stay bound within the material of the sample. The estimated saturation activity based on the previous fitted curve is 1/a ≈ 1600 Bq/m³.

The total activity of Ra-226 in the zirconium-silicate sample is A ≈ 3290 Bq. If Rn-222 completely emanated from the zirconium silicate, then its activity would be equal to the activity of Rn-226 after achieving radioactive equilibrium in a closed container (t ≈ 10 $T_{1/2}$(Rn-222) ≈ 40 days), i.e., inside a glass container, the Rn-222 activity would be 3290 Bq, corresponding to the activity concentration of A/V = 600 000 Bq/m³. Therefore, the factor of emanation of Rn-222 from the zirconium silicate is very low: 1600 Bq/m³ / 600 000 Bq/m³ ≈ 0.3%.

The results of the measurements of radon emanation from siporex, red brick, ceramic pads, and sand for three different granulations are shown in Figures 8, 9, 10 and 11, respectively, and in Tables 4 and 5.

For siporex it is evident that differences between increments of radon concentrations do exist during the measurement period for different granulations.

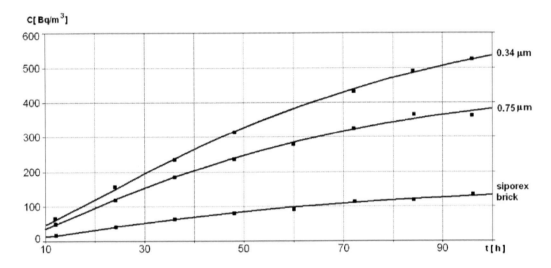

Figure 8. Emanation of radon from siporex.

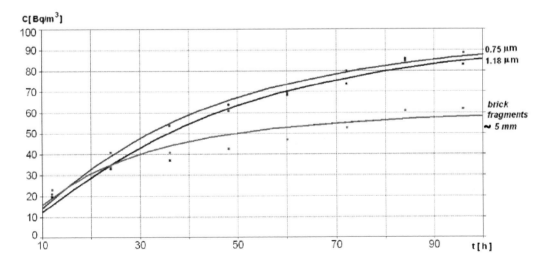

Figure 9. Emanation of radon from red brick.

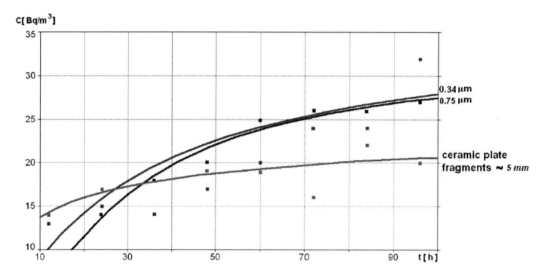

Figure 10. Emanation of radon from ceramic pads.

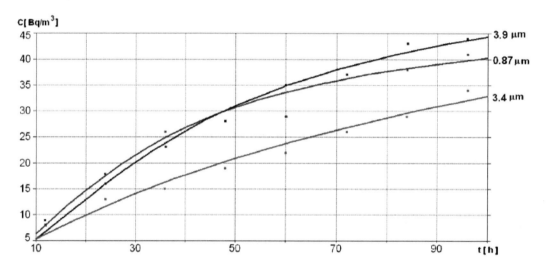

Figure 11. Emanation of radon from sand.

For red brick, the same functional dependency that was used for siporex was applied to fit the results. The results of the fitting are very similar for the 0.75 μm and 1.18 μm granulations, but a weaker emanation was detected from the broken brick fragments.

For ceramic pads, the functional dependency of radon activity concentration in relation to time is given in Figure 10. The weakest emanation is from the fragments of the ceramic pads, which is similar to the results obtained for the brick.

The dependency of the emanation rate for sand on the granulation was the weakest (Figure 11). There was no tendency for a greater degree of emanation with the reduction of the size of the particle. This suggests that the structure, or rather the composition, is the dominant aspect of radon emanation for sand.

The relative measurement uncertainties of the displayed results of the experiment range from 10%–20%.

The Analysis of Radon Diffusion, Emanation and Adsorption ...

Table 4. The results of the measurements of radon emanation from siporex, red brick, ceramic pads, and sand for three different granulations

Sample Average size of particles	Function	Parameter a	Parameter b	Expected saturated concentration Rn-222 [Bq/m^3]	Radon emanation coefficient [%]
Zirconium-silicate 0.39 µm	$C(t)=\dfrac{1}{a+\dfrac{b}{t^{0.5}}}$	0.0006355	0.001704	1600	0.3
Siporex	$C(t)=\dfrac{1}{a+\dfrac{b}{t^{1.5}}}$	--------------	---------------	-----------------	-----------------
0.34 µm		0.0012253	0.64089	815	27
0.75 µm		0.0018443	0.76676	540	18
komad		0.0052515	2.22871	190	6.4
Sand	$C(t)=\dfrac{1}{a+\dfrac{b}{t^{1.5}}}$	--------------	---------------	-----------------	------------------
0.87 µm		0.020619	4.22209		
3.89 µm		0.017269	5.29733	~ 60	~ 6
3.39 µm	$C(t)=\dfrac{1}{a+\dfrac{b}{t}}$	0.013042	1.73638		
Ceramic pads	$C(t)=\dfrac{1}{a+\dfrac{b}{t^{1.5}}}$	----------------	---------------	-----------------	--------------------
0.75 µm		0.031468	4.87489	32	0.43
0.34 µm	$C(t)=\dfrac{1}{a+\dfrac{b}{t^{0.5}}}$	0.027218	0.85636	37	0.5
Peaces (~ 5 mm)	$C(t)=\dfrac{1}{a+\dfrac{b}{t}}$	0.036906	0.11452	27	0.36
Red brick	$C(t)=\dfrac{1}{a+\dfrac{b}{t^{1.5}}}$	---------------	---------------	------------------	--------------------
0.75		0.0095187	1.88110	105	2.8
1.18		0.0093617	2.26310	107	2.9
peaces (~ 5 mm)		0.015563	1.53593	65	1.7

The concentrations of radon depending on time inside the glass chamber, measured by alpha-spectrometer RAD7, during all exposures of zeolite samples to radon, are presented in Figure 12.

The results of gamma-spectrometry measurements by method described in [12], performed in order to precisely quantify the adsorption characteristics of various zeolite granulations, are presented in Table 1 [13].

Relative uncertainties of presented net intensities of 352 keV gamma ray line, associated with lead-214, a short-lived decay product associated with radon-222, expressed at 95% confidence level, are about 10%.

Table 5. The calculated factors of radon emanation

Sample	Siporex			Sand			Ceramic pads			Red brick			Zirconium silicate
Averaged size of particles [μm]	0.34	0.75	bulk	0.87	3.89	3.39	0.34	0.75	pieces	0.75	1.18	pieces	0.39
Radon emanation coefficient [%]	27	18	6.4	~ 6	~ 6	~ 6	0.5	0.43	0.36	2.8	2.9	1.7	0.3

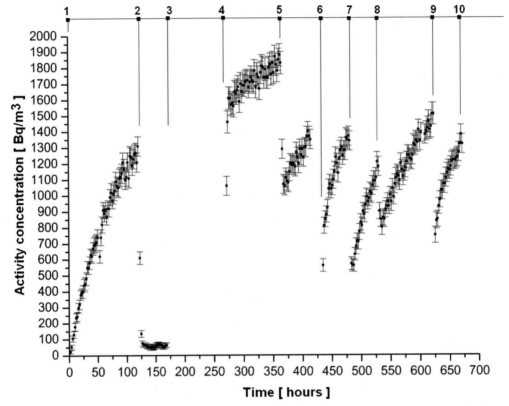

Explanation for corresponding time intervals: 1-2 - only zirconium-oxide closed inside glass chamber; 2-3 – activated charcoal + zeolite, not milled (grain size 5-10 mm); 3-4 - without measurement (only zirconium-oxide closed inside glass chamber); 4-5 - only zirconium-oxide closed inside glass chamber; 5-6 – test of radon loss by short opening (5 seconds) of glass chamber; 6-7 - exposure of the finest zeolite fraction (40 minute milling); 7-8 - exposure of zeolite fraction, 20 minutes milled; 8-9 - only zirconium-oxide closed inside glass chamber; 9-10 - exposure of zeolite fraction, 10 minutes milled. Error bars are presented for each single measurement.

Figure 12. Time dependence of radon concentration inside the glass chamber.

Although relatively simple, our experimental setup provided the possibility for estimation of adsorption coefficients of Rn-222 gas on on zeolite.

The Analysis of Radon Diffusion, Emanation and Adsorption ...

Table 6. Gamma-spectrometry measurements of samples

Sample	Intensity of post-radon 352 keV line before sample exposure to radon [c/(s kg)]	Intensity of post-radon 352 keV line after sample exposure to radon [c/(s kg)]	Net intensity of post-radon 352 keV line due to radon adsorption by sample [c/(s kg)]
Zeolite (not milled)	0.23	0.78	0.55
Zeolite (40 min milling)	0.23	0.70	0.47
Zeolite (20 min milling)	0.23	1.18	0.95
Zeolite (10 min milling)	0.23	0.75	0.52

This coefficient, k (m^3/kg), characterizes the capacity that sorbent material has in adsorbing of radon, and can be determined in the following way [14]:

$$k = Q/C \tag{8}$$

where Q is radon activity adsorbed by unit of mass of sorbent material (Bq/kg), and C is concentration of Rn-222 in air during exposure of sorbent material (Bq/m^3). In our case the time interval of exposure to radon of each zeolite sample was 48h. The temperature of air inside the glass chamber was in interval $(25 - 30)$ °C, while the relative humidity was about 7%.

Knowing the net intensity of 352 keV post-radon line for activated charcoal after exposure (Table 6), gamma emission probability for this line ($p_\gamma = 0.371$), and detection efficiency at this energy of our HPGe detector for cylindrical geometry of measured sample ($\varepsilon = 0.03$), we found $Q = 42$ Bq/kg for zeolite (40 min milling). The average radon concentration inside glass chamber for 48h exposure interval for this zeolite sample (time interval 6 -7 in Figure12) was relatively high, $C \approx 1100$ Bq/m^3. From these values of Q and C, we found $k_{zeolite}$ (40 min. milling) = 0.038 m^3/kg.

Based on values given in Table 6 and radon, the adsorption coefficients of zeolite samples with 20 min milling and 10 min milling are found to be 0.11 m^3/kg and 0.046 m^3/kg, respectively. The relative uncertainties of derived adsorption coefficients reach up to 20%.

CONCLUSION

For materials used in this experiment, the aluminum sheet has lowest value of diffusion coefficient. Slightly higher value (1.35 times compared to aluminum) has a tire, and follow the condor material and Tarkett linoleum. As the measured values of diffusion coefficients for these materials $\sim 10^{-10}$ m^2/s, they are very good insulators of radon and recommended them to use in residential care facilities. For Speer plate and Tarkett laminate significantly higher values of diffusion coefficient of radon ($13(17) \cdot 10^{-7}$ m^2/s and $25(18) \cdot 10^{-10}$ m^2/s) were obtained, and these materials are not good insulators of radon due to their relatively high porosity. Based on the results of experiments described in this paper can be noticed significant differences in the quality of certain material and their possible use to improve the protection of the interior of the existing facilities of the radon, as well as the design of new buildings.

In this chapter, the influence of the granulation of building materials on radon emanation has been considered. Apart from the zirconium-silicate, the concentration of Ra-226 in these materials was relatively low, which lead to relatively large measurement uncertainties during the measurements of the concentration of Rn-222 inside the sealed glass chamber using the RAD7 device.

Even though the zirconium-silicate was in the form of a fine powder with an average particle size of 0.39 µm, the factor of radon emanation from this material was very low (below 0.5%), which indicates a weak dependency between the factor of emanation and the granulation. It was also found that in the case of sand, granulation did not play an important role in the radon emanation rate. It can be concluded that the degree of emanation was not increased with the reduction of the size of the sand particles. For ceramic pads and red bricks, a certain reduction of the factor of emanation was observed when these materials were in the form of broken fragments rather than in powder form. The most prominent influence of the granulation on the factor of radon emanation was detected in the sample of the siporex. The factor of emanation was the highest (27%) for the finest powder granulation (average particle size of 0.34 µm), while this factor was approximately 30% lower for a rougher granulation (0.75 µm). The factor of emanation from a piece of siporex block was four times lower than the factor for the finest granulation of siporex. This can be attributed to the longer mean free path of radon within siporex (due to less crystalline character of this material) than in other analyzed materials, so the finer siporex granulation leads to the higher radon escape.

In this chapter it is shown that simple experimental setup could be used for radon adsorption in materials investigation. The adsorption coefficients obtained in our experiment for natural zeolite samples were in the range (0.038 m^3/kg - 0.11 m^3/kg).

For the optimal grain size of the zeolite sample (achieved by 20 minute milling), the exposure of the zeolite to the average radon concentration of about 800 Bq/m^3, during 48h, resulted in an activity increase of the 352 keV post-radon line of about 1 c·s-1·kg-1 (35% nominal efficiency HPGe).

ACKNOWLEDGMENTS

The authors acknowledge the financial support of the Provincial Secretariat for Science and Technological Development within the project Development and application low-background alpha, beta spectroscopy for investigating of radionuclides in the nature, City Administration for Environmental Protection, City of Novi Sad, Serbia and Ministry of Education and Science of Serbia, within the projects Nuclear Methods Investigations of Rare Processes and Cosmic Rays No.171002 and Biosensing Technologies and Global System for Continuous Research and Integrated Management of ecosystems No.43002.

REFERENCES

[1] Colle, R., Rubin, R. J., Knab, L. I., Hutchinson, J. M. R., *Radon transport through and exhalation from building materials,* US Department of Commerce, 1981.

[2] S. Stoulos, M. Manolopoulou, C. Papastefanou, Measurement of radon emanation factor from granular samples: effects of additives in cement, *Applied Radiation and Isotopes* 60, pp. 49-54(2004).

[3] S. C. Lee, C. K. Kim, D. M. Lee, and H. D. Kang, Natural Radionuclides Contents and Radon Exhalation Rates in Building Materials used in South Korea, *Radiation Protection Dosimetry*, Vol. 94, No. 3, pp. 269-274 (2001).

[4] International Commission On Radiological Protection, *Limits on Inhalation of Radon Daughters by Workers, Publication 32*, Pergamon Press, Oxford (1981).

[5] P. Bossew, The radon emanation power of building materials, soils and rocks, *Applied Radiation and Isotopes* 59, pp. 389-392 (2003).

[6] Bikit, J., Slivka, D., Mrdja, N., Žikić-Todorović, S., Ćurčić, E., Varga, M., Vesković, Lj. Čonkić: Simple Method For Depleted Uranium Determination, *Jpn. J. Appl. Phys.*, 2003, No. 42, str. 5269-5273.

[7] Bikit, I., Mrdja, D., Todorovic, N., Varga, E., Forkapic, S., Veskovic, M., Slivka, J., Conkic, Lj.: Possibility Of Prompt U-238 Activity Concentration Determination By Gamma-Ray Spectroscopy, *Jpn. J. Appl. Phys.*, 2005, Vol. 44, No. 1A, str. 377-379.

[8] N. Todorović, S. Grujić, U. Kozmidis-Luburić, I. Bikit, D. Mrđa, S. Forkapić, Radon Diffusion Through Various Building Materials, *6th Conference on protection Against Radon at Home and at Work*, 13-17 September 2010, Prague, Czech Republic.

[9] Bikit, I., Mrđa, D., Grujić, S., Kozmidis-Luburić, U., Granulation Effects On The Radon Emanation Rate, *Radiation Protection Dosimetry*, (2011), vol. 145, br. 2-3, str. 184-188.

[10] M. Jiranek, K. Rovenska, A. Fronka, Radon diffusion coefficient – a material property determining the applicability of waterproof membranes as radon barriers, *Proceedings of the American Association of Radon Scientist and Technologist, AARST 2008*, Las Vegas.

[11] G. Keller, B. Hoffmann, T. Feigenspan, Radon permeabilty and radon exhalation of building materials, *The Science of the Total Environment* 272, pp. 85-89 (2001).

[12] I. Bikit, D. Mrdja, J. Nikolov, K. Bikit, and S. Forkapic, Methods of low level gamma spectroscopy, *Gamma Rays: Technology, Applications and Health Implications* (2013), Nova Science publishers Inc., pp. 41-73.

[13] Bikit, I., Grujić, S., Mrđa, D., Bikit, K., Forkapić, S., Kozmidis-Luburić, U.: Radon Absorption by Zeolite, *7th International Conference on Protection Against Radon at Home and at Work*, Prague, 2-6 September 2013.

[14] López, F. O. and Canoba, A. C. 222Rn gas diffusion and determination of its adsorption coefficient on activated charcoal. *Journal of Radioanalytical and Nuclear Chemistry* 252, No. 3 (2002), 515-521.

INDEX

#

20th century, viii, 99
^{222}Rn activity, viii, 44, 60, 78, 79, 81, 82, 85, 87, 119, 193, 217, 221, 222

A

accelerator, 19
accommodation, 150
accounting, 68, 179, 221
acetaldehyde, 141
acid, 100, 126, 128, 137, 139, 141, 166, 167
acidic, 100, 122
acidity, 100
ACTH, 136
activated carbon, 231
activity level, 153
activity rate, 138
additives, 267
adjustment, 197
adrenocorticotropic hormone, 136
adsorption, xii, xiii, 189, 193, 197, 198, 206, 208, 215, 231, 232, 233, 251, 252, 254, 258, 259, 263, 264, 265, 266, 267
adults, 171, 179, 180, 181, 183
aerosols, 121, 124, 177, 246
age, xi, 96, 122, 123, 164, 176, 235
age-related diseases, 123
air quality, 113, 116
air temperature, 242
airborne particles, 177
airways, 177
alanine, 138, 146
alanine aminotransferase, 138, 146
alcoholic liver disease, x, 133, 136, 138, 144, 145, 146
Algeria, 40, 184

algorithm, 246
alkylation, 141
alpha activity, 25, 191, 221
alpha particles, vii, 1, 119, 123, 124, 153, 176, 179, 191, 213, 215, 217, 236
ALT, 138
aluminium, 253
ambient air, 120
amplitude, 221
analgesic, 121, 126, 127, 134, 136, 143, 148
angstroms, 191
ankylosing spondylitis, 121, 127, 135, 144
antioxidant, x, 121, 123, 133, 134, 135, 137, 138, 141, 143, 144, 145
antioxidative activity, 141
apoptosis, 142
aquatic life, vii, 43, 44
aquatic systems, 130
aquifers, 44, 46, 47, 48, 50, 53, 68, 71, 77, 92, 94, 96, 120, 122, 125, 175, 176
argon, 150
arithmetic, 153, 205, 237, 241, 242
arterial hypertension, 128
arteries, 128
arthritis, 121, 144
ascorbic acid, 137, 141, 145
assessment, viii, 31, 35, 43, 44, 45, 94, 95, 101, 116, 144, 164, 183, 187, 208, 236, 245, 248
asthma, 121, 135
atherosclerosis, 128
atmosphere, viii, ix, xii, 10, 23, 44, 45, 59, 60, 67, 82, 109, 110, 150, 171, 190, 194, 196, 209
atmospheric pressure, xii, 55, 209, 229
atoms, x, 117, 219, 246
audit, 199
Austria, 94, 122, 126, 167, 185, 207, 208, 250
authorities, ix, 109, 110, 248

B

background radiation, vii, 164, 183
Balkans, 166
balneotherapy, vii, x, 118, 120, 126, 127, 171
banks, viii, 44, 46, 50
barometric pressure, 148
barriers, 166, 267
base, 44, 54, 58, 71, 91, 128, 206, 229
basic research, 16
baths, 118, 119, 122, 124, 126, 127, 135, 183
batteries, 2
bauxite, 253
behaviors, 143
beneficial effect, x, 126, 133, 134, 137, 145
beta particles, 220
bias, 19
bicarbonate, 168, 173
bioavailability, 123
biological activity, 123
biological systems, 144
bismuth, 213, 228
blood, 119, 121, 127, 128, 134, 176, 179, 217
blood circulation, 119
blood flow, 176
blood pressure, 119, 121, 127, 134
blood supply, 128
blood vessels, 121
bloodstream, 164
body weight, 141
boreholes, 168, 170, 171, 176, 178
brain, 134, 137, 138, 142, 144, 145, 146, 148
brain stem, 148
Brazil, 118, 120, 121, 122, 123, 124, 125, 126, 128
breakdown, 3, 4, 36
breathing, 178
bronchial asthma, 121, 135, 144
bronchial epithelium, 176, 180
bronchiectasis, 121
bronchitis, 121
Bulgaria, 151, 159, 208
by-products, 229

C

cadmium, 221
Cairo, 94
calcium, 93, 257, 258
calibration, xii, 3, 6, 9, 12, 14, 16, 17, 18, 21, 23, 29, 31, 54, 189, 193, 194, 195, 196, 197, 198, 199, 206, 207, 220, 222, 232, 233

campaigns, viii, 28, 44, 55, 59, 61, 62, 68, 71, 72, 73, 75, 78, 80, 82
Campania region, viii, 43, 47, 54, 69, 74, 82
cancer, 116, 128, 150, 176, 177, 184, 207, 222
cancer death, 222
carbon, 2, 3, 4, 27, 32, 134, 144, 145, 147
carbon dioxide, 32
carbon tetrachloride, 134, 144, 145, 147
carcinogen, 171, 183
case study, 56, 60, 87, 91, 93, 94
catchments, 46, 48, 52
cation, 167
cell cycle, 144
Central Europe, viii, 99, 100
central nervous system, 127
ceramic, xiii, 251, 256, 261, 262, 263, 266
cerebral edema, 145
CFR, 161
challenges, 95
charge density, 4
chemical(s), viii, x, 44, 46, 55, 58, 71, 75, 76, 77, 91, 94, 117, 119, 120, 122, 128, 135, 167, 170, 174, 177, 191, 193, 194, 216, 236
chemical characteristics, 122
chemical etching, 216, 236
children, 114, 115, 128, 171, 179, 183
China, 122, 124
chromosome, 183
chronic diseases, 126
chronic illness, 128
circulation, xiii, 45, 50, 69, 119, 165, 173, 174, 213, 251
cirrhosis, 146
cities, 152
citizens, 182, 235
city, 149, 161, 186, 266
classes, 47, 103, 257
classification, 91, 95, 120, 123
classroom, 202
cleaning, 151, 178
clinical trials, 121, 126, 144
closure, 72
CO2, 126, 165, 166, 168, 173
coal, xii, 189, 194, 202, 253
colitis, 142, 148
collaboration, 242
Colombia, 39
colon, 142, 177
color, 150, 174
commercial, 26
communication, 101, 230
communities, 44
community, 128

compilation, 123
complex interactions, vii, 43, 69, 91, 92
complexity, 47, 50
compliance, 87
composition, 120, 121, 167, 170, 174, 193, 211, 262
computation, 16
computer, 216
conceptual model, viii, 44, 63, 68
conceptualization, 68, 69
condensation, 28
conditioning, 47, 50, 69
condor, 252, 265
conductivity, 4, 17, 38, 46, 71, 73
conductor, 153
congress, 92, 96, 207
conjugation, 141
constipation, 121
constituents, 118, 167
construction, 3, 151, 194, 211, 221, 235, 242, 248, 252
constructional materials, 116
consumption, xi, 118, 126, 127, 128, 161, 163, 179, 181, 190, 217
containers, 153, 211, 234, 253
contaminant, 167
contamination, viii, 43, 44, 178, 199
continental, 50, 71
Continue Natural Ventilation (CNV), ix, 109, 110, 111, 113, 114, 115
control group, 135
conversion rate, 8
cooling, 245, 246, 247
copper, 221
correction factors, 18, 38, 194
correlation, viii, 11, 22, 57, 99, 100, 104, 106, 107, 167, 238, 239
correlation coefficient, 107
cosmetics, xiii, 251
cost, 36, 115, 184, 191, 215, 241
cost benefit analysis, 241
covering, 2, 21, 30
cracks, 101, 239, 252
CRM, 20, 211, 212
Croatia, 208
crops, 208
crude oil, 175, 231
crust, 22, 50, 119, 125, 164
crystalline, 165, 169, 170, 171, 185, 266
crystals, 191
CT, 67, 84, 85, 86
Cuba, 126
curcumin, 148
cure(s), 118, 119, 126, 143

cycles, 214, 219, 244, 254
cystitis, 126
cytoplasm, 138
Czech Republic, ix, 99, 100, 101, 119, 120, 126, 178, 267

D

damages, 141
data analysis, 25, 77
database, 54, 120, 122, 242
deaths, 210, 222
decay, vii, viii, x, xii, 1, 3, 12, 18, 20, 21, 24, 25, 27, 28, 31, 33, 38, 45, 60, 94, 99, 100, 117, 118, 120, 122, 123, 125, 127, 150, 153, 164, 165, 167, 171, 176, 177, 179, 180, 183, 187, 189, 190, 191, 193, 194, 195, 196, 197, 206, 208, 209, 210, 213, 215,216, 217, 218, 224, 228, 229, 231, 233, 246, 263
decongestant, 121
decontamination, 159
deficiency, 18
degenerative joint disease, 121
deoxyribonucleic acid, 137
Department of Energy, 26
deposition, viii, x, 3, 25, 99, 118, 178
deposition rate, 178
deposits, viii, 68, 71, 77, 99, 120, 122, 165, 174
depression, 115, 134
depth, viii, 99, 101, 119, 164, 168, 170, 174, 175, 215, 234
dermatology, 127
desorption, 181, 193, 215, 220
detectable, 16, 65, 127, 214, 218
detection, xii, 12, 16, 22, 33, 35, 36, 210, 213, 218, 222, 231, 233, 245, 247, 254, 265
detection system, xii, 210
detoxification, 137, 142
deviation, 15, 153, 203, 204, 205, 241, 242
diabetes, 122, 128, 134, 139, 141, 144, 147
dielectric constant, 4, 7
diet, 126
differential equations, 246, 247
diffusion, viii, xiii, 23, 24, 44, 60, 82, 213, 216, 221, 228, 235, 236, 251, 252, 253, 254, 259, 265, 267
diffusion time, 23, 24
dilation, 127
dipoles, 3, 4
directives, 229
directors, 204
discharges, 17, 19
discrimination, 220, 221, 222
discs, 255

disease activity, 142

diseases, x, 121, 126, 127, 128, 133, 134, 135, 136, 137, 138, 139, 141, 143

dislocation, 170

dispersion, 120

disposition, 239

dissolved oxygen, 46, 55, 71

distilled water, 153, 218, 220

distribution, x, 17, 38, 68, 110, 117, 118, 154, 155, 158, 165, 183, 190, 230, 237, 245, 256, 257, 258

diversity, 191

DNA, 137, 141, 176

DNA damage, 176

doctors, 135

DOI, 161

dosage, 121

dosing, 146

draft, 15, 16

drainage, 45, 50, 69, 70

drawing, 112, 213

drinking water, vii, x, xi, 43, 44, 50, 82, 94, 118, 119, 125, 126, 129, 149, 150, 151, 154, 157, 158, 159, 160, 161, 164, 167, 177, 178, 179, 180, 181, 182, 183, 184, 187, 211, 217, 220, 222, 224, 225, 248

drugs, 136, 146

drying, 218

dynamic viscosity, 252

E

ecology, 161

edema, 121, 137, 142, 148

electret, vii, 1, 2, 3, 4, 5, 6, 7, 8, 9, 10, 11, 12, 13, 14, 15, 16, 17, 18, 19, 21, 23, 24, 25, 27, 28, 30, 36, 37, 38, 39, 40, 191, 211

Electret Ion Chambers (EICs), vii, 1, 5, 7, 8, 9, 10, 11, 13, 14, 15, 16, 17, 18, 19, 20, 22, 23, 29, 31, 36, 40, 41, 42

electric field, 2, 3, 4, 9, 10, 15, 16, 17

electrical conductivity, 55, 71, 75, 123

electricity, 193

electrodes, 4

electron, 19, 144

electrostatic field, vii, 1, 2, 11, 19

elementary school, 202, 204, 205

e-mail, 227, 251

emission, ix, 19, 42, 84, 100, 110, 167, 247, 265

emitters, 191

emphysema, 121

encoding, 138

endocrine, 127

endocrine system, 127

energy, 25, 119, 124, 127, 150, 176, 198, 199, 220, 241, 265

energy transfer, 176

engineering, 96, 164, 212, 230

England, 116

environment, xi, 2, 10, 15, 18, 23, 27, 28, 44, 45, 46, 68, 91, 93, 95, 111, 120, 121, 122, 124, 163, 171, 176, 179, 180, 183, 185, 206, 211, 213, 217, 221, 248, 249, 252

environmental aspects, 130

environmental conditions, xii, 28, 122, 209, 245

environmental impact, vii

environmental protection, 248

Environmental Protection Agency (EPA), xii, 21, 22, 39, 42, 151, 153, 161, 167, 177, 181, 184, 187, 189, 191, 192, 193, 194, 195, 196, 201, 202, 205, 206, 207, 208, 214, 215, 220, 223, 225, 231, 232

enzymatic activity, 121, 126

enzymes, 135, 136, 137, 138, 140, 141, 144

E-PERM®, vii, 1, 2, 14

epidemiologic, 107, 124

epithelium, 177, 178

equilibrium, xii, 18, 26, 27, 34, 39, 124, 177, 180, 189, 193, 195, 215, 218, 220, 234, 241, 245, 248, 250, 259, 261

equipment, xii, 3, 22, 189, 199, 213, 216, 235, 236

erosion, 174

ERS, 211

etching, 191, 216, 236

ethanol, 137, 145, 147

Europe, vii, x, 1, 93, 119, 133, 134, 135, 165, 179, 191, 202, 207, 221, 237, 248, 249

European Commission, 126, 187, 206, 222, 223, 242, 249

European Community, 93, 179

European Parliament, 93

European Union (EU), 93, 95, 178, 182, 192, 221, 249

European Water Framework Directive (EWFD), viii, 43, 44, 93

evapotranspiration, 124

evidence, 198, 221

excitation, 127

excretion, 121, 127

exercise, 144

exploitation, 123

exposure, ix, xi, xii, 6, 11, 14, 17, 18, 23, 34, 87, 89, 90, 109, 110, 116, 119, 122, 123, 124, 126, 127, 128, 134, 138, 139, 146, 163, 171, 176, 177, 178, 179, 180, 182, 183, 184, 186, 187, 189, 190, 191, 192, 193, 194, 195, 196, 197, 201, 202, 205, 206, 208, 210, 212, 215, 216, 217, 222, 223, 224, 229,

230, 231, 232, 233, 234, 235, 236, 242, 248, 249, 250, 252, 258, 264, 265, 266

extraction, ix, 110, 111, 217

extracts, 218

F

facial pain, 148

families, 63

Federal Register, 161, 184

fiber, 246

films, 215, 236

filters, xii, 28, 189, 191, 213

filtration, 45

financial, 184, 222, 248, 266

financial support, 184, 222, 248, 266

Finland, 178

fish, 45

flavonoids, 137

flaws, xii, 190

flexibility, 135

fluctuations, 45, 60, 159, 211, 232

fluid, x, 3, 117, 118, 126, 165

fluvial eco-systems, vii, 43

foils, 36

food, xiii, 136, 251

force, 123

Ford, 45, 69, 93

formation, 58, 168

formula, 179, 180, 232, 241, 245, 259, 260

foundations, ix, 99

fractures, 45, 58, 165, 166, 169

fragments, 256, 262, 266

France, ix, 93, 95, 109, 110, 111, 113, 115, 116, 122, 126, 206, 207

free radicals, x, 133, 134, 137, 138, 139, 141, 143

frequency distribution, 237

G

gamma dose rates, viii, 99

gamma radiation, xii, 3, 8, 9, 11, 13, 14, 15, 16, 17, 19, 40, 79, 189, 190, 229

gamma rays, 218

gas diffusion, 267

gastrointestinal tract, 164, 176

genes, 138

geology, vii, 22, 34, 36, 115, 164, 168, 170, 171, 183, 184, 241

geometry, xii, 189, 195, 199, 205, 246, 259, 265

geophysical prospecting, vii, 1

Georgia, xi, 149, 151, 159, 160

germanium, 194

Germany, ix, 16, 99, 101, 119, 121, 126, 178, 200, 211, 212, 258

glucose, 136

glutathione, 136, 138, 139, 145, 146, 147

glycogen, 134

gout, 121, 127

GPS, 71

grain size, 256, 264, 266

graph, 79, 80

Greece, 127

grounding, 3

groundwater, vii, x, xi, 43, 44, 45, 46, 47, 48, 50, 51, 53, 55, 57, 58, 65, 67, 68, 69, 71, 72, 74, 77, 79, 80, 82, 91, 92, 93, 94, 95, 96, 97, 117, 118, 119, 122, 123, 125, 163, 164, 165, 166, 167, 168, 169, 173, 175, 178, 183, 216, 220, 224, 239

groundwater assessment, viii, 43

growth, 121, 150

guidance, 80, 180, 217

guidelines, xii, 192, 209

H

half-life, x, 45, 46, 117, 118, 119, 122, 126, 195, 210, 213, 219, 228, 233

hardness, 215

hazards, x, 118, 207

healing, 119, 128

health, vii, x, xi, xiii, 110, 118, 119, 128, 150, 163, 164, 169, 171, 177, 179, 183, 184, 190, 210, 211, 214, 217, 221, 227, 248

health effects, xiii, 164, 183, 190, 227

health risks, x, xi, 118, 119, 163, 217

heart disease, 126

height, 170, 196, 214, 216, 236, 259

helium, 30, 95, 150, 217

hepatitis, 138, 145, 146

hepatopathy, x, 133, 134, 139, 140, 141, 143, 144, 147

hepatotoxicity, 137, 145, 147

high blood pressure, 121

high school, 202, 204, 205, 235

highlands, 47, 48, 50

highways, 151, 159

hippocampus, 142

histogram, 80, 158, 254

historical data, 55

history, viii, 99, 153

holocene, 82

homeowners, 201

homes, 10, 11, 18, 22, 27, 34, 86, 116, 164, 178, 179, 191, 192, 201, 202, 206, 211, 222, 230

hot springs, 122, 124, 175
hotel(s), 118, 124, 173, 244, 245
house, 56
housing, 111, 152, 158, 192
human, vii, x, xii, xiii, 44, 81, 82, 91, 118, 119, 127, 128, 142, 164, 179, 180, 183, 190, 209, 210, 217, 222, 251
human body, 127
human exposure, 119, 179, 183
human health, vii, x, xii, 81, 82, 91, 118, 119, 164, 183, 209, 210, 222
humidity, xii, 2, 3, 4, 10, 11, 13, 15, 17, 22, 27, 34, 38, 134, 193, 194, 195, 196, 209, 212, 215, 218, 219, 229, 233, 239, 245, 258, 265
Hungary, 96, 122, 185, 187, 214, 216, 236, 249, 250
hydrogen, 136
hydrogen peroxide, 136
hydro-geomorphological system, viii, 44, 68
hydrological behaviour, viii, 43
hydroxide, 168, 174
hydroxyl, 136
hypertension, 122, 134, 144
hyperthermia, 135
hypothesis, 101, 102, 121
hypothesis test, 121

I

ID, 72, 144, 147, 148
ideal, 215, 254
identification, 91, 152
immersion, 126, 127
immune function, 134
immune system, 119, 127
immunity, 128
improvements, 126
impulses, 150
in vivo, 135, 147
incidence, 179
India, 16, 120, 121
individuals, 230
inducer, 139
induction, 146
industries, 229
industry, ix, 99, 127, 253
infarction, 145
infection, 128
inflammation, 128, 148
inflammatory cells, 135
inflammatory responses, 142
ingestion, x, xii, 118, 119, 126, 127, 128, 161, 164, 176, 177, 179, 180, 182, 183, 184, 209, 210, 222, 229

inhibition, 121, 134, 137, 141, 142
inhibitor, 148
injury, x, 133, 139, 142, 143, 147, 148
inner ear, 148
institutions, 100, 106, 207
in-stream springs, viii, 43
Insufflating Mechanical Ventilation (IMV), ix, 109, 110, 111, 113, 114, 115
insulation, ix, 99, 202
insulators, 265
insulin, 134, 136, 141
integration, 95
interface, 45
interference, 23, 24, 40
interferon, 145, 146
International Atomic Energy Agency, 94, 187, 192, 225
international standards, 16
intervention, 135, 192, 230, 236
intrusions, 122
ion collection, vii, 1
ionising radiation, 230, 249
ionization, vii, 1, 2, 3, 8, 9, 10, 11, 13, 19, 22, 37, 38, 127, 136, 191, 212, 213, 230
ionizing radiation, xii, 9, 10, 19, 38, 89, 127, 208, 209, 229, 230, 249
ions, vii, 1, 2, 4, 6, 10, 12, 15, 17, 25, 120, 167
Iowa, 186
Iran, 130, 150, 159, 161
Ireland, 178
iron, 137, 139, 168, 174, 210, 231, 234, 259
irradiation, x, 17, 133, 136, 138, 139, 140, 141, 146, 147, 164
irrigation, viii, xi, 44, 82, 163
ISC, 72
ischemia, 137, 142, 147, 148
ischemia-reperfusion injury, 137
isolation, 228, 237, 242, 248
isotope, 100, 228
Israel, 94, 178
issues, 44, 217
Italy, viii, 43, 47, 69, 71, 74, 82, 92, 93, 96, 119, 121, 122, 126, 127, 131, 212

J

Japan, x, 23, 46, 121, 122, 123, 126, 133, 134, 135, 186, 214
joints, 50, 128, 166, 169, 173, 183

Index

K

karst systems, viii, 44, 93
Kartli, 151
kidney(s), 126, 134, 138, 141, 142
kindergartens, 202, 203, 204, 205, 206, 207, 208
Korea, 122

L

lakes, 30, 44, 178
landscape(s), viii, 43, 45, 50, 94, 100
languages, 135
laryngitis, 121
laws, 45
leakage, 173
legislation, 87, 91, 95, 123, 127
leisure, 118, 123
lifetime, 165
light, 44, 48, 49, 81, 113, 213, 257
limestone, 49, 50, 51, 52, 55, 69, 71, 77, 96, 165, 168, 171, 173
lipid peroxidation, 136, 137
lipid peroxides, 122
liquid phase, 34
liquids, xiii, 251, 254
lithology, 165
Lithuania, 178
liver, x, 133, 134, 136, 137, 138, 139, 140, 141, 142, 143, 144, 145, 146, 147, 177
liver cirrhosis, 145
liver damage, 136, 137, 140, 145, 147
liver disease, 136, 138, 143, 145
low risk, 106
low temperatures, 126
lung cancer, 82, 100, 101, 110, 116, 128, 150, 164, 177, 179, 184, 190, 210, 221, 229
lung disease, 121

M

Macedonia, 161, 202, 208
magnesium, 174
magnet, 3
magnetic field, 15, 19
magnetic resonance, 145
magnetic resonance imaging, 145
magnets, 2
magnitude, 45, 58, 69, 86, 90, 94, 95, 104, 197, 217, 247
majority, 123, 128, 159, 167, 170, 178
management, viii, ix, 43, 44, 92, 93, 94, 96, 109, 110

manufacturing, 21
mapping, xiii, 25, 186, 222, 223, 227, 242
Maryland, 42
mass, 66, 124, 164, 193, 194, 195, 196, 234, 254, 258, 265
materials, xii, xiii, 2, 3, 21, 27, 32, 33, 36, 79, 87, 89, 123, 151, 190, 192, 196, 199, 209, 221, 228, 229, 231, 247, 248, 251, 252, 253, 254, 255, 256, 259, 265, 266, 267
matrix, xii, 166, 189, 195, 199, 205
Mechanical Ventilation with Double flow (MVD), ix, 110, 111, 114, 115
median, 22
medical, xiii, 17, 119, 126, 128, 129, 135, 167, 171, 183, 211, 227
medical reason, 171
medicine, 38
Mediterranean, viii, 43, 44, 45, 91, 92, 93, 95
melatonin, 137, 145
mellitus, 123
membranes, 24, 25, 267
meta-analysis, 135, 144
metabolism, 141, 147
meter, viii, 5, 13, 20, 27, 42, 99, 120
methodology, 30, 31, 46, 93, 198
Mexico, 22
mice, x, 133, 134, 137, 138, 139, 140, 141, 142, 143, 144, 145, 146, 147, 148
microscope, 191, 216
microstructure, 254
migration, 46, 164, 165, 166, 230
mineral water, 81, 118, 119, 120, 122, 123, 124, 127, 128, 130, 166, 167
mineralization, 82, 168, 169, 185
miniature, 2
Ministry of Education, 184, 222, 266
Miocene, 171, 173, 175, 176
mitochondria, 136
mitogen, 136
mixing, 46, 66, 67, 77, 93, 96, 174, 257, 258
modelling, viii, 43, 54, 69
models, 17, 18, 45, 96, 148, 200
modifications, 113, 114, 115, 191
moisture, 232
molecules, vii, 1, 145
Montana, 135, 144
Montenegro, 249
morbidity, 101
Morocco, 122
morphine, 134, 136, 148
morphometric, 70
mRNAs, 146
mucous membrane, 120, 127

mucous membranes, 120
mutant, 147

N

National Academy of Sciences, 161
national policy, 111
National Research Council, 161, 249
NATO, 161
natural gas, 175, 231
natural resources, 119
Nd, 152, 155, 247
nephritis, 126
nephrosis, 126
nerve, 121, 143, 148
neuralgia, 121
neuronal apoptosis, 148
neurons, 142, 147, 148
neuropathic pain, 143, 148
neutrons, 17, 19
neutrophils, 136
nitric oxide, 142, 148
nitric oxide synthase, 148
nitrogen, 217
NMR, 130, 144, 147
noble gases, 10
normal distribution, 236, 237, 238, 239, 240
Norway, 178
NRC, 249
Nrf2, 138, 146
nuclei, 171, 228, 247
nucleus, xiii, 138, 148, 227
nuclides, 164
nursery school, ix, 109, 111, 114, 115

O

oil, 221
operations, 11, 128, 199
opportunities, 95
organism, 128
organs, 119, 138, 139, 141, 144, 146, 147, 150, 164, 176, 177, 179, 187, 217, 224
osteoarthritis, 134, 135, 144
overproduction, 136
oxidation, 126, 136, 141
oxidative damage, 134, 136, 139, 140, 141, 142, 144, 147
oxidative stress, 134, 136, 137, 138, 140, 144, 145, 147
oxygen, 127, 147, 168, 174

P

pain, x, 119, 127, 133, 134, 135, 141, 143, 144
Pakistan, 150, 159, 161
pancreas, 138, 142, 147
parallel, 25, 151
paralysis, 121
parents, 167
participants, 135
passive type, 29
pathogenesis, 136
pathophysiology, 135
pathways, 46, 113, 121
peat, 125
percolation, 69
peristalsis, 126
permeability, ix, 51, 53, 68, 95, 99, 100, 101, 102, 103, 104, 105, 107, 165
permeation, 126
permit, 128
permittivity, 7
peroxidation, 134
peroxide, 135, 136, 140, 146
peroxynitrite, 148
pH, 46, 55, 71, 125, 167
Philadelphia, 42
phosphate, 136, 165, 253
photons, 17, 194
physics, 235
physiological, 130
pigs, 121
pilot study, 127
placebo, 145, 146
plastics, 191
plexus, 126
Pliocene, 176
PLS, 31
Poland, 127, 167, 179, 185
polarity, 11
polarization, 3
policy, 120, 123
politics, 229
pollutants, 136
pollution, viii, 43, 44, 123
polonium, 213, 214, 219, 228
polymers, 2, 36
polypeptide, 134, 136
polypropylene, 3, 7, 13
pools, 173, 244, 245
population, vii, xi, 33, 43, 44, 87, 89, 90, 91, 122, 150, 152, 183, 190, 192, 208, 229, 230, 242, 248
porosity, 125, 165, 169, 175, 176, 183, 214, 229, 231, 239, 265

positive correlation, 77
potassium, 120, 174
precipitation, 65, 126, 168, 229
preparation, 164, 185, 194, 221, 222, 234, 244
preservation, viii, 43
President, 1
pressure gradient, 252
prevention, 134, 230
principles, 127, 184
probability, 150, 176, 216, 247, 265
probe, 5, 13, 234
professionals, 18, 28
progenitor cells, 179, 217
project, 27, 108, 160, 184, 222, 266
prostatitis, 126
protected areas, viii, 43, 45
protection, viii, 42, 43, 81, 82, 89, 137, 150, 187, 190, 192, 206, 207, 208, 222, 224, 225, 230, 239, 249, 265, 267
proteins, 146, 148
PTFE, 3, 5
public health, 164, 249
public water supplies, x, 118, 119, 151, 161
Puerto Rico, 46, 93
pumps, 191
Punctual Natural Ventilation (PNV), ix, 109, 110, 111, 112, 113
purification, xi, 163, 216
purity, 194

Q

qualifications, 101
quality assurance, 198, 206
quality control, ix, 99, 198, 199, 208
quantification, 144
quantitative estimation, 176
quartile, 101
quartz, 171
quercetin, 137

R

radiation, vii, 1, 2, 3, 6, 8, 9, 10, 11, 12, 16, 17, 19, 25, 36, 37, 38, 39, 40, 42, 89, 95, 101, 119, 120, 121, 122, 124, 125, 126, 127, 128, 130, 136, 144, 150, 153, 160, 161, 165, 171, 172, 176, 177, 178, 179, 182, 183, 184, 185, 186, 187, 189, 190, 199, 200, 206, 207, 208, 211, 212, 214, 216, 217, 220, 221, 223, 224, 225, 229, 230, 236, 239, 242, 248, 249, 250, 252, 267
radicals, x, 133, 136, 139

radio, 150
radioactive, vii, x, xi, xii, xiii, 12, 17, 28, 81, 95, 96, 100, 110, 117, 118, 119, 120, 121, 122, 123, 124, 125, 126, 127, 135, 149, 150, 164, 165, 166, 167, 168, 183, 185, 190, 191, 192, 193, 199, 209, 210, 215, 216, 220, 227, 229, 230, 252, 254, 259, 261
radioactive decay, vii, x, 12, 28, 117, 118, 123, 164, 165, 183, 190, 191, 193, 210, 215
radioactive waste, 95
radioisotopes, vii, 95, 164, 183, 190
radium, vii, viii, xi, xiii, 18, 20, 31, 32, 94, 99, 100, 102, 119, 120, 124, 125, 163, 164, 165, 167, 172, 174, 181, 182, 183, 185, 195, 210, 218, 221, 222, 228, 229, 231, 238, 239, 251, 253, 254
radius, 120
radon activity, 46, 47, 60, 61, 63, 71, 73, 79, 82, 84, 196, 245
radon detectors, vii, 1, 17, 193, 194, 215, 216, 236
radon flux, vii, 1, 21, 27, 28, 29, 30, 31, 32, 33, 39, 40, 105, 107, 252, 253
radon inhalation, x, 120, 133, 134, 136, 139, 140, 141, 142, 143, 144, 146, 147, 148
radon progeny, vii, xiii, 1, 2, 21, 24, 25, 26, 27, 39, 42, 101, 116, 177, 184, 190, 199, 210, 211, 245, 246, 248, 250
Radon-222 activity, viii, 43, 58
rainfall, 55, 65, 95
ramp, 18
randomized controlled clinical trials, 135
reactions, 119, 124, 127, 134, 144
reactive oxygen, x, 133, 134, 144
reading, 6, 10, 13, 36, 216, 245
real time, 191
reality, 95
receptacle, 5, 13
reception, 199
recession, 55, 73, 76
recombination, 17
recommendations, 15, 119, 122, 182, 201, 205, 207, 212, 222, 225, 229, 230
recovery, 140, 147
red blood cells, 121
regeneration, 119, 128, 136
regionalization, 95
regions of the world, 159
regression, 14, 15
regression equation, 14, 15
regulations, 87, 201, 207
rehabilitation, 126, 148
rehabilitation program, 126
relaxation, 134
relevance, 68, 71, 119, 128
reliability, viii, 44, 200

relief, 127, 141, 143
REM, 242
remediation, 116, 212, 230
remission, 143
repair, 128, 211
reparation, xii, 190, 206, 248
requirements, ix, 46, 99, 207
researchers, 46, 123, 127
residential, 41, 210
resolution, viii, 43, 194, 231, 234, 248, 254
resources, vii, 43, 44, 66, 96, 119, 125, 126, 158, 198
response, xii, 3, 8, 9, 11, 12, 13, 17, 19, 23, 27, 40, 45, 46, 77, 95, 136, 138, 142, 143, 147, 210, 216
rheumatic diseases, 127, 135, 144
rheumatoid arthritis, 123, 126, 135, 148
rhinitis, 121
risk, vii, ix, xi, xiii, 82, 90, 91, 99, 100, 106, 107, 109, 110, 111, 116, 123, 126, 128, 150, 163, 164, 176, 177, 178, 183, 184, 192, 207, 210, 214, 227, 229, 230, 248
river flows, 49, 50, 174
River Monitoring System, viii, 44
Rn222 activity, viii, 44, 68
RNA, 146
ROI, 196, 197, 198, 221
Romania, 202, 205, 206
room temperature, 134
root, 198
ROS, x, 133, 134, 135, 136, 137, 138, 139, 140, 141, 142, 143
Royal Society, 95
rubber, 31, 252
runoff, 69, 71, 95
Russia, 126, 134, 135, 179

S

safety, 124, 208, 230, 249
salinity, 82
salts, 119, 123
saturation, 216, 253, 254, 259, 260
scarcity, 44, 95, 103, 122
scatter, 103
schema, 3, 13
scholarship, 128
school, xii, 96, 114, 115, 116, 174, 183, 190, 192, 202, 204, 205, 206, 207, 208
science, 248
scope, 236
sea level, 18
secretion, 142
sedative, 121
sediments, 83, 100, 165, 168, 171, 173, 175

seismicity, x, 118
semiconductor, 191, 211, 212, 214
sensing, 13
sensitivity, 16, 22, 23, 25, 27, 29, 211, 213, 216, 236
septum, 121
Serbia, 122, 163, 166, 167, 168, 169, 170, 171, 172, 184, 185, 186, 189, 192, 195, 201, 207, 208, 209, 222, 223, 224, 225, 227, 229, 230, 231, 239, 243, 248, 249, 250, 251, 266
serum, 126, 139, 140, 142, 146
services, 44, 86, 230
settlements, xi, 149, 151, 152, 153, 154, 155, 156, 157, 158, 159, 160, 237
sexual activity, 121
sham, 139, 142
shape, 171, 183, 199, 221
shear, 53
showing, 60, 199
signals, 30
signs, 2
silicon, 211, 212, 214
SiO2, 166
skin, 124, 126, 127
slag, 253
small intestine, 138, 176
smooth muscle, 134, 136
sodium, 142, 148
software, vii, 1, 25, 27, 220, 233, 234, 236
soil particles, 165
soil type, 122, 210
solid phase, x, 117
solid state, 101, 153, 214
solubility, xiii, 119, 124, 125, 167, 176, 183, 251, 254
solution, x, 20, 45, 67, 117, 120, 123, 168, 173, 216, 236
South Korea, 267
SPA, 40
space-time, viii, 43, 45, 47
Spain, 119, 167, 185
specialists, 12, 36
species, x, 133, 134, 144, 193, 194
specific surface, 166
specifications, 12, 18
spectroscopy, xiii, 184, 218, 222, 227, 244, 245, 247, 249, 254, 266, 267
spine, 128
spleen, 138
Spring, 48, 53, 56, 60, 62, 63, 76, 78, 82, 84, 87, 89, 120, 125, 185, 244
sputum, 183
SSI, 223, 249
stability, 2, 11, 19, 153, 154, 199, 200, 206

staff members, 29

standard deviation, 63, 153, 154, 156, 202, 237, 238, 239, 241, 242

standardization, vii, 2, 12, 220

state, xiii, 20, 21, 93, 123, 151, 159, 190, 214, 218, 227, 228

statistics, xii, 96, 189, 197, 198, 206, 241, 242

steel, 3, 11, 31, 234, 258

stimulation, 143, 148

stomach, 128, 138, 150, 176, 177, 178, 179, 182, 184, 217

storage, xi, 2, 4, 18, 25, 68, 69, 76, 77, 87, 89, 95, 149, 179, 194

streamflow, viii, 43, 44, 45, 46, 50, 52, 54, 55, 58, 64, 65, 67, 68, 69, 72, 73, 91, 92, 93, 94

stress, x, 118, 137, 144, 145, 169

stroke, x, 133, 137, 145

structure, ix, 55, 69, 99, 118, 171, 230, 252, 262

style, 30

substrate(s), 68, 125

subterranean pathway, vii, 43

succession, 48

sulfate, 142, 148

Sun, 148

supplementation, 145

surface area, x, 6, 30, 33, 117, 118, 166, 228

surface charge, vii, 1, 2, 3, 4, 7, 12, 13, 17, 38

surface component, 65

surface layer, 102, 107, 234

sustainable development, 96

Sweden, 179, 253

Switzerland, 122, 179, 207

symptoms, 121

syndrome, 121

synthesis, 134, 138, 144, 146

T

T cell, 136

tap water, vii, xi, 118, 149, 150, 151, 152, 154, 155, 156, 157, 158, 159, 160, 171

target, 27, 158, 176, 200, 235

Tbilisi, xi, 149, 150, 151, 152, 153, 154, 155, 156, 157, 158, 159, 160, 161

teachers, 235

techniques, ix, xii, xiii, 27, 29, 68, 91, 109, 110, 111, 120, 123, 191, 207, 209, 210, 211, 218, 222, 227, 243

technologies, 36

technology, vii, 2, 9, 11, 30, 31, 36, 248

temperature, ix, xii, 2, 3, 4, 15, 38, 46, 55, 71, 75, 109, 111, 112, 113, 115, 120, 123, 124, 125, 135, 159, 165, 167, 169, 170, 171, 174, 183, 193, 194,

209, 216, 217, 218, 228, 236, 253, 254, 256, 258, 265

temporal variation, 78

terraces, 56, 103

territorial, xi, 149, 151, 154, 157

territory, xi, 150, 151, 154, 155, 157, 158, 159, 160, 222

testing, xii, 45, 46, 59, 189, 207, 212, 256

therapeutic agents, 138, 145

therapeutic use, 123

therapy, x, 18, 121, 128, 133, 134, 135, 137, 138, 139, 141, 143, 144, 145, 148, 183, 244, 245

thermal treatment, 136

thorium, 22, 23, 119, 122, 125, 129, 164, 180, 190, 210, 217

thoron, x, 10, 15, 21, 22, 23, 24, 25, 27, 28, 36, 117, 118, 120, 122, 123, 124, 172, 179, 190, 213, 214, 221, 223, 231

thymus, 138

thyroid, 128

TID, 180, 217

time periods, 3, 18

time series, 54, 96

time use, 45

tissue, 119, 121, 127, 150

Title V, 223, 249

TNF-α, 142

total cholesterol, 136

toxicity, vii

toxicology, 121

Toyota, 129, 143, 144, 147

trace elements, 126

tracks, 176, 191, 236

trade, 1

transformation(s), 65, 176

translocation, 138, 146

transmission, 252

transparency, 24

transport, xi, 93, 94, 95, 100, 107, 144, 149, 194, 245, 252, 266

transport processes, 93

transportation, 164

treatment, xi, 122, 126, 127, 128, 134, 135, 137, 140, 141, 143, 144, 146, 147, 163, 183, 217, 250

trial, 126, 135, 138, 144, 145, 146, 148

tuff, 253

tumor(s), 142, 176, 177

tumor necrosis factor (TNF), 142

tumor progression, 176

turbulence, 61

Turkey, 120, 127, 166

U

U.S. Department of the Interior, 184
U.S. Geological Survey (USGS), 95, 161, 184
Ukraine, 160
UNESCO, 96, 185
uniform, 17, 151, 158, 190
unique features, 21
United Kingdom (UK), 93, 111, 179
United Nations, 165, 184, 187, 208, 248, 249
United States (USA), viii, 21, 44, 76, 92, 96, 125, 134, 135, 144, 151, 159, 178, 179, 195, 206, 207, 208, 211, 212, 223, 232
uranium, vii, viii, x, xi, 1, 18, 21, 22, 25, 27, 28, 29, 30, 31, 33, 34, 99, 100, 101, 110, 117, 119, 120, 122, 125, 129, 163, 164, 165, 167, 168, 178, 180, 183, 185, 190, 207, 208, 210, 217, 222, 223, 228, 231, 248, 253
urban, xi, 149, 248
urban population, xi, 149
urethritis, 126
uric acid, 121
urinary tract, 126
urine, 126
US Department of Commerce, 266

V

validation, 96, 198
valve, 254
variables, 55, 67, 124
variations, x, 2, 10, 22, 45, 47, 57, 68, 96, 116, 118, 124, 176, 183, 191, 217, 248, 254
vector, 147
vein, 128
velocity, 75, 101, 106, 165, 246, 252, 258
ventilation, ix, 28, 109, 110, 112, 113, 114, 115, 183, 201, 211, 212, 228, 235, 242, 245, 248
vessels, 118, 128, 254
virus infection, 138
viscosity, 121
visualization, 236
vitamin C, 137, 141, 145
vitamin E, 141, 145, 146
vitamins, 141, 144
volatility, 45, 178
volatilization, 67, 82, 84

W

Wales, 116
walking, viii, 99
Washington, 161, 184, 223, 225, 249
waste, 18, 229, 253
water policy, 93
water quality, viii, 43
water resources, viii, xi, 43, 44, 65, 67, 94, 149, 151, 157, 158, 160
water supplies, x, 117, 119, 151, 161, 164, 178, 222
water vapor, 193, 210, 228
watershed, 50, 92, 93
wear, 18
web, 21, 31
well-being, 146
wellness, 129
wells, x, xi, 33, 107, 117, 120, 149, 150, 163, 164, 178, 216, 222
white blood cells, 121
wildlife, viii, 43
wind speed, 83, 229
windows, ix, 110, 112, 113, 115, 196, 201, 205, 212, 236
wood, 243
workers, 11, 18, 29, 87, 88, 127, 183
working conditions, 89
working hours, 87
workplace, 87
World Health Organization (WHO), 129, 185, 187, 192, 224, 241, 248, 249
worldwide, x, xiii, 118, 227, 245

X

X-irradiation, 138

Y

yield, 165, 174, 190, 215
Yugoslavia, 185

Z

zinc, 213
zirconium, 254, 258, 259, 260, 261, 264, 266